Praise for
The Other Brain

"A compelling, entertaining and information-rich narrative that explains why the brain's glial cells—traditionally considered mere 'packing materials' separating nerve cells—may be living a secret life of their own. Fields persuasively argues that this 'other brain' may hold the key to curing brain diseases such as multiple sclerosis, migraine and stroke, as well as enhancing our understanding of the mind."

—Richard Restak, M.D., author of *Mozart's Brain and The Fighter Pilot*
and *Think Smart*

"A brilliant and indispensable guide to the brave new frontier of brain science. With a storyteller's heart and a scientist's keen eye, R. Douglas Fields serves as our neural Jacques Cousteau, roving the depths of this thrilling realm to show us the vital discoveries that are being made today, and the breakthroughs that may come tomorrow. Read this book—your brain will thank you for it."

—Daniel Coyle, author of *The Talent Code*

"*The Other Brain* offers an insightful, complex, and nuanced picture of the most interesting substance on earth: the matter inside our heads."

—Anthony Doerr, *The Boston Globe*

"*The Other Brain* is a wonderful illustration of a paradox that lies at the heart of modern science. While the truths of science are independent of the personal beliefs of any particular scientist, the convictions of scientists and the social settings within which they conduct their work influence the direction and pace of discovery."

—John Strawn, *The Oregonian*

"The work described in this book, and the smart, funny scientist describing them, are key to the growing recognition that glia, which vastly outnumber neurons, are the 'brilliant glue' that holds our brains, thoughts, and acts together, coordinating not just the work of our neurons but the tremendous plasticity that accounts for our learning, our resilience, even our evolutionary success. Dig in. It will make your glia happy."

—David Dobbs, author of *Reef Madness: Charles Darwin, Alexander Agassiz, and the Meaning of Coral*

"A detailed exploration of a major part of the brain that has been ignored for decades. . . . With lucid prose—and a helpful ten-page glossary—the author reveals [the] vital role [of glial cells]. . . . Fields makes a convincing case that understanding this 'other brain' opens the door to dazzling possibilities."

—*Kirkus Reviews*

"A provocative new look at the brain."

—Mark Roth, *Pittsburgh Post-Gazette*

"[A] fascinating overview of contemporary neuroscience. . . . Fields gives life to a potentially dry medical topic by eavesdropping on the work of other neuroscientists, past and present, and shows how penetrating glia's secrets offers hope for breakthroughs in healing Alzheimer's, brain tumors, and even spinal cord injuries. Highly recommended."

—*Booklist*

The Other Brain

THE SCIENTIFIC AND MEDICAL
BREAKTHROUGHS THAT WILL HEAL OUR BRAINS
AND REVOLUTIONIZE OUR HEALTH

R. DOUGLAS FIELDS, PH.D.

2009
SIMON & SCHUSTER PAPERBACKS
NEW YORK   LONDON   TORONTO   SYDNEY

Simon & Schuster Paperbacks
A Division of Simon & Schuster, Inc.
1230 Avenue of the Americas
New York, NY 10020

First Simon & Schuster trade paperback edition January 2011

SIMON & SCHUSTER PAPERBACKS and colophon are
registered trademarks of Simon & Schuster, Inc.

Figure credits are listed after the Index.

For information about special discounts for bulk purchases, please contact
Simon & Schuster Special Sales at 1-866-506-1949 or business@simonandschuster.com.

The Simon & Schuster Speakers Bureau can bring authors to your live event.
For more information or to book an event, contact the Simon & Schuster Speakers
Bureau at 1-866-248-3049 or visit our website at www.simonspeakers.com.

Designed by Kyoko Watanabe

Manufactured in the United States of America

3 5 7 9 10 8 6 4 2

The Library of Congress has cataloged the hardcover edition as follows:

Fields, R. Douglas.
The other brain : from dementia to schizophrenia, how new discoveries
about the brain are revolutionizing medicine and science / R. Douglas Fields.
    p.   cm.
Includes bibliographical references and index.
1. Brain—Physiology.   2. Brain chemistry.   3. Cognitive neuroscience.   I. Title.
            QP376.F46   2009
         612.8—dc22   2009034287

ISBN 978-0-7432-9141-5
ISBN 978-0-7432-9142-2 (pbk)
ISBN 978-1-4391-6043-5 (ebook)

CONTENTS

PART III

Glia in Thought and Memory

PREFACE

How do the 100 billion neurons in our brain allow us to remember who we are; to learn, think, and dream; to be stirred by passion or rage; to ride a bike or conjure meaning from inked patterns on paper; or to pluck out instantly a mother's voice from the muddle of a noisy crowd? What goes wrong with neural circuits in schizophrenia or depression, or in dreadful diseases like Alzheimer's, multiple sclerosis, chronic pain, or paralysis?

We are on the cusp of a new understanding of the brain that transforms a century of conventional thinking about the brain, specifically the role of the brain's neurons. Crowding around the computer screen in a darkened room in 1990, scientists watched information passing through peculiar brain cells, bypassing neurons and communicating without using electrical impulses. Until this discovery scientists had presumed that information in the brain flowed only through neurons by using electricity. In fact, a mere 15 percent of the cells in our brain are neurons. The rest of our brain cells—called glia—have been overlooked as little more than packing material stuffed between the electric neurons. "Housekeeping cells" they were called. Dismissed as cellular domestic servants, glia were neglected for more than a century after they were discovered.

Now scientists are shocked to learn that these odd brain cells communicate among themselves. Scientists' understanding of the brain has been shaken to its foundation by the discovery that these cells not only sense electrical activity flowing through neural circuits—they can control it.

How did scientists miss half the brain until now? Glia do not fire electrical impulses, and so the probing electrodes neuroscientists use to monitor neurons were deaf to glial transmissions. Glia are not connected through synapses into circuits the way neurons are. Rather than passing messages sequentially like a falling line of dominos, glia broadcast their messages broadly throughout the brain.

How will this new discovery change our understanding of the mind? Will the mysteries of how the mind fails in mental illness or disease be solved as we explore this new dimension of the brain? Will the search uncover answers to how the brain can be repaired after disease or injury?

The discovery of glia—the other brain—is a dawning that illuminates every aspect of brain science, touching simultaneously all researchers working on the brain. This is a story of science in action, with twists and turns, insight and confusion, controversy and consensus. Along the way you will meet some fascinating scientists who are real people, each one different and sometimes peculiar, but each engaged in the most collaborative of all human activities—science.

The information here is so new it has not yet found its way into textbooks. This information will change your understanding of the brain. It will also provide you essential knowledge that will benefit you or your loved ones in your own health. The book is packed with the latest information on neuroscience and medicine, and it brings you inside for an eyewitness view through the eyes of one of the researchers involved.

ACKNOWLEDGMENTS

Although I had written many academic articles on this subject, this book originated when the editors at *Scientific American,* Ricki Rusting and John Rennie, were intrigued enough to publish an article submitted by an unknown author. At a time when the information was new and strange, and still discounted by many, they appreciated the importance of this breakthrough and published the first article on glia for a popular audience as the cover story of the April 2004 issue, "The Other Half of the Brain." I owe much of the success of that article to the editor who worked with me on it, Mark Fischetti.

It might have ended there had not Jeff Kellogg, a literary agent who read the *Scientific American* article, contacted me with the suggestion that I write a book on the subject. Knowing nothing of literary agents or book writing for the popular market, I promptly deleted his e-mail—and his next one as well. Fortunately he persisted, and soon I learned from Jeff how to write a book proposal. I could not have had a better person helping me and working as my agent. When Simon & Schuster agreed to publish the book, the difficult task of guiding a research scientist through the complex process of writing a book that the public might want to read fell to Senior Editor Bob Bender. Bob's expert guidance has been crucial in bringing this book to completion, and I am grateful for his help and commitment to the project. I am indebted to Loretta Denner, to Amy Ryan for her discerning copyediting, to Morgan Ryan for careful fact-checking, and to Johanna Li.

To all my scientific colleagues and coworkers who permitted me to share their stories and comments, I offer my sincere thanks. I hope that the reading will provide for them, as the writing did for me, a renewed appreciation of the joy and privilege of what we do as scientists. I owe a special thanks to Dr. Susanne Zimmermann and Prof. Christoph Redies of the University of Jena for their help in researching Hans Berger, and to Lydia Kibiuk, Alan Hoofring, and Jen Christiansen for their outstanding illustrations.

One of the delights of this project has been working with and learning from some of the top editors and authors bringing science to the public. In particular I acknowledge Mariette DiChristina, Christine Soares, and David Dobbs at *Scientific American*. Author Dan Coyle read the entire book and offered helpful suggestions. Tom Cecil, Tony Barnes, Pam Hines, and Beth Stevens all graciously read parts of the manuscript and offered encouragement and advice. Above all, I thank my wife, Melanie, and family for tolerating the many impositions the prolonged project of writing a book demands. They were the strongest supporters and most willing critics of early drafts of each chapter; this includes my son, Dylan, and daughters Morgan and Kelly, my brother and sister, Kyle and Peggy, and my parents, Marjorie and Richard. I thank my dad, an electrical engineer who had me drawing atoms and electrons in kindergarten when my friends were drawing cars, for my love of science, and my mom, a farm girl who rode bareback into the fields to write stories on the back of her horse, for my love of letters. And to Melanie I owe thanks for a bit more than this book, for love, life, and family.

PART I

Discovering the Other Brain

CHAPTER I

Bubble Wrap or Brilliant Glue?

EINSTEIN'S BRAIN

With a final slice he dropped the scalpel into the stainless steel tray, and reaching into the open skull with both hands, he carefully scooped out the brain. Cradling a human brain in his hands always released a torrent of thoughts and emotions over mortality, individuality, biology, spirituality, and the mystery of one's own place in the world. Only hours before, everything that was this unique human being had been embodied in these three pounds of convoluted tissue. Although the pathologist had felt these emotions countless times before, this time was different. The corpse laid out on the stainless steel table before him was Albert Einstein, and in his hands he held Einstein's brain.

Scrutinizing the brain under the bright examination lights he stared with profound wonder at how this brain, slumping under its own weight like Jell-O and looking identical to any other human brain, could have created one of the most extraordinary minds of the last century. Suddenly, Dr. Thomas Harvey saw in this brain his own destiny and purpose. It was meant for *him*.

Rinsing blood from the brain carefully in saline solution, he weighed and measured it and then placed it in a freshly made 10 percent solution of formaldehyde, the toxic fumes stinging his nose and eyes. When this great man's body was laid to rest, his phenomenal brain lay sunken in a jar of preservative, like a curious museum specimen, hidden away for the next forty years by a pathologist who felt an overwhelming compulsion to keep it for himself. It was an unethical and illegal desecration, but Harvey felt it was his fate and duty to science and humanity to unlock the secrets that had enabled this brain to give birth to such an extraordinary scientific mind.

The task was far beyond the ability of this pathologist, who saw his role as guardian of this priceless scientific treasure. Over the next four decades, Harvey doled out small slices of Einstein's brain to scientists and pseudoscientists around the world to probe in different ways for clues to Einstein's genius.[1]

Here was a mind so extraordinary it conceived thoughts beyond the ability of any other mind to imagine, and beyond the capacity of many minds to understand even after the theory of relativity was fully formulated and articulated to them. A mind that could conceive the idea that time itself was flexible. Time and space, matter and energy lost their identity and freely morphed from one to the other, and time contracted or dilated to frame events fluidly. And to reach that revelation through no other means than the power of thought—a mind imagining itself riding a beam of light.

Thirty years after Einstein's brain was stolen, four pieces of it reached a distinguished neuroanatomist at the University of California, Berkeley. She now held in her hands vials containing four bits of tissue selected from carefully chosen regions of Einstein's cerebral cortex. Dr. Marian Diamond reasoned that since Einstein's genius related to extraordinary abilities of imagination, abstraction, and higher-level cognitive function, any physical basis for Einstein's genius would be found in the regions of cerebral cortex serving these cognitive functions, rather than regions of cortex handling other functions such as hearing, sight, or motor control, which were not extraordinarily different in Einstein. Harvey had cut up Einstein's cortex, numbered the blocks, and embedded them in celloidin, a nitrocellulose compound that, when hardened, encased the tissue like insects in amber. Diamond wanted to examine two samples of association cortex, parts of the cerebral cortex where information is brought together for analysis and synthesis. She requested that Harvey send her samples from the prefrontal region, which lies just under the forehead, and one from the inferior parietal region, located slightly behind and above the ears. It was important to have samples from both left and right sides of the brain, because in most people, the left and right hemispheres tend to dominate in different cognitive functions, just as we engage the world differently with our left and right hands. The prefrontal cortex is involved in planning, recent memory, abstracting, and categorizing information. The infamous prefrontal lobotomy procedure severs this region from the brain, leaving basic mental functions unimpaired but rendering patients docile by the loss of higher level cognitive abilities to make mental abstractions and syntheses from their experiences. Diamond also requested samples from Einstein's inferior parietal cortex, because this region is associated with imagery, memory, and attention. People with damage to this area, especially on the dominant (usually left) side of the brain, lose the ability to recognize words and letters and cannot spell or calculate. Medical literature documents the story of a mathematician who found it difficult to formulate mathematical problems after damage to this area of his cortex.

A career of studying human cortical anatomy could not diminish the wonder, excitement, and anticipation Diamond now felt as she held these four cream-colored bits of human brain, the size of sugar cubes, up to the light. These were different—at least the mind that had emerged from them was. If she could discover the secrets that had enabled this brain tissue to produce Einstein's genius, that discovery could give insight into the cellular mechanisms linking mind and brain. It could tell us how our own brain operates and how the diseased minds of those less fortunate fail.

Diamond would need to compare these samples with appropriate control samples. Her excitement was tempered by doubt, for while she was surrounded in her lab by boxes of microscope slides containing tissue samples from many different human brains, there is but one of Einstein's brain. The truly extraordinary nature of Einstein's brain meant that no matter what result she obtained, the experiment could never be replicated. The gnawing uncertainty that faces every scientist at the conclusion of an experiment would be harder to vanquish without the possibility of replication. Conclusions from any data can be fallible, but science progresses through observation, data collection, and assembly of the facts. Would it be better not to look?

Scientists deal with the uncertainty of experimental results by making measurements and calculating mathematically the odds that the difference between the data from the control group and the experimental group could have come about due to chance alone. Similarly the significance of a single blond hair found at a crime scene can be partly evaluated knowing the probability of finding a blond hair in the population.

Diamond and her associates prepared to study the cellular structure of the samples. To do this, the brain tissue had to be sliced thinner than the diameter of a cell and stained with dyes, so that individual neurons could be discerned in detail among the jumble of cells forming the tissue. So thin are these slices that a stack of fifteen would only equal the thickness of a human hair. Laid out before her was a series of glass dishes filled with brilliantly colored solutions, ranging from deep purple to a shimmering pink that changed to green depending on how light struck it, like an oil slick on water. After she had collected a series of tissue sections, she transferred each one carefully with an artist's fine paintbrush to a small glass dish of staining solution.

The next day, when she studied the sections under the microscope, shadows appeared through the formless fog; suddenly, as from an airplane descending through the clouds, the image came into sharp focus and detail like the panorama of a city. The cell she was viewing was a neuron from a region of Albert Einstein's cortex. Perhaps this very neuron had imagined riding a beam of light. What was the difference between it and an ordinary neuron in another region of Einstein's cortex that sparked commands to Einstein's fingers, perhaps to write out on paper the mathematical symbols that brought this imagination to a tangible reality? How similar was this neuron to one in the same part of *her* cortex now sparking images and thoughts through circuits in her own mind as she contemplated the priceless treasure and mystery before her? How could a microscopic cell have so radically changed the world? How would this cell compare to a neuron in the same part of Isaac Newton's brain? Science and technology progress through the combined action of thousands of small steps, but scientific advance is sometimes punctuated by great conceptual leaps, like the Copernican view of the solar system, Newton's laws of gravitation and motion, Darwin's theory of evolution of species, and Einstein's theory of relativity. The number of such leaps can be counted on the fingers of both hands, and this neuron had come from a mind that had changed the world.

After days of carefully measuring and counting cells, Diamond added up the data and compared it with the identical regions from eleven control brains, from men ranging in age from forty-seven to eighty. There was no difference.

A neuron from the brain of a genius was indistinguishable from one taken from a typical brain. And on average, there were just as many neurons in Einstein's creative cerebral cortex as in the cortex of men not noted for being unusually creative. But there *was* one difference in the data. The number of cells that were *not* neurons was off the charts in all four areas of Einstein's brain. On average, the samples from normal brain tissue had one cell that was not a neuron for every two neurons counted, but the samples from Einstein's brain had nearly twice as many nonneuronal cells, about one for every neuron. The biggest difference was seen in the sample of parietal cortex from the dominant side of Einstein's brain, the region where abstract concepts, visual imagery, and complex thinking take place. Was this a fluke? Diamond calculated the mathematical odds that this difference could have happened by chance, considering the range of variation in the control tissue samples. In all the regions sampled from Einstein's brain the odds that the difference could have occurred by chance were small.

The only difference Diamond could see between Einstein's brain and an average brain was in these nonneuronal cells. Could this be the cellular basis of genius? How? What were these nonneuronal cells—called "glia"—doing? For decades these glial cells had been considered little more than mental bubble wrap, connective tissue that physically and perhaps nutritionally supported the neurons, but Einstein's brain had more. Speculating that glia could be involved in mental function was well outside the conceptual box of most neuroscientists. Their very name sealed this box for a century: neuroglia—Latin for "neuro glue."

INTELLECTUAL BLIND SPOT: GLIA HIDDEN IN PLAIN SIGHT

To appreciate the implications of Dr. Diamond's findings, it is important to understand some basic facts about glia and to consider the origins of our current view of the way the brain works. The image most people have of the nervous system resembles something like the jumble of wires in a telephone network. This image has changed little in the last one hundred years. So deeply ingrained is this concept that it is difficult to imagine the nervous system working in any other way or to imagine that when this view was first conceived, it was seen as radical, and the controversy and debate about it would rage for a quarter century.

Santiago was an artistic boy, born in 1852, the son of a medical doctor in Spain. He excelled in drawing and enjoyed the new art of photography, but these pursuits did not lend themselves readily to a prosperous profession. Studying for his medical degree, he spent hours making anatomical drawings of the cadavers his father carefully dissected.

At the age of thirty-three, Santiago Ramón y Cajal (pronounced *Ca-hall*) held the position of professor of anatomy at Zaragoza, Spain. In 1887 on a visit to Madrid, Ramón y Cajal saw a microscope slide of nervous tissue stained with a procedure developed by the Italian anatomist Camillo Golgi fourteen years earlier. That image transformed Ramón y Cajal's life. Abandoning his previous and highly regarded research in bacteriology, Ramón y Cajal assumed the chair of Normal and Pathological Histology in Barcelona and applied himself to improving and exploiting the Golgi method of staining to uncover the cellular structure of the brain. The Golgi method had not attracted attention for fourteen years because it was capricious. Often the staining failed, but when it worked properly, the results were stunning.

The staining method shares the same chemical reaction used in the brand-new method of black and white photography that so interested Ramón y Cajal. For reasons that are still a mystery, only a small number of neurons take up the stain, perhaps one in a hundred. But those that do become stained in their entirety, causing them to stand out in fine detail like the black silhouette of an oak tree against a yellow winter sunset. If the method stained all the neurons in the sample, it would be useless, because the branches of nerve cells packed together in any slice of brain tissue would form a thicket of incomprehensible tangles. Instead Ramón y Cajal saw individual neurons exposed bare and in their entirety like fossils cleaved from stone.

Today we tend to think of the brain by analogy with computers and electronics, but before the electronic age a different model prevailed. Nineteenth-century mills and factories harnessed the power of water diverted from rivers and streams to waterwheels, then channeled it back through canals and streams to the river that was the original source. Hydraulics were the most advanced mechanism for transmitting force over distance. Power could be applied where needed by control valves connecting the plumbing lines and hoses to direct the force. At the time, this was the analogous view to how the nervous system functioned. The nerves in our bodies were thought to be plumbed together so that their force could be applied to any muscle. Through a microscope lens one could see hundreds of tiny tubes inside nerves, presumably all interconnected with control valves and leading to the master cylinder in the brain. Inside the brain, thousands of tangled microscopic tubes, called axons, could also be seen under the microscope, coursing through tracts of white matter streaking through brain tissue.

The headwater for these tracts was in the grey matter, which formed a thick rind over the convoluted surface of the brain, like the stem of a broccoli plant dividing into finer branches to terminate among the green florets. The Golgi staining method revealed the individual nerve cells, called neurons, in exquisite detail, but these details were interpreted differently by two groups of scientists. Golgi saw the axon, the slender tube emerging from the body of each neuron, projecting great distances and branching to interconnect with other axons through countless connections. This network of interconnected fibers would allow tremendous facility in directing nervous commands or routing incoming messages from our sense organs. At the opposite end of the nerve cell, Golgi saw the highly branched and tapering rootlets called dendrites, because they resembled trees. These, he surmised, were for extracting nutrients to sustain the nerve cell and power the flow of nervous energy, now understood to be electrical current, through the network of axons. Ramón y Cajal looked at the same material, using the very staining method Golgi invented, and saw something completely different. Ramón y Cajal proposed a new theory, which came to be known as the "neuron doctrine."

Ramón y Cajal worked feverishly, sixteen hours a day, seven days a week, examining pieces of brain tissue from animals of all kinds and of all ages, and in all regions of the brain and body. He looked at samples from humans, rabbits, dogs, guinea pigs, rats, chicks, fish, frogs, mice, and fetal animals. With an artist's precise observation he drew out the silhouetted neurons, and studying them, he began to see a logic in the structure. Although the axons, those wire-like extensions of nerve cells, could travel over tremendous distances in the brain, they always terminated among fields of dendrites, the finely branched rootlets of neurons. In a conceptual leap, Ramón y Cajal realized that a neuron was not a node in a net, it was an independent unit! Moreover, the neuron had a functional polarity. He perceived that signals were not radiated through neural networks in all directions, like vibrations through a spider's web. Instead the signals were conducted through each neuron in one direction, like horse-drawn buggies on a one-way street. Information came into a neuron through its root-like dendrites and commands exited through the axon, emerging from the opposite side of the cell. The axon did not connect into a meshwork of other axons, but instead ended at the dendrite of another neuron. Somehow, nervous signals were passed across the threshold between axon and dendrite into the next neuron, like boxes left on the doorstep of the recipient neuron to pick up. The cellular contents (protoplasm) of the two nerve cells were not plumbed together like fluid in hydraulic couplings.

This separation between an axon and a dendrite is called the synapse. By passing or not passing a message from axon to dendrite at each synapse, the brain directs information flow with great complexity, just as switchboards direct telephone calls.

Ramón y Cajal became the most renowned neuroanatomist of the twentieth century, receiving the Nobel Prize in Physiology or Medicine in 1906 together with Camillo Golgi, his rival, who had developed the essential staining method, but who disagreed with Ramón y Cajal's neuron doctrine. Working prodigiously, Ramón y Cajal made discovery after discovery and published volumes of scientific papers and books on the cellular structure of the brain that are still a rich and valuable source of accurate information today. But what he left out of his drawings is an equal measure of his genius.

The wilderness of cellular structure in each microscope slide is refracted through the discerning lens of an artist's eye in each of Ramón y Cajal's drawings to isolate the essential information. Pioneering his way through this wilderness of complex brain structure, Ramón y Cajal never failed by drawing something that was not really there— no Martian canals, no homunculus inside the head of a sperm. He was deliberate and careful never to mix things that did not seem to belong together.

Among the things he saw clearly and always left out of his neuron drawings were glia. These he drew separately, filling volumes of notebooks over years of research with the strange-looking cells. These cells fascinated him, but their structure as revealed by the Golgi stain gave no clues to their function. Glial cells lacked both wire-like axons and root-like dendrites. Through a microscope most of them resembled a bullet hole shot through glass: a circular middle with fine fracture-like extensions radiating outward in a halo. Ramón y Cajal called them "spider cells" because of the many protoplasmic legs extending in all directions from their corpulent cell body. Other scientists thought these cells resembled stars and called them "astrocytes." That name prevails today for one of four major types of glial cells now recognized. But Ramón y Cajal saw that glial cells came in an endless variety of bizarre forms. Some looked like grotesque fan corals, and others like sausages strung up on axons. Most authorities considered these nonneuronal cells some form of cerebral connective tissue, filling the space between neurons. If these peculiar brain cells had a higher function, Ramón y Cajal knew that their secrets would not be cracked with the primitive tools at his disposal. So he wisely drew these cells separately, and in dividing them out in this way in his notebooks, he called implicitly to neurobiologists of the future to answer the question, What is this other half of the brain?

GLIA ARE LISTENING:
SHINING LIGHT ON RAMÓN Y CAJAL'S MYSTERY

Ninety years later I sit in a tiny room, my face illuminated by the cold blue glow of computer monitors. To my left is a massive pool-table-sized stainless steel table, eight inches thick, floating on air pistons inside massive steel legs to provide a precise, optically flat, and vibration-free surface. The table is covered with electronic instruments linked together and bolted firmly to the table. The whirr of cooling fans inside the equipment fills the air, punctuated by clicks of automated valves and shutters coming from inside the mass of instruments. Thick black hoses circulate chilled water to a washing-machine-sized device in the adjacent room to cool the ultraviolet laser that is the heart of the instrument. A corrugated hose resembling the duct from a clothes dryer sucks toxic ozone fumes out of the small room.

In the center of the table sits a box the size of a large cabinet, constructed of brilliant orange translucent Plexiglas to shield me from the ultraviolet light. Inside the box is the only object in the room Santiago Ramón y Cajal would have recognized: a microscope. He would have marveled at this instrument. Massive and precisely built, it is about three times the size of the microscope he had used to peer at brain tissue samples he had hand cut with a razor. Nevertheless, he would have recognized the microscope's basic structure and components—the movable specimen stage and two eyepieces like those he had stared through with childlike wonder. Now these are used for little more than a hasty glance to position the specimen beneath the objective lens before shuttering them closed and diverting the light path to the digital camera or photomultiplier tubes that intensify the dim image and display the microscopic scene brilliantly on a computer screen. I navigate through the microscopic terrain with a joystick control like a helicopter pilot.

Turning a knob like a radio dial, I section optically through the cell from top to bottom, peeling away cellular structure one visual "slice" at a time. First a spot of cell membrane, like the smudge left by a ball bounced off a window pane. Then a ring as though I had sliced the top off the ball, and sliced deeper with successive cuts all the way through it to the other side, recording along the way every minute structure inside the cell from the cell membrane through the nucleus. More startling yet, the cells I am now exploring with this advanced light microscope are alive. They had been removed from a fetal mouse, isolated individually, and grown for more than a month inside a laboratory incubator, serving as an artificial womb to keep them warm and oxygenated.

Ramón y Cajal would have recognized instantly the image on the screen as neurons. In fact he would have identified the exact type of neuron by their unique spherical shape as those that carry the sensations of touch, heat, and pain from our skin to our spine. These are called dorsal root ganglion (DRG) neurons. But he would have been bewildered at the image of these cells seen in crisp focus as though they had been sliced into impossibly thin sections.

Donning special glasses, I see the computer screen transform into an open window, and suspended inside it this single DRG neuron floats in three-dimensional space like a dangling Christmas ornament. Ramón y Cajal would surely have been astonished as with a touch of my finger on the mouse, I spin the cell on any axis to examine its finest internal structural details.

But the best is yet to come. This instrument is a laser-scanning confocal microscope, the first one in our institute at the National Institutes of Health. Distinct from a common light microscope, this instrument not only reveals cellular structure in crisp optical sections, but it also can reveal cellular biochemistry and physiology in action as molecules move through and between living cells carrying messages and commands from the electrical signals at their surface to the heart of each cell's nucleus. At the time, 1994, this was one of only a handful of such instruments in the country, but today no research department at any major university is without one, and most have several.

REVELATIONS FROM THE DEEP

Twenty years before, when I was a marine biologist at Scripps Institution of Oceanography in San Diego, I spent summers conducting research on obscure deep-sea fish called chimaeras that rise from the deep along cold-water currents that come close to the surface in the North Pacific region of the San Juan Islands in Washington, where Friday Harbor Marine Laboratory is located. Scientists from around the world assemble, forming a summer camp community engaged without distraction in around-the-clock scientific research. As we prepared to ship out on a fishing boat to collect specimens with a bottom trawl, I noticed several students running up and down the pier with butterfly nets excitedly dipping them into the water to collect tiny silver-dollar-sized jellyfish known as *Aequorea victoria*. These beautiful transparent creatures are common summer residents of Friday Harbor waters, but I could not quite understand the collectors' enthusiasm. In answer to my question, one of the students told me they were interested in the jellyfish's bioluminescence, the ability a number of sea creatures have to emit cold light, usually green-blue phosphorescence. "Why? Are you trying to figure out how they make light?" I asked.

"No, we know that. We're extracting the protein that makes the light when it binds to calcium. We're injecting the protein, aequorin, into cells to study calcium currents."

Suddenly I understood. Electrophysiologists study nerve activity by making extremely fine electrodes, which they jab into nerve cells using micromanipulators to position them precisely under a microscope. The flow of ions inside a nerve cell generates a biological electric current, which, when greatly amplified with electronic instruments and displayed on the phosphorescent screen of an oscilloscope, enables scientists to see nerve impulses traveling thorough neural circuits, much as doctors track heartbeats on an operating room monitor. Electrophysiologists want to know how these electric currents are formed and regulated and which of the many ions inside cells contribute to the electrical current. This is a difficult task, usually approached by substituting different ions in the solution bathing nerve cells, or by applying drugs that block certain protein channels in the cell membrane that admit different ions—such as sodium, potassium, or calcium—into the cell.

If these scientists could get this technique to work, they would inject this fluorescent jellyfish protein into a nerve cell and watch it under a microscope. If a current of calcium flowed into or through a cell, they would see its tracks by the glow of green phosphorescence left behind, like the contrail of an airliner across a cloudless sky. They would be seeing with their own eyes the biochemical reactions and physiological activity in the living cell. And they could watch these events in real time and in three-dimensional space, not just as green streaks and blips of electrical events flashed across the screen of an oscilloscope.

Now, twenty years later, I was processing DRG neurons through a series of solutions so that they would absorb a synthetic calcium-sensitive fluorescent dye much like that extracted from jellyfish. I had grown these cells in a dish equipped with platinum electrodes that allowed me to deliver weak electric shocks to make the DRG neurons fire impulses. When a neuron fires, the change in voltage on the cell membrane opens protein channels on the membrane that allow calcium ions to enter the cell. I was delighted to see that every time the neuron fired, the cell flashed as the calcium ions entered the cell and bound with the fluorescent dye. Ten years earlier, the first time my colleagues and I saw this happen, our shouts of joy echoed through the halls.

This experiment was different. In addition to the DRG neurons, I had asked Beth Stevens, my lab technician, who would later become my graduate student (now a new professor at Harvard University), to add glial cells to the cultures. The type of glial cells we added are called Schwann cells. Found in nerves, Schwann cells attach to axons and form a layer of electrical insulation, called myelin, around large-diameter axons, or they embed several small axons inside themselves, like a bun wrapped around several hotdogs. These glial cells provide structural and possibly physiological support for the axons. Neural impulses, of course, are carried only by axons. Heretofore it was presumed that glial cells insulated the axons like plastic coating around a wire but that they could not detect impulse activity flowing through the axons. We wanted to test this assumption.

After months of preparation it had finally come to the moment all scientists live for—when with the flick of a switch, your hypothesis will or will not be proven. The computer screen transformed the intensity of fluorescent light in the cell into a colored scale like a TV weather map showing rainfall in the local region. The more calcium, the brighter the light, and the brighter the light, the warmer the color. The DRG neurons and Schwann cells were deep blue, indicating little calcium inside. The instant I flicked the switch to stimulate the neurons, the neurons changed from blue to green to red to white, indicating a flood of calcium into their cytoplasm. The Schwann cells, unable to fire electrical impulses or detect the weak electric shock that causes axons to do so, remained blue. Then after a long fifteen seconds of disappointment, Beth and I were elated as the Schwann cells suddenly began to light up like a Christmas tree. These glial cells had somehow detected the impulse firing in nerve axons and responded by increasing the calcium concentration inside their own cell bodies. Glia, which for so long had been regarded as little more than bubble wrap for the brain, were a party to information sent between neurons. Now new questions emerged. How were these cells tapping into the electrical signals flowing inside nerve cell axons? And more importantly, why?

CHAPTER 2

A Look Inside the Brain: The Cellular Components of the Brain

DISSECTING THE BRAIN

With eager hands of a ten-year-old boy I sliced the heart in two with a butcher knife. All was revealed—four chambers separated by moist, gristly valves that suck blood into auricles and squeeze it out the aorta and pulmonary artery. Fascinated, I asked Mom if next time, could she bring me a brain? When she returned from the butcher shop with a calf brain my excitement welled as I sliced, cleaving it into two halves. But inside there was nothing. Just a hollow cavity at the core of a fleshy mush.[1]

How did it work? Books offered names for its various bumps and folds—"cerebellum", "pons", "medulla", "lateral ventricle"—but this information failed to provide the slightest inkling into how this organ, the most supreme of all bodily organs, might function. My parents, teachers—no one really seemed to have the answers.

Today I know that the brain's power comes from miniaturizing and concentrating its components to such an extreme that its working parts are invisible. Like the working parts of a computer miniaturized beyond the resolution of the human eye, cellular components of the brain are invisible unless they are magnified hundreds or thousands of times larger by powerful microscopes. It is natural now to think of the nerve cell, or neuron, as the microprocessor of the brain, but recall that scientists thought differently about the brain before the electronic age. How sure can we be that our microprocessor analogy of the brain is accurate? Thoughtful neuroscientists are beginning to wonder—could our fundamental concept of how the brain works be naive?

To try to understand why glia might be tapping into lines of communication between neurons, we must begin by examining brain structure and these shadowy cells more closely. By far most of the cells in your brain are glia, not neurons. Unlike neurons, glia cannot fire electrical impulses. Consequently, glia do not have the characteristic features of neurons, with their wire-like axon for sending electrical impulses over long distances and the bushy dendrites for receiving electrical signals through thousands of synapses. To find answers to what these brain cells are doing, we must trace the trail of discovery back to its roots when pioneering scientists first reported seeing neurons in microscopic sections of nervous tissue. Ironically, glia were there in abundance, right before their eyes, but the dazzle of neurons blinded scientists intellectually to the glia, which constitute the bulk of the brain.

GREY MATTER

In contrast to the descriptive names anatomists give every bump and minute feature of the body (usually in Latin), the best they could do when it came to the brain was "grey matter." Inept even as a physical description, for the brain's true color is pink, the name is a poetic allusion to our murky understanding of this enigmatic tissue. To the early anatomists, there was nothing black and white about it.

Imagine the frustration of nineteenth-century anatomists scrutinizing this mushy tissue for any sense of its structure or clue as to how it worked. When the microscope was invented, scientists rushed to look inside grey matter, but they were bewildered by the world they saw, so unlike any other tissue in the body. So finely miniaturized is the cellular structure of the nervous system that the length of a wave of visible light is too blunt to probe it. The wavelength of green light is ten times longer than the size of a synaptic vesicle, the fundamental apparatus for communication between neurons. It was not until physicists developed ways to focus beams of electrons in place of beams of light in the electron microscope that there was a tool sharp enough to see synaptic vesicles and resolve the fine structure of the brain. The electron microscope was invented in the mid-twentieth century, but it required decades of technical development to perfect and years to understand the complex images it produced. More than a century after the cellular structure of the rest of the body was mapped out in fine detail, the cellular structure of the nervous system is still an area of vigorous research today.

WHITE MATTER

The rest of the brain is white matter. This glistening-white brain tissue is a mass of millions of tightly bundled communication lines connecting neurons between distant points in the brain. These vital communication lines are packed beneath the grey matter cortex, much like tightly wound fibers beneath the leather skin of a baseball. White matter in the brain, like white space on paper, is easily dismissed as something defining the areas between the functional components, but recently this naive view has been changing. Unraveling this part of the brain is such a daunting task that only in the last few years have new brain imaging techniques allowed scientists to venture into the white matter realm. As we will see later, these new findings are changing fundamental concepts about how the brain processes and stores information—how we learn. Here inside the blank white regions of brain, glia are the heart of the mechanism.

A revolution in our understanding of how the brain is built, how it functions, how it fails in mental illness and disease, and how it is repaired has been ignited with the recent exploration of these long-neglected brain cells. Glia are the key to understanding this new view of the brain. There is little or no information available about these cells to nonscientists, so we can begin our inquiry with about as much knowledge as the pioneering scientists who discovered these various odd brain cells. Since the answers are known to only a few specialists, we can experience the same puzzles, clues, and revelations as the scientists who sleuthed out these peculiar cells in the brain. When these clues are assembled, will they reveal another brain working in parallel with our neuronal brain?

NEURONS:
A USER'S GUIDE TO HOW THE BRAIN WORKS

Before venturing further, it is essential that we proceed from a common base of knowledge about how the brain operates at the level of cells and circuits. The nervous system works by sending electrical impulses down a wire-like axon at top speeds of 200 miles per hour. Impulses travel through some axons, such as pain fibers, much more slowly—only 2 miles per hour, the pace of our footsteps in a slow walk. This explains the build-up to the full painful sensation when you accidentally hit your thumb with a hammer. The reason for the hundred-fold increase in transmission speed through our high-speed nerve fibers is that they are wrapped with electrical insulation, called myelin. In contrast, pain fibers are uninsulated thread-like axons.

As Ramón y Cajal surmised, neurons are not fused to one another like copper wires soldered in a circuit; instead, each neuron in your brain is an island unto itself. Each of these neuronal islands communicates by sending a message to another neuron across a tiny gulf of the saltwater that bathes every cell in your body. Because of this gulf of separation, information is not passed on to the next neuron in a circuit by electricity. Instead, the neuron floats chemical messages across the gulf to reach the neuron on the other side. This gulf is the synapse, and the neurons on either side are called presynaptic or postsynaptic neurons, depending on whether they are the sending or receiving shore of the gulf. The presynaptic neuron is always the one sending the message from its axon tip; the postsynaptic neuron receives messages across the synapse through its root-like dendrites.

The messages are sent in the form of a chemical substance called a neurotransmitter. Microscopic "bottles" inside neurons, called synaptic vesicles, are filled with neurotransmitter molecules. Each synaptic vesicle is a tiny sphere too small to see with a light microscope; they are visible only under the high-power magnification of an electron microscope. The messages are not floated across the synaptic gulf in these spherical bottles, as you might expect; instead, their contents are dumped into the gulf and diffuse across to the opposite shore. Synaptic vesicles accumulate inside the axon right next to the cell membrane at the tip.

Like cellular water balloons, one or more synaptic vesicles are smashed against the cell membrane of the axon by the force of the electrical impulse when it arrives, releasing the vesicle's contents into the cellular sea. The neurotransmitter then flows across the synaptic gulf to reach the postsynaptic neuron on the other side.

Sentinel molecules along the shore of the postsynaptic neuron are specially designed to detect the neurotransmitter substance in the synaptic gulf. These neurotransmitter receptors are large protein molecules acting as biological nanomachines. In each neurotransmitter receptor there is a passageway that can open into the dendrite of the receiving neuron when neurotransmitter is detected. When the tunnel through the receptor opens briefly, charged ions floating in solution leak out, reducing the voltage inside the postsynaptic neuron. This brief drop in voltage in the postsynaptic neuron is the receiving signal, called the postsynaptic potential. If the synaptic voltage change is big enough, the voltage drop triggers the postsynaptic neuron to fire an impulse out its own axon to signal the next neuron in the circuit. This may seem an awkward way to design a nervous system, but consider the engineering challenge facing Nature: to build a powerful, high-speed biological computer using nothing other than cells—tiny bags of saltwater.

So the nerve impulse speeds down an axon, releasing neurotransmitter when it reaches the end. The neurotransmitter flows across the synaptic gulf and activates neurotransmitter receptors on the postsynaptic neuron, causing a voltage drop in the recipient neuron that will make it fire an electric impulse down its own axon to release neurotransmitter onto dendrites of the next neuron in the circuit in relay fashion. To reduce the time it takes for neurotransmitter to diffuse across the synapse, the gulf of separation is infinitesimally narrow (25 billionths of a meter). The synaptic cleft is so narrow, in fact, it is impossible to see the separation through the most powerful light microscopes. This fact caused decades of controversy in the field of neuroscience until the electron microscope proved that every synapse in the body has a gulf of separation between the pre- and postsynaptic neurons. A message passes across the synapse in about one-tenth of an eye blink, but compared with the two hundred mile per hour speed of the neural impulse, the synapse slows information flow much like a toll booth on a turnpike.

This explains why neuronal circuits have as few connections as possible: the delay in neurotransmission at each synapse slows information flow through the circuit. This limitation is the reason that neurons are the largest cells in your body—incredibly large cells, in fact. Neurons are too fine to see with the naked eye, but some can be monstrous: three feet long in the case of the neuron from your spine to your big toe. This size is necessary because sending chemical messages across the watery synapse separating each neuron places severe engineering limitations on the circuit. For greatest efficiency, it's best to have as few synapses as possible in a circuit carrying out any function in your nervous system.

When your doctor taps his rubber mallet below your knee to test your knee-jerk reflex, you are seeing a circuit in action that controls the vital coordination crucial for you to walk. Should you stub your toe as you are walking, this misstep will jerk the tendon below your kneecap, just as your doctor does with his mallet. To avoid stumbling, you must now quickly swing your lower leg forward to catch your fall mid-stride. It is vital that this entire sensory-motor reflex is executed in a split second of time, otherwise you *will* trip and fall.

To execute this lightning-speed response, there is only one synapse in the entire circuitry controlling the vital reflex that keeps you on your toes. When nerve endings in your kneecap tendon sense a sudden tug from a stubbed toe (or doctor's mallet), they shoot impulses at two hundred miles per hour up the axon from the nerve endings into your spine. There is no time to send signals to your brain; instead, there in your spinal cord a single synapse separates this sensory neuron (bringing information about your leg motion into your spine) from a motor neuron that will fire electrical impulses down to your leg muscle to jerk your lower leg forward in a flash—one synapse separating us from falling on our face. (The messages will be relayed by other nerve circuits to your brain, but they arrive after your leg muscles have already responded, and you have no conscious control over what has happened. This is why the doctor's mallet triggering the knee-jerk reflex always delights us with surprise as we watch our leg react automatically.)

This system of neuronal islands creates a severe constraint for rapid communication, but there is a silver lining to this awkward setup: the synapse becomes a control point for directing the flow of information through neural circuits. Like switches and volume controls, these synaptic control points greatly expand the computational power and information-processing ability of the nervous system far beyond what could be provided by hardwired points of connection between neurons. By regulating the flow of information across a synapse, circuits can be strengthened or weakened, in effect allowing the circuits to change their behavior from experience—that is, *to learn*. Our memories are not bottled up inside neurons, but rather they are stored in the connections between neurons linked by synapses. With new experiences, new connections are made between neurons and others are lost. In a sense, memories are not stored *inside* matter; they are stored in the spaces between.

Synapses do much more than connect neurons; they enable flexibility of information processing. Synapses permit adjustments in functional connections based on experience. The process of learning is more finely regulated than simply making and breaking synapses: the strength of a synaptic connection can be finely tuned in a process called synaptic plasticity. How? The molecular changes that strengthen or weaken a synaptic connection are intensely studied by neuroscientists interested in memory and learning, but in principle, the mechanisms are quite simple. Either by releasing a bit more neurotransmitter from the pre-synaptic ending when an impulse arrives, or by adjusting the sensitivity of the postsynaptic neuron receiving the neurotransmitter signal, the same input to a synapse can produce greater or lesser voltage change in the postsynaptic neuron, thereby weakening or strengthening the connection.

But there is one additional crucial aspect to this process of synaptic transmission: cleanup. Communication across a synapse would fail if the synaptic gulf were not cleared of neurotransmitter quickly to permit another message to be sent. It was long understood that glia bordering the synaptic cleft carried out this cleanup operation. Protein molecules in the glial membrane pump the neurotransmitter out of the synaptic cleft and into the astrocyte—one of the four major kinds of glial cells—where it is reprocessed. After filtering out the neurotransmitter and recycling it into an inert form that cannot be confused as a signal, the astrocyte surrounding a synapse delivers the reprocessed substance back to the presynaptic nerve terminal. The neuron then carries out a simple chemical reaction to convert the inert neurotransmitter back into active neurotransmitter and repackages it into synaptic vesicles.

Astrocytes also provide the energy source for neurons, lactate—the same substance that gives yogurt its tang. The astrocytes deliver the fuel in proportion to neuronal demand.

These subservient custodial functions of glia held little interest for most neurobiologists. Yet, some now see that the dependence of the neuronal synapse on glial housekeeping could empower glial cells with the means to completely control the synapse.

If neurotransmitter is not taken up efficiently, communication across a synapse will fail because the gulf will become saturated with stale messages. If neurotransmitter is taken up too quickly, the message will appear too briefly to have full effect on the postsynaptic cell. If the energy requirements of the neuron cannot be met by the nutrients supplied by astrocytes, the neuron will run out of gas. Astrocytes are thus in a position of control.

As long as scientists probed glia with the same electrodes they used to probe neurons, they were destined to fail to discover what glia are doing in our brain and nerves. Because glia do not generate electrical impulses, understanding how glia communicate and interact with neurons required a new technique: calcium imaging. This is the technique that had revealed Schwann cells responding to electrical impulses in axons. We will explore the breakthrough calcium imaging experiments shortly, but first we need to learn a bit more about glial cells in all their variety.

CELLULAR COMPONENTS OF THE "OTHER BRAIN"

Today we know that in addition to neurons there are four main types of glia in nervous tissue. Two glial cells, Schwann cells in nerves and oligodendrocytes in the brain and spinal cord, form myelin insulation on axons. Throughout the brain and spinal cord there are also astrocytes and microglia. The microglia protect the brain from injury and disease, making them central to recovery from brain and spinal cord injury. Tantalizing clues point to the possibility that all types of glia may sense and respond to electrical activity in neurons.

Imagine the implications. Electrical activity in the brain conveys our perceptions, experiences, thoughts, and moods. Glial cells perform such diverse functions in our nervous system that a vast range of brain functions might be influenced by glia if they could sense nerve impulse activity. Everything from immune system responses to infection, to insulating axons, to wiring up and rewiring the brain, to recovery from brain disease and injury might be influenced by impulse activity acting through glia.

Glia outnumber neurons six to one, but the exact ratio differs in different parts of the nervous system. Just as the ratio of men to women is one to one on average, the exact ratio of men to women ranges widely in different areas. For example, the sex ratio may be ten men to one woman in barbershops and just the opposite in fabric stores. Along nerves or in white matter tracts in the brain, the ratio of glia to neurons can be one hundred to one, because one axon can be ensheathed by myelin-forming glial cells spaced roughly one millimeter apart along the full length of the axon. In the human frontal cortex, the ratio of astrocytes to neurons is four to one, but whales and dolphins have seven astrocytes for every neuron in their gigantic forebrains. This glia to neuron ratio is larger than seen in the frontal cortex of any other mammal. No one knows why this is the case. Whales and dolphins are highly social creatures and very intelligent. Perhaps, as with Einstein's cortex, the larger proportion of glia somehow contributes to the animal's obvious intelligence. But whales may also need more abundant glial cells to sustain their neurons in a healthy state during their long breath-holding dives to the ocean depths.

Outside the brain, in the nerves of the body, there is a different type of glial cell packed around nerve axons all along the length of every nerve. These Schwann cells were the glia that I first studied in detail in the experiment I described earlier.

Schwann Cells

My thoughts raced back to Theodore Schwann, the man whose name is given to these glia cells Beth and I had just seen light up when the axon fired. What would Theodore Schwann have thought if he could have seen what we had just seen—glia detecting neural impulses? Most people imagine that the moment of discovery explodes with the elation of reaching the summit of a mountain and peering down on the world below, or the excitement of winning a car race or a grand prize. There is certainly an element of that elation in the unique mix of emotions un-leashed by a scientific discovery, but the overwhelming feeling for me is one of gratitude and amazement. Nature has revealed for the first time in history a secret long hidden. You feel a sense of gratitude and fellow-ship with the many scientists across space and time who labored to bring you to this new insight into Nature. Scientists sharing your own curios-ity left clues written in books and journals, sensing they had found frag-ments of a larger puzzle they could not possibly piece together, hoping that someone in the future would pick them up, put them together with fresh insights, and discover the secret they sensed Nature had hidden here.

As a child Theodore Schwann displayed a gifted intellect, and as a scientist he was far ahead of his time—too far ahead. Before the mid-1800s, scientists pondered the substance of living things, wondering how life was structured at its most basic level. What made animate things so distinct from the inanimate? This was the era when chemistry had ascended from alchemy to explain the transformations of matter as the result of the fundamental properties of atoms from which all things are made. Theodore Schwann not only discovered one of the four major types of glia, he also gave us the very concept of the cell itself.[2]

Schwann was a sensitive person, devout and modest by nature. He obtained an M.D. degree at the age of twenty-four in Berlin, under the guidance of a renowned scientist, Johannes Müller. By the age of twenty-nine he had conceived the theory that every living thing was composed of cells. He defined a cell as a membrane-bound structure surrounding a nucleus. Cells, he proposed, could reproduce themselves and change from simple shapes into special forms. They could assemble into groups, form layers, or hollow out masses of cells into organs and thus build a complete body. He proposed that all plants and animals—every sub-stance of the body, from bone and sinew to skin and blood—was formed of cells, each one bounding a single nucleus.[3]

Shattering accepted theological doctrine, Schwann reasoned that the body was not infused by a mysterious force, but instead operated ac-cording to blind natural laws, just as physical laws govern inorganic matter. Even though he was a devout Catholic, he came to believe that the life force emerged from the action of the fundamental properties of natural forces in the inorganic world, combining into action to give rise to life. He imagined that living cells arose from biological substances, much like crystallization in the inorganic world, but that once formed, cells could change by the guiding forces of physical chemistry and physics. Philosophically he had stolen the vital force from the Creator and transferred it to the chemist and physicist.

His short and brilliant scientific career spanned only the years from 1834 to 1839. In that final year, he suffered a particularly vicious personal attack on his scientific work by the renowned German chemists Justus Liebig and Friedrich Wöhler, who ridiculed Schwann's ideas in a prominent scientific journal. They mocked his idea that alcoholic fermentation was the result of cells (yeast) acting on sugar, insultingly depicting the cells excreting gas from their imaginary anuses and pissing alcoholic urine from their wine-bottle-shaped bladders. The prevailing theory at the time was that sugar was transformed into alcohol through a chemical reaction involving air and the nitrogenous substances in fruit juices. In comparison to this reasonable chemistry, Schwann's theory that microscopic bugs (yeast) consumed the sugar and released carbon dioxide gas and alcohol as metabolic byproducts appeared laughable, and the most highly respected authorities in science laughed openly. Humiliated, Schwann spent the rest of his life in relative isolation, suffering episodes of depression and anxiety, denied the academic promotions and funds he needed to continue scientific research.

As if the cell theory and the discovery of glia in nerves were not enough, Schwann made other fundamental discoveries during his brief but brilliant scientific career. Suspecting that there was more to digestion than hydrochloric acid, he discovered one of the principal enzymes of digestion, pepsin, which breaks down proteins in our diet. He also proved the vital role of bile in digestion.

Before long the scientific world would see that Schwann had shown brilliant insight. In 1847 the English translation of his fundamental scientific work, *Microskopische Untersuchungen,* was heralded as "worthy to be ranked among the most important steps by which the science of physiology has ever been advanced."[4] No one was laughing now, but by this time Schwann had ceased to work in the laboratory. He became an inventor of machines for the mining industry, designing pumps for removing water from coal mines and a respiratory apparatus for rescue operations that would one day allow divers to walk on the floor of the ocean.

Today many people pay tribute to Schwann on a regular basis with small tokens of change. In the 1890s Caleb Bradham, a druggist in New Bern, North Carolina, developed an aid to digestion claimed to contain pepsin as its active ingredient. The name for this popular tonic is "Pepsi Cola."

But what are these Schwann cells that cling to the axon like flattened pearls on a string? What do they do? Where do they come from?

Schwann and other anatomists of his day used fine glass needles to carefully tease nerves apart and examine the splayed fibers under a microscope. Seen this way, a nerve was a bundle of hundreds of microscopically slender fibers—nerve axons—each one clearly a conduit for the nervous energy of sensation and motion. Each axon was dotted along its entire length with a chain of cells, looking like droplets of dew on the strand of a spider's web.

Just as a plumber fits segments of pipe together to span from sink to drain, Schwann imagined that the axon leading from each nerve cell must be formed in the fetus by a chain of tiny cells joining together to make a long cylindrical axon tube by fusion. He speculated that the cells clinging along the length of the axon must be the remnants of these fetal cells. Schwann cells, as they are now known, might simply be vestigial remnants that had no more function in the adult than our navel, a mark left by a discarded embryonic structure.

Alternatively, these glial cells might still support or nourish the axon in adult nerves, a reasonable speculation offered by other scientists considering the extreme distance between the tip of an axon and the cell body of the neuron up to a yard away. Rather than shipping nutrients from the cell body of the neuron all the way to the axon tip, Schwann cells might deliver the required supplies locally, supplying every segment of the axon from the cell body to the tip. The answer to these questions about Schwann cell glia would not come for another sixty years, but the answer to the question where these glial cells originated was provided by another pioneering scientist who was quite different from Theodore Schwann.

Mariner of the Mind—Fridtjof Nansen

If Theodore Schwann represents the sensitive, shy personality in science, Fridtjof Nansen represents the exact opposite. Nansen, who lived from 1861 to 1930, is famous for his Arctic explorations, but few people know that this Norwegian adventurer began his career as an explorer of the nervous system. From an early age he was an avid outdoorsman, and when an opportunity to join an expedition on a ship sailing north to Greenland arose, Nansen, then a zoology student from the University of Christiana in the city now known as Oslo, was eager to join. His studies of small parasitic worms from this expedition, published in 1885, remain a classic in the field.

Nansen's athleticism and fascination with the unknown led him to attempt an adventurous crossing of Greenland on skis. In 1888, a Norwegian sealer dropped off Nansen and his five-man party on the frozen sea ice off Greenland. He and his men scaled mountains, endured frigid −50 degree centigrade temperatures, and skied through horrendous fog and snowstorms to reach the west coast three months later. There they survived for the winter, living with Eskimos (Inuit) and learning their ways.[5]

Later Nansen developed a theory that there was an ocean current flowing from Siberia under the frozen Arctic Sea to Greenland. To prove it, he proposed to trap a specially designed ship in ice and ride this current to its destination. At 78 degrees 50 minutes north latitude this ship, the *Fram,* was intentionally stuck in the ice and drifted with the ice sheet northward nearly to the pole. Nansen and shipmate Hjalmar Johansen elected to leave the shelter of the *Fram* and make a dash by dogsled for the North Pole from a latitude of 85 degrees 55 minutes north on March 14, 1895. With no assured plan for reuniting, their shipmates bade them farewell on an expedition that they feared would lead their captain and comrade to a frozen, lonely death. The two explorers struggled intrepidly through the frigid Arctic alone toward their destination at the very top of the globe, but they were turned back by impassible blocks of ice only 268 miles from their goal. Still their achievement was the farthest north any known explorer had reached at the time. Through the winter they survived the barren polar isolation by killing their sled dogs one at a time to feed the remaining dog team and themselves. Alone in the frozen Arctic and sharing a single sleeping bag for warmth, the two survived the next nine months living in a small hut they built of whale bones and the skins of polar bears and walrus they shot for food. They endured by adopting the ways of the Eskimos that Nansen had learned that winter years before when he skied across Greenland.

The *Fram,* in the meantime, continued its tedious drift locked in the frozen ice, its timbers groaning constant threats of shattering under the enormous pressure. Finally their vessel crept with the glacial pace of the ice floe to near the edge of open water, where the crew dynamited the ship free of its icy encasement and sailed their leaky ship home safely, but without their captain and crewmate.

At the end of yet another bitter winter and after several near-death battles with polar bears, Nansen and Johansen were at last rescued in May after walking the ice floes and paddling the frigid waters between them in makeshift kayaks until they reached the outpost of an English expedition. After nearly three years alone in the Arctic, Nansen and Johansen finally rejoiced in reuniting with their comrades.

Nansen made many discoveries in the field of oceanography, including inventing special devices for exploring the ocean depths. The Nansen bottle, which I still used as a marine biologist in the 1980s to collect water samples, was his invention. Nansen received the Nobel Prize in 1922—not for science, but for peace. As a Norwegian delegate to the League of Nations he was awarded the prize for his humanitarian work with war refugees.

This explorer of the Arctic also crossed paths with famous explorers of the nervous system. In 1888 Nansen received a Ph.D. degree for his studies of the nervous system, and he then traveled from Oslo to Italy, where he visited Camillo Golgi's laboratory in Pavia. There he learned the silver impregnation staining technique that revealed nerve cells so clearly. Nansen went on to describe both nerve and glial cells of the hagfish, a grotesque, mucus-covered eel-like fish with a sucking, jawless mouth. This fish is important biologically because it represents the most ancient of all living species of fish-like creature, and its ancestors are the forerunners of all animals with backbones (amphibians, reptiles, birds, and mammals).

Nansen the explorer set out to track the route of sensation into the central nervous system. Studying this most primitive vertebrate and comparing the anatomy of its nervous system with that of other animals, Nansen made the fundamental discovery that all nerve fibers upon entering the spinal cord immediately divide into two branches. One runs up the spinal cord toward the brain, and the other branch runs down to the tail. This structure—governing the route of entry of all sensation into the nervous system—holds true for all animals, from fish to humans.

What is even more remarkable than Nansen's sharp perception was what he could not see. Nansen simply could not see what his renowned mentor, Camillo Golgi, endeavored to show him: the connections fusing nerve cells into a network. To Nansen, each nerve cell stained by the Golgi method appeared like an isolated island. Just as he had nearly reached the North Pole, Nansen came closer than anyone before to reaching a true understanding of the neuron as an individual cell. Had he trusted his own observation that nerve cells were not fused into a network as his mentor insisted, he might have formulated the neuron doctrine before Ramón y Cajal, who began his own studies in the field the year Nansen received his Ph.D.

The prevailing view in Nansen's day was that Schwann cells were akin to the connective tissue that binds all cells together. It was believed, therefore, that Schwann cells derived from embryological cells that generate connective tissue rather than from the specialized embryonic cells that ultimately form nervous tissue. This presumed lowly pedigree was one of the reasons Schwann cells were so easily dismissed by neuroscientists as cellular glue. But from his experimental explorations of Schwann cells, Nansen concluded that the prevailing wisdom was false and that Schwann cells derived from the same regal cellular parentage that gave birth to neurons. The famous embryologist Ross G. Harrison later proved Nansen's theory for the origin of Schwann cells by removing the primordial tissue from frog embryos that gives rise to neurons. When he did, there were also no Schwann cells after the embryo matured.

The prevailing wisdom prevented researchers from asking the obvious question: if these glial cells were formed from the same cellular line as neurons, what might that mean in terms of the functions that these cells might be capable of performing? Nansen, however, after carefully exploring the nervous systems of animals, observed in 1886 that glia might be "the seat of intelligence, as [their number] increase in size from the lower to the higher forms of animals."[6] This law of increasing ratio of glia to neurons ascending the ladder from lower to higher vertebrates holds today. Nansen, the pioneer, may have glimpsed "the other brain" a century before nearly everyone else.

Schwann cells coat the nerve fibers all along their length right up to the point where the nerve enters the spinal cord or brain, but they do not cross that threshold. Why the Schwann cells stop there was a major puzzle. If Schwann cells are so important, why are there none inside the brain and spinal cord?

In addition, anatomists recognized immediately that not all axons in nerves are coated with Schwann cells. These cells entwine around only large-diameter axons. On a large-diameter axon, hundreds of Schwann cells are attached in a series all along its full length like pearls on a necklace, but thin axons lack even a single pearl. If these cells are absent from the brain and missing from many axons in nerves, how important could they be?

Today we know that Schwann cells in our nerves actually exist in three different forms: nonmyelinating and terminal, as well as the myelinating Schwann cells just described. It is evidence of the general neglect of glia that all three go by a single name. Such would never have been the case for neurons. Each type of Schwann cell has an entirely different structure and unique function.

Small-diameter axons are not studded with Schwann cell "pearls," yet they are not naked. These tiny axons are cabled together by huge globular cells grasping bunches of slender axons like a fistful of spaghetti. The anatomists called these fist-like cells "nonmyelinating" Schwann cells, to distinguish them from the pearl-type "myelinating" Schwann cell. These protective nonmyelinating Schwann cells assure that none of the most fragile slender axons in nerves are ever left bare. The nonmyelinating Schwann cells also undermine the clever idea that axons are formed in embryonic development by connecting together Schwann cells to form the axon tube, because one nonmyelinating Schwann cell engulfs a dozen or more small-diameter axons inside itself. Some pioneering neuroscientists suspected that these glial cells in our nerves must have a hidden function, but what the function might be was unclear.

When an axon reaches its target—for example, the synapse onto a muscle fiber that will make the muscle twitch—the entire tip of the axon is completely engulfed by another glial cell that seals off the nerve junction like shrink wrap. This cell is called the "terminal" Schwann cell or "perisynaptic" Schwann cell ("perisynaptic" meaning "surrounding the synapse"). Until recently, this was essentially the function most scientists presumed it served: sealing off the nerve ending. In recent years, that naive view has crumbled with the discovery that these terminal Schwann cells can sense and control information flow from nerve to muscle.

For now we should understand that Schwann cells come in three basic types: (1) myelinating, (2) nonmyelinating, and (3) terminal. Although these cells look completely different, they are all called Schwann cells simply because early anatomists recognized that none of them was a type of nerve cell. As will become apparent, each of these Schwann cells performs entirely different functions, and our nerves will fail to work properly if any one of them is defective. This static picture belies the dynamic nature of Schwann cells: they react with rapid changes in their structure and undergo cell division in response to nerve injury. Schwann cells must perform all the functions of the various specialized glia found in the central nervous system (CNS).

Schwann cells were ignored for decades because there was no reason to imagine that they could have any function in information flow through our nerves, but the mystery of what I had just seen was before me on the computer screen: Schwann cells all along the axon in our experiment had somehow detected impulses flowing through the nerve fiber. How were the Schwann cells picking up the signals from electrical impulses in the axons? An even more intriguing question was, why would Schwann cells all along the axon need to tap into the information flowing through the nerve cell? And what would they do with the information they gleaned? These questions lay ahead of us as I flipped off the switch and watched the Schwann cell lights dim slowly, returning the screen to the shadowy darkness of silent neurons.

OLIGODENDROCYTES: OCTOPUS'S GARDEN

How important can Schwann cells be when there are no comparable glial cells inside our brain or spinal cord? The axons that pierce the brain leave their glial partners behind as they weave through neural networks in our central nervous system. The early anatomists looked closely for cells resembling Schwann cells inside the brain and spinal cord, but without success. Ultimately, however, the search led to the discovery of oligodendrocytes. These were the last glial cells discovered, and these odd brain cells were a great puzzle to anatomists. Like astrocytes, these glial cells are found only inside the brain and spinal cord, never in the nerves of our body. When the mystery of oligodendrocytes was finally solved, the most widely appreciated and intricate form of neuron-glia interaction was revealed—an elegant partnership between axon and glia that is absolutely essential for high-speed impulse conduction. This is myelin.

The name "oligodendrocyte" means "stubby dendrites" or "short branches." Anatomists could recognize these cells by their small cell body and several short branches radiating out like a cocklebur. They floated freely throughout the cellular terrain of the brain, unattached to neurons or any other cellular structure, an isolation that left no clues to their possible function. Ramón y Cajal left these glia to his student Pío del Río-Hortega to ponder while he explored neurons with furious passion.

Oligodendrocytes are seen almost everywhere in the brain, but they are especially numerous in white matter tracts. White matter streaks through the core of the brain of animals with backbones (fish, amphibians, reptiles, birds, mammals, and humans). This white matter consists of the information trunk lines formed by thousands of axons bundled together to carry information between distant parts of the brain. Under a microscope, anatomists could easily see why the trunk lines were sparkling white. Each axon was coated with a substance that reflected light brilliantly. In the focused beams of the light microscope, an axon looks like the branch of a tree encased in a crystalline sheath of ice deposited in a winter storm.

Simple tests showed that the white coating was a fatty substance, because oily dyes would stain it, but water-soluble dyes were shed like water off oilskin. Strangely, the axon was never coated evenly, but instead a series of droplets of the oily sheath punctuated the entire length of the axon, leaving minute bare spots of axon between each droplet. Was this the natural structure of the axon sheathing, or had the fragile coating been damaged by scientists teasing the bundled fibers apart with fine glass needles to see the individual axons under the microscope?

This droplet sheathing on brain axons resembled the string of flat-tened pearls formed by Schwann cells on large axons in nerves, but the sheathing droplets lacked a nucleus, so this oily coating was apparently not made up of cells at all. The mysterious cocklebur oligodendrocytes drifted freely like flotsam throughout the brain. How was this oily mate-rial deposited on the axon? Scientists could watch the process of axons becoming coated during development, because this action begins in the late stages of fetal life and the coating continues to accumulate in young animals after they are born or hatched. It was impossible with the mi-croscopes of the time to tell whether the coating was deposited outside the axon or stuffed just under its cell membrane. If it was on the outside, some other cell might have plastered it there, but what cell?

The early anatomists were acutely aware of the limitations of their staining methods. As the Golgi method had shown, a new stain could reveal a completely new cell, previously invisible to the world of science. Río-Hortega continued to tinker with the Golgi silver staining method, using different combinations of metal salts and chemical treatments on brain tissue. Suddenly everything made sense! One of the staining vari-ations he tried illuminated the true structure of oligodendrocytes. As if teased by Nature, Río-Hortega saw that the name "oligodendrocyte" would become ironic. The supposedly stubby cell branches (or "pro-cesses") described by the prefix *oligo-* did not stop where the old weak stains had failed. This new stain showed that the cellular processes ex-tended great distances, and each slender process ended like a long ten-tacle of an octopus wrapping around an axon. The cocklebur had morphed into a monster octopus with dozens of long tentacles, each one grasping a different axon.

The fatty droplets on axons were now understood to be the places where these cellular octopuses gripped the axon with several wraps of a tentacle. Countless tendrils from many oligodendrocytes form these droplets, sheathing the full length of each axon, like the stacked hands of children grasping the length of a bat to determine which team bats first. This discovery explained why there were tiny bare spots on the axon between each droplet; they were the gaps between adjacent tenta-cles wrapping the axon. This discovery also explained why there was no cell nucleus inside the myelin sheath on axons in the brain and spinal cord. In nerves of the trunk and limbs, myelin sheath was formed by chains of individual Schwann cells clinging to the axon, like railroad cars on a train track, but in the central nervous system (brain and spinal cord), the myelin sheath was formed by the scores of tentacles extending from oligodendrocytes, while their cell body and nucleus remained far away from the axon.

Now it was obvious that this glial grip on nerve axons must be important: axons of both nerves and brain are encased by glia in a myelin sheath in a most intricate and complex manner. In people with multiple sclerosis and other demyelinating disorders, destruction of the myelin sheath allows electricity to leak out, blocking the flow of impulses along axons in the central nervous system. This causes paralysis, loss of vision, and impairment of other bodily functions. Without these octopus glia, the nervous impulse will fail and the brain circuits become powerless.

Today there is no question that oligodendrocytes make the myelin sheath on axons in the brain. Research in my lab has revealed that these glial cells in the brain are also "listening" to axonal firing, just as Schwann cell glia "listen" in nerves. It now seems possible that neuron-glia communication may take place between every type of glial cell in our brain and nerves, including between axons and oligodendrocytes. What would it mean to our current understanding of the mind and medicine if information flowed not only through neural circuits, but through glia as well?

ASTROCYTES:
CELLULAR STARS AND MYSTERIOUS SICKNESS

Autumn 1947 in London was developing into one of the driest on record. On an unseasonably warm October 30 morning, a fifteen-month-old boy was admitted to the London Hospital screaming. Soon after his birth the boy's parents faced the heartbreak of realizing that their child was not developing normally, and they coped with the struggle of caring for a baby who was always ill—and now, as was obvious to everyone, severely mentally retarded. At the age of seven months, the child's head started growing enormously. In the past month the parents had watched their baby's head balloon another inch in circumference. Now the baby boy's forehead bulged grotesquely, and he suffered fretful sleepless nights, screaming inconsolably night and day, wracked with attacks of copious vomiting. After admittance to the hospital, the child's condition had deteriorated to the point where he could not sit up or even raise his bulbous head.

The baby had a high fever, and Dr. Stewart Alexander treated him for this, but penicillin and sulphamethazine failed to lower the fever. Alexander could see that despite the high fever, the boy exhibited none of the telltale signs of infection. The baby's fever raged for three more weeks, but none of the modern equipment and resources of medicine were of any use, because the real cause of the child's suffering was not to be found on the pages of any medical book. After several days in the hospital, seizures started on the right side of the child's body. His right eye, arm, leg, and mouth twitched convulsively, and when the convulsions released him, he would slip into a five-minute coma. The tiny body became wasted and dehydrated from the diarrhea and vomiting, and finally on November 19, 1947, the fever rose above 106 degrees and he died.

Alexander could see plainly that the cause of this child's fever and hydrocephalus (enlargement of the head due to fluid pressure buildup in the brain) was not the result of an infection. Neither of the child's parents had any history that would suggest a genetic cause. An autopsy was ordered. Alexander found a severe degeneration of the child's brain tissue, unlike anything ever reported in the medical literature.

There was one prominent and especially curious finding: the support cells filling the space between neurons—the astrocytes—were packed full of foreign bodies looking like aggregates of tiny rods. The most similar microscopic deposit known to brain research was the fibrillary tangles seen inside neurons in Alzheimer's disease. But that is a disease of old age. As far as the doctor could discern from different chemical stains applied to the microscopic brain section, these foreign bodies were different in structure and chemical composition from those in Alzheimer's. More importantly, there was no sign of the curious rods infecting any neurons, which is where the neurofibrillary tangles are found in Alzheimer's disease.

The white matter in the frontal lobes of the boy's brain was also severely affected. There were far fewer oligodendrocytes than normal, and the myelin was threadbare or missing. Although the neurons were not affected by the rod-like foreign bodies, there was severe degeneration of the cerebral cortex. Alexander concluded that the primary cause of this disease, which had produced such severe mental retardation, lack of myelin, neuron degeneration, and death, was a disorder in the astrocytes. This disease now bears the name of the doctor who first described it.

Alexander disease shows tragically and dramatically that astrocytes do more than act as passive padding or connective tissue encasing the neurons. What was it about these sick astrocytes that could dissolve a young child's brain?

A neurochemist, Lawrence Eng, isolated a protein from brain tissue taken from adult patients afflicted with multiple sclerosis, a wasting disease that causes paralysis, blindness, and other sensory-motor and cognitive disabilities. Multiple sclerosis is caused by defective electrical communication between neurons that results when the body's immune system attacks the myelin sheath, stripping the electrical insulation from axons.

After making an antibody to recognize the purified protein he had extracted from the diseased brain of multiple sclerosis patients, Eng was able to use it to stain microscope sections of normal and diseased brain.

When Eng treated normal brain tissue with the green fluorescent-labeled antibody, he saw the bright green light, identifying the protein, glowing inside many brain cells. Curiously, the protein was present throughout the brain but in only one type of brain cell. That cell was not a neuron. The protein Eng had purified was filamentous and tightly bundled inside astrocytes and nowhere else. It was quite abundant in astrocytes in normal brain tissue, but in the damaged areas in the brains of patients with multiple sclerosis and other mental and neurological illnesses, it was markedly increased in intensity and abundant in many more astrocytes. He called this astrocyte protein glial fibrillary acidic protein (GFAP). Today we know that this protein is part of the cellular skeleton of astrocytes, but why it increases in astrocytes as a result of so many different brain stresses and disorders, including multiple sclerosis, is unknown. What the GFAP protein does for the astrocyte's cellular skeleton and how increasing the amount of it may change an astrocyte's function are still mysteries.

Years later, Michael Brenner, a neurobiologist now at the University of Alabama at Birmingham, used molecular biological methods to explore what GFAP did in astrocytes. He discovered that too much GFAP caused the astrocytes to become choked with Rosenthal fibers—the same rod-shaped filaments that Alexander had seen in his young patient in 1947 and that are now recognized as the hallmark of Alexander disease. Moreover, mice with excess GFAP in their astrocytes developed all the symptoms of Alexander disease and soon died. This series of discoveries proved that mental retardation was not exclusively a disorder of neurons. It could be caused by a defect in astrocytes.

Despite this powerful evidence of a connection between astrocytes and mental retardation in Alexander disease, it was not until many years later that scientists began to realize that an obvious question had been overlooked. If astrocytes were connected with the deteriorating mental function seen in Alexander disease, might not they also have a role in normal mental function in a healthy brain? It seemed clear that the survival of neurons, the formation of the myelin sheath, and the normal integrity of brain tissue and brain structure were all somehow supported by these tiny microscopic protein filaments inside astrocytes.

THE MAGIC OF MYELIN

There is a great divide in the natural world between animals like dogs and dinosaurs, fish and birds, porpoises and people, and other creatures like starfish and slugs, butterflies and bees, roaches and snails. The vast difference in intelligence and speed of mental computation between animals with backbones (vertebrates) and those without (invertebrates) is due to a fundamental difference in nervous systems of all animal life. This essential difference has little to do with neurons.

The invertebrate nervous system is to the vertebrate's what a schoolchild's calculator is to a NASA supercomputer. But astonishingly, the neurons in a fly's brain operate exactly the same way as neurons in your brain, often right down to using the same neurotransmitter chemicals. The essential difference between these two great groups of animals is in their glia: specifically the glia that form myelin (oligodendrocytes and Schwann cells). Vertebrates have myelin-forming glia; invertebrates do not. The evolution of this amazing glial contribution to nervous system function could not have had more profound consequences.

Vertebrates have a far more complex nervous system than invertebrates. The vertebrate nervous system is also centralized, that is, concentrated into a brain and spinal cord. In lower animals like crabs or slugs, neurons are bunched together like grapes wherever they are needed. There are clusters of neurons at each segment of the articulated tail of a lobster and knots of neurons near the mouth parts of slugs to operate structures for feeding, for example. But in animals with backbones, the brain is concentrated into one massive supercomputer encased inside a thick armor of bone. The backbones of vertebrate animals protect their vital spinal cord. This cerebral concentration of brain power and complexity could not have occurred without this fundamental difference in glia separating vertebrates from invertebrates. Glia—not neurons—are responsible for this biological revolution.

Myelinating glia wrap dozens of layers of membrane around axons, insulating each one like electrical tape wrapped around a wire. Without this insulation, invertebrates must cope with slow communication lines that leak electricity. This signal loss limits how far the electrical impulse can reach down the invertebrate axon before it fades away, like water pressure dropping through a leaky hose the farther it flows from the spigot.

Some invertebrates, such as squid, which can move quickly, have developed an ingenious method to overcome the absence of myelin. Through the course of evolution, squid and some other invertebrates have greatly enlarged the diameter of critical axons that are essential for the life-saving reflexes needed to escape predators. The principle is simple: you can get more water through a fire hose than a garden hose, even if both are leaky. The giant axons in squid are so enormous that they can be seen with the naked eye as you clean squid in preparing calamari for dinner. They look like damp cotton strings about a millimeter in diameter, stuck to the underside of the squid mantle, which is the fleshy part of the squid. The mantle is designed to squeeze quickly, like the rubber bulb of a turkey baster. The spurt of water squirting out a small opening propels the squid suddenly to escape a predator. The giant axons offer less resistance to electric current, enabling more rapid flow of nerve impulses to trigger this quick escape from a predator's jaws.

This brute-force solution comes with a serious disadvantage, however. Consider that there are thousands of myelinated axons connecting the light-sensing retina in your eyes to your brain, and each axon is so slender that it is invisible. (Myelinated axons in the human brain are as small as one-thousandth of a millimeter in diameter.) If the axons from your eyes were unmyelinated and instead increased in diameter like the squid's to achieve comparable conduction speed, each of your optic nerves carrying information from your eyes to your brain would grow to three feet in diameter. Trying to extend this inelegant engineering solution to the enormous complexity of the vertebrate brain would be as impossible for practical reasons as trying to build a supercomputer with vacuum tubes instead of integrated circuits. In engineering better electronic brains, as in evolving better biological ones, the solution is the same: miniaturization. The insulating glia allow this miniaturization by in effect plugging the electrical leaks in the axon hose with multiple wraps of myelin. This allows extremely slender axons to transmit information quickly and over long distances in the vertebrate brain and nerves of its body.

How this biological revolution came about is a great mystery, because myelin leaves no fossil record. The most primitive living animals in which myelin appears are sharks and ancient fish known as chimaeras. More primitive eel-like fish that have sucking mouths like leeches but no jaws (lampreys and hagfish) have no myelin. There are no animals showing any obvious intermediate steps, such as axons that are partly myelinated or poorly myelinated. We do not know what cells the first myelinating glia evolved from or whether they evolved first in the brain or in the body. This is especially curious because, as discussed, myelin is formed by entirely different types of cells in the brain (oligodendrocytes) than in the rest of the body (Schwann cells). If there is a living myelin missing link, no one has yet found it. When myelin appears in these primitive fish, it is fully formed in both the brain and the body. Moreover, the structure of this myelin in these ancient species is in every way as elaborate as the myelin in our bodies, right down to the electron-microscopic level. Considering what an exquisite and complex interaction between axon and glia myelin represents, this observation is even more astonishing. How did such an intricate cellular choreography necessary for glia to wrap layers of its cell membrane around axons come about? This cellular dance involves cellular recognition, motility, and synthesis of huge amounts of cell membrane lipids and proteins compressed into thin layers as seen nowhere else in any other cell.

Not only is this neuron-glia interrelationship one of the most complex interactions between two cells anywhere in the body, the process of myelination is also regulated with extreme precision. Glia wrap myelin only around axons, never around dendrites, cell bodies, or other cells in the brain such as blood vessels. No one knows how they do this. Not all axons are myelinated, only those that must carry information at high speeds and over long distances. How do oligodendrocytes and Schwann cells know which axons to myelinate? How do they know when in development to start doing it?

It is hard to understand how an evolutionary process of gradual change from random variation could explain myelination. Of what advantage is a partly myelinated axon? It would be as useful as half a wing. Mysteriously, myelination seems to have occurred suddenly and simultaneously in the central nervous system and elsewhere in the body. How this sudden revolution occurred cannot be answered at present. But this glial revolution was the breakthrough that elevated life on Earth beyond the level of slugs and worms.

When you see an egret take flight, soaring with grace from a marshy shore, or a stallion galloping in an open field, you are seeing what glia have enabled vertebrates to accomplish: swiftness and grace of motion. In ourselves, we can see the importance of myelinating glia in the growing ability of a baby to hold up his head and in the severe debilitation in people suffering from multiple sclerosis. With this increased speed and complexity of information processing provided by myelinating glia came elevation of behavior beyond the simple reflexes of invertebrates to what we call intellect and wonder in human beings. We owe this to glia.

MICROGLIA: THE MIND'S MILITARY

The fourth major type of glial cell is the microglia. These glial cells were actually discovered before oligodendrocytes, but they were mistakenly believed to be cells infiltrating the brain from the bloodstream.

The central nervous system reaches every part of the body, but like a king, it is guarded and completely isolated. Encased entirely inside a thick shield of bone—the brain in its skull and the spinal cord inside articulating armor of vertebrae—our central nervous system is isolated even from the blood and fluids bathing all other tissues in the body. This barrier between brain and blood is called, appropriately, the blood-brain barrier. The cells forming the walls of blood vessels in the central nervous system are sealed together so tightly that cells and molecules in the bloodstream, which freely pass into tissue elsewhere in the body, are unable to cross into brain tissue.

The brain's isolation behind the blood-brain barrier deprives it of access to the vital immune system that fights infection and disease, for the immune cells circulating in our blood and lymph do not penetrate the tightly sealed walls of blood vessels in the brain and spinal cord. How then does the brain resist attack by microorganisms and toxins?

The answer is that the brain has its own exclusive guard, a special class of glial cells called microglia, the smallest and most dynamic of all glia. Each microglia can transform from a latent multibranched solitary cell into a highly mobile amoeboid cell when it detects the danger of infection or injury. Squeezing between tangles of dendrites and axons as they rush to kill the invader, microglia attack and devour any harmful organism. These cells are no doubt tunneling through your brain at this very moment like tiny worms through fertile garden soil. Their mission accomplished, they transform back again into stationary multibranched cells, camouflaged like the apple-throwing guard trees in *The Wizard of Oz*, looking like just another part of the landscape.

These microglia, "microglue cells," constitute 5 to 20 percent of the entire glial population in the brain. This means that there is nearly one microglial cell for every neuron. Each neuron has, in effect, its own private bodyguard. Some of these cells wrap themselves around a particular neuron, protecting it like a Secret Service agent shielding the president from a bullet. So stealthy in their disguise are these quick-change artists that fifteen years ago, scientists were still debating whether they existed. Dr. Alois Alzheimer, who described the degenerative disease that now bears his name, encountered these cells in his studies of diseased brains. He studied them with intense interest because they accumulated in large numbers around the senile plaques in brain tissue that are the hallmark of Alzheimer's disease. He and most other authorities at the time speculated that these cells were simply invaders from the bloodstream venturing into neural tissue in a diseased state. (He was partially correct, because after damage to the brain's blood-brain barrier, certain immune system cells in the blood can enter the damaged brain tissue and transform into microglia.)

From the beginning, anatomists supposed that microglia must derive from an embryonic stock that is different from the cells generating neural tissue. Ramón y Cajal agreed, calling these cells the "third element," distinct from both neurons and the other glia he recognized. Today we know that microglia are indeed derived from a different embryonic stock than that giving rise to neurons and other glia. Microglia originate from the same embryonic line that gives rise to other immune system cells in the body. Microglia do not invade the brain, they grow up with it. The formative type of microglial cells enter the part of the embryo that will become the brain at a very early stage in development. They continue their development along with the brain in fetal and early postnatal life. The resident microglia in the brain do not originate from the blood, because microglia arrive in brain tissue before the blood vessels even form in the embryonic brain. Thus microglia are not the same as the blood-borne immune cells, the white blood cell monocytes that protect the rest of the body, but they are distantly related.

At about the time of birth many of these microglial cells assemble at specific parts of the baby's brain, and they begin to build their ranks by rapid cell division. Río-Hortega called these regions of the brain "fountains of microglia." Over the next few weeks these cells fan out all over the brain to take up positions on the cerebral cortex, transforming there into their highly branched disguise, lying in wait for invaders. These cellular soldiers are equipped with a powerful ability to resupply their ranks by rapid cell division, something neurons cannot do.

Microglia are everywhere in the brain, but they fan out to guard individual territories in nonoverlapping domains. They do not contact one another physically or tangle their cellular branches together, although they remain in close biochemical communication with their comrades. As if to camouflage themselves, they adapt the physical arrangement of their cellular branches to the cellular terrain. In grey matter they are bushy, radiating their cell extensions symmetrically in every direction, constantly moving their cellular fingers over the dendrites and synaptic connections between neurons looking for signs of injury or disease. In white matter fiber tracts they align their cellular fingers along and perpendicular to the axons in a protective grid.

Microglia will track down and pounce on a bacterium, virus, or cellular debris and devour it, but they also attack using chemical weapons. Some of the chemical agents they release—for example the excitatory neurotransmitter glutamate, cytokines, reactive oxygen, and nitrogen species—are particularly harmful to neurons in high concentrations. Like all defending armies and soldiers, microglia are both saviors and potential enemies. Collateral damage caused by microglia is the source of many neurological disorders. They also carry out mercy missions, bringing aid to neurons by dispensing neuroprotective chemicals to injured nerve cells.

The many branches of these bushy cells are encrusted with an array of cell sensors always on the lookout for signals of danger and disease. They have receptors for immunological recognition molecules, beacons of self and nonself that identify foreign cells invading the brain. They also have certain sensors like those on neurons (neurotransmitter receptors and ion channels) that allow microglia not only to detect invading cells and toxic conditions, but also to monitor neuronal function and remain alert to possible neuronal distress.

Considering their armament of toxic weapons, their array of sensors that can respond to disease and monitor neuronal states, and their ability to secrete healing proteins that repair neurons, microglia deserve closer attention. As we've seen, microglia are equipped with powerful enzymes that enable them to cut through the matrix of proteins that bind cells into a tissue as they rush to attack an invading organism. There is evidence that they can also apply these weapons to strip synaptic connections from neurons—not only in disease, but also in rewiring circuits in learning. Microglia, it appears, may be able to unplug the connections between neurons.

How are microglia involved in maintaining a healthy brain? How do they damage it? Do they work alone or ally with astrocytes or oligodendrocytes or both? Do microglia decline as we age? Can drugs target microglia to help them cure disease? Finally, can microglia affect mental abilities such as cognition and memory? These and other questions of glial involvement in health and disease will be considered in the next part of this book.

But for now we will return to the glial cells in Ramón y Cajal's other notebooks, the astrocytes.

ASTROCYTES: THE SOURCE OF BRAIN POWER

The first clues that there might be another, nonelectrical part of the brain interacting with neurons came from studies of astrocytes. Remember that astrocytes were the brain cells that perplexed Ramón y Cajal and were the first major glial cell to be discovered. The most obvious things about them are their enormous variety and that they have neither axon nor dendrites. Astrocytes are two to ten times more plentiful than neurons, depending on which region of the brain is examined. Many investigators now think that there must be at least as many different types of astrocytes in our brain as there are different types of neurons. Yet in our ignorance, these cells are all given the same general name, with a few notable exceptions.

Astrocytes are everywhere in the brain and spinal cord, but they are not present in the nerves of the peripheral nervous system. They are found in the optic nerve because the eye forms during embryonic development as a swelling growing out from the brain, and it is in fact part of the brain.

Astrocytes support neurons in several ways. They provide a physical matrix for structural support, they deliver energy to neurons and remove their waste products, and they react to brain injury by forming scars. Like all living cells, astrocytes have an electrical voltage, but they do not fire nerve impulses. However, their constant battery-like voltage can strengthen or weaken slowly in some interesting circumstances.

Potassium Rechargeable Battery

The brain is an electrical device, but where does the electricity in a nerve cell come from? Every neuron has its own cellular battery. All of the electrical currents in our brain are carried by charged ions dissolved in the salty fluids in our body, not by free electrons as in a wire. The battery that powers the nervous impulse is fueled by these charged ions. Like all batteries, a neuron develops its voltage simply because there is a net imbalance in the number of positive and negative charges on opposite sides of a barrier separating two poles. Positive and negative charges attract, so any time there is an imbalance in distribution of charges, the ions will move toward one another to neutralize the imbalance. This flow of charges is an electric current.

The cell membrane is the barrier between battery poles in a neuron, separating the inside from the outside of the cell. Inside the nerve cell there is an excess of negative charges, giving the nerve cell a voltage of −0.1 volt. If this imbalance of ions across the neuronal membrane ever depletes to zero, the neuron battery is dead and it will be electrically silent, unable to fire an electrical impulse. This is where astrocytes come into play, for they are vital in maintaining the proper balance of ions in the space between cells in our brain. By controlling these charged ions outside the neuron, glia recharge the battery and help control the power source for neurons.

Like sponges, astrocytes absorb discarded potassium ions from the space around neurons, sucking them into their own cytoplasm. Potassium ions are released by neurons when they fire an electrical impulse. Accumulating these excess positive charges inside astrocytes does not create a problem for glial function, because glia do not communicate by firing electrical impulses. Removing the excess potassium is essential for recharging neurons.

How do astrocytes collect and dispose of the excess potassium ions? Astrocytes are connected to one another in a vast multicellular network through protein channels called gap junctions. Gap junctions not only couple astrocytes together like snaps on a jacket, they allow potassium to flow freely through the channels and between adjacent glial cells. These gap-junction connections between astrocytes allow them to siphon off potassium from around a neuron that is actively dumping potassium ions as it fires impulses, dispersing the excess positive ions into the network of astrocytes. The community of glial cells, coupled by gap junctions, works cooperatively to maintain the proper potassium ion concentration outside neurons. To dispose of the excess potassium, specialized astrocytes have structures called end feet. These cellular extensions grasp small blood vessels like the clinging feet of a bat. Through these end feet, astrocytes dump the accumulated potassium into the bloodstream, as if it were ridding the brain of the waste generated by neuronal activity.

The consequences of glia failing to maintain potassium ions at the proper concentration outside neurons are obvious. During high states of neuronal activity—the extreme being a brain seizure—the potassium concentration builds up around neurons quickly, and its removal by astrocytes is crucial. Without astrocytes sopping up potassium ions, the brain will run out of electrical power. When it is unable to fully recharge its neuronal batteries, the brain waves go flat. In comparison to normal brain waves, these flattened brain waves, called spreading depression, are much like the dim flash of a camera strobe, too feeble to work properly. These depressed waves are seen in EEG recordings accompanying many pathological conditions.

For the brain to function properly, astrocytes must tend to the various and constantly changing demands of neurons. Doesn't this imply that astrocytes somehow monitor neuronal operations? Doesn't the vast intercellular network of astrocytes coupled together through gap junctions imply the possibility of another cellular network of communication in the brain?

CHAPTER 3

Transmissions from the Other Brain: Glia Know and Control Your Mind

In 1966, Richard Orkand, John Nicholls, and Stephen Kuffler, pioneering electrophysiologists (neuroscientists who study the electrical signals in neurons), slowly drove an electrode through the optic nerve of an anesthetized fish and watched their electronic instruments, designed to detect voltages generated by neurons, swing suddenly from zero volts to −100 millivolts, and then back to zero again. As they pushed the electrode deeper into the nerve the same thing happened again, like a divining rod mysteriously dipping toward unseen water deep in the ground. This response occurred repeatedly all the way through the optic nerve and back out again as they withdrew their electrode.

The scientists were puzzled, because they knew that the slender axons in the optic nerve were much too small for their electrode to penetrate, so the voltages they were recording could not come from optic nerve axons. The only possible conclusion was that they were penetrating the optic nerve glia, astrocytes more specifically, that lie between the cables of axons. Could the astrocytes somehow be participating in impulse conduction through nerve axons? Using their electronic instruments, the scientists began to explore the question in depth, providing the first evidence that this other side of the brain might do quite a bit more than glue neurons together.

They shined a light on the fish's retina and the voltage dropped suddenly inside the astrocytes. Glia had responded to the light! How? After considerable experimentation they determined that these glia in the optic nerve were responding to positively charged potassium ions released from nerve axons when light on the retina stimulated the axons to fire electrical impulses to the brain. Recall that positively charged potassium ions are always released by axons when they fire nerve impulses. The researchers surmised that when astrocytes sucked up the potassium as part of their cleanup function, the positive charges they accumulated weakened the negative voltage that astrocytes carry at rest. Light stimulating the retina had made the optic nerve axons fire, releasing potassium ions, which the astrocytes absorbed, making the voltage inside the astrocyte more positive.

This was the first clue that glial cells could respond to impulse activity, but it was dismissed as little more than a curiosity. The voltage drop in astrocytes was simply a consequence of the well-understood cleanup role of glia in keeping the cellular environment pristine for neurons. The impulse impacted astrocytes in this way, but there was no foreseeable consequence from the interaction. Decades later, new methods of imaging chemical changes inside living cells are toppling that conclusion, by revealing that astrocytes use a nonelectrical means of communication in addition to slow voltage changes. Electrodes, designed to detect the electrical impulses of neurons, were incapable of sensing this entirely different chemical form of communication used by this part of the brain. Using new tools neuroscientists now see that astrocytes not only respond to visual stimulation, they participate in vision by controlling neurons.

VACUUM TUBES AND ELECTRON BEAMS: A NEW FRONTIER

The brain is often likened to an electronic device. Neurons are like integrated circuits, synapses are transistors, and axons are wires, but what are glia? The metaphor of glia as solder no longer works; some glia, like Schwann cells, act as insulation, but could other glia function analogously to different types of electronic components? Relays? Batteries? Parallel processors? Glia may fracture the worn-out electronic metaphor of the brain.

When I was a boy the innards of a television set were a cityscape of glass tubes glowing warm with life. When something went wrong with the picture, which was a sporadic but expected occurrence, you looked inside the box and felt for the cold glass coffin of the tube that had died. Every supermarket had a kiosk in the front of the store with a maze of receptacles for you to match to the particular pattern of plug pins on the bottom of the suspect vacuum tube. Planted now on the stage, the tube awaited the verdict you unleashed with the push of a black button. The needle of a meter swung smartly into the green wedge "good," or fell sullenly into the red zone of death: "bad." If the tube was bad, you opened the cabinet beneath the test table arena and selected the proper replacement.

After you returned home, you inserted the fresh tube into the empty socket inside the cavernous mahogany chassis of the ailing TV. You flipped the switch powering a warm orange glow of life into the happy new tube as it hummed in concert with the others, and through the snowy rolling blizzard and grating static burst the clear voice and moving image of Huntley and Brinkley sharing the day's news.

Today the explosive advance in the science of electronics has showered society with an endless and wondrous variety of televisions and countless other consumer devices, but no one who buys them really knows how they work. This is deeply unsettling. In contrast to the early days of electronics, inside the plastic box of today's television sets we can see small black wafers stuck to shiny mint-green cards dotted with soldered pimples and pits, the surface embossed with a maze of fine copper lines resembling a map of the Tokyo subway system. In sickness or in health, all is cold and dark. The innards are lifeless and inanimate. There is no clue to how the device works, what might be wrong, or how to fix it when it breaks. Peering inside a broken TV today, I feel helpless. No one can fix this. The very idea that one might go into the corner supermarket to diagnose and purchase a replacement for the defective component is inconceivable.

It is troubling to live in a world we cannot hope to comprehend. When it comes to the universe of biology—the science of life—and especially the human body, the desire to know is urgent, acute, and personal. How does it work? Why do I feel or do what I do? The mysteries of growth, digestion, reproduction, healing, and illness propel our fascination and compelling need to know.

In this sense, the ultimate engine of humanity emerges from one bodily organ, a bodily organ that has grown so powerful in abstracting and analyzing the world, it has broken through to ask the question of how it asks a question. How does it perceive, dream, think, control the body, dare to break the binding logic of begging the question: to understand the inner workings of how it knows?

Now is a unique moment in the history of brain science. Nature offers an endlessly vast terrain of complexity, but at certain points in the history of scientific progress, we find ourselves on the crest of a new summit that reveals a new frontier of discovery. From this perspective and at this point in time we can see it all in one fresh vista. Soon, as we descend into the morass and become engulfed in tangled complexity, we reach a point where no one person can grasp it all. But at this moment, we are on that summit of brain research looking out on the world of glia with new eyes.

This makes the subject exciting and comprehensible to everyone. Soon scientific progress will expand to where only specialists can perceive and understand their small bit of the whole. But at this moment you know just as much, in fact far more, than the pioneering scientists who ventured into this new frontier of brain science with the simplest of tools, a razor blade and microscope, at the turn of the twentieth century. With no theory or chart to guide them, they dared to proceed by dead reckoning, using their wits. They had to invent a new vocabulary, including names for the microscopic marvels they found, and they tried to make sense of it all by thinking hard and sharing what they found with others on the same journey.

This is a journey that you are free to join. You will need to learn some new language, and at times think hard and differently, but right now is a moment of opportunity. It is all within your grasp.

VIDEO GAMES AND LASER SWORDS:
SECRETS OF GLIA REVEALED
IN GLOWING TERMS

The synapse is the transistor of the brain, the fundamental switch connecting neurons into circuits that allows us to think and feel, remember and hope. But is this the *only* way information flows through our brain? If there were another way for information to flow through our brain, sticking wires into glia the way neuroscientists plug their electronic amplifiers into neurons was not going to reveal it. Glia do not communicate with electrical impulses. A far more refined method had to be invented to detect glial communication. As with the sophisticated advance of a CD read by laser beam surpassing the primitive vinyl record probed with a needle, the secrets of glia were literally illuminated with beams of light.

Rapid technological advances in the 1980s driven by the explosive growth of the video game market were the key to the scientific revolution of calcium imaging. The electronic toy market fueled development and production of new and affordable home computers with improved color graphics for displaying video games. A new class of anatomists tinkered with this new technology, grafting it onto their microscopes to build a new instrument: the video microscope. Attached to computers, their microscopes now harnessed the new power of image processing, which surpassed the antique methods of staining tissue with colored dyes to reveal cellular structure. For the first time, video microscopy permitted scientists to see structure in unstained living cells and even watch the flow of ions engaged in physiological processes moving inside living cells, exposing the other brain to science.

Calcium Signaling: Replacing Electrodes with Light

Outside the lab he favors Hawaiian shirts and plays classic rock and roll on his Gibson J-200, but in his scientific career at Yale and Stanford, Stephen J. Smith specializes in building and using video microscopes and advanced laser microscopes to visualize the rise and fall of calcium ions inside living cells. These calcium signals are now understood to be the primary means that all cells use to transmit information from outside the cell, across the cell membrane, and into the cell's interior fluids (its cytoplasm). Calcium ions are the means by which information about the world outside is transmitted in coded form into cells.

A dazzling variety of sensors on the surface of cells constantly monitor their chemical environment. When a specialized receptor detects the specific chemical that it was designed to sense, it signals a cell-wide alert by opening up a pore in the cell membrane to allow calcium ions to flood briefly into the cell. This calcium signal is the equivalent of "The British are coming!" alerting all parts of the cell to the event this molecular sentinel had been watching for and marshaling the appropriate cellular response. By soaking cells in a dye that fluoresces with an eerie green glow in the presence of calcium, Stephen Smith and others enthusiastically exploited this new method to explore how stimuli of all kinds trigger calcium signals inside living cells.

Our cells use calcium as a signal because all cells in our body live in a virtual sea of calcium. The situation inside cells is quite different; cellular membrane pumps, much like those holding back the sea in a levee system, constantly pump calcium out, so that there is 10 million times less calcium inside our cells than outside. This sets up the perfect condition for calcium ions to function as potent messengers inside the cell. Some of the calcium is also pumped into cytoplasmic storage tanks called the endoplasmic reticulum. This pumping activity not only helps clear residual calcium ions from the cytoplasm, it provides a reservoir that can release a powerful flood of calcium to activate processes downstream that will initiate the cell's response.

Calcium is the main currency of information inside all cells, coordinating cellular responses to the ever-changing conditions in their environment. Neurons also rely on this intracellular currency of information. Specialized ion channels in the cell membrane of neurons sense the voltage change produced when an electrical impulse fires down its axon. These protein channels snap open briefly in response to electrical impulses, allowing a spurt of calcium ions into the neuron. By monitoring these wakes of calcium rippling through the cytoplasm from electrical events at the surface, cellular operations deep within the cell constantly monitor electrical events outside. Using the new calcium imaging methods, scientists could now literally see neurons fire.

The revolution in live cell calcium imaging had such an impact because it opened a new window for scientists to monitor living cells at work in real time. Dyes that fluoresce when they bind calcium ions exposed cellular messages that were previously carried out in secret. No longer were anatomists in search of clues to the cell's function in life limited to a forensic analysis of dead tissue pickled with preservatives and stained to reveal their dead structure. Now scientists could study the inner workings of living cells at work by watching them through these new video microscopes.

This new breed of anatomists was distinguished not only by their revolutionary imaging tools, but also by temperament. Traditional anatomists, working at the slow methodical pace of a museum curator carefully examining and comparing minute structural differences among collections of specimens, gave way to the quick-thinking reflexes of the electrophysiologist, collecting and analyzing data on the fly from living cells. Using all their wits, these new anatomists/physiologists had to make critical observations rapidly and form quick judgments, before the cell died. These were scientific explorers daring enough to follow an inspiration in an instant and improvise a new experiment based on what they were seeing. Much like electrophysiologists, but more so because their imaging techniques were entirely new, these new anatomists had to quickly distinguish the important and true from the trivial and false responses as they confronted things never before seen, and had to debug and fix technical difficulties and equipment failures rapidly and ingeniously while the cell remained alive, theirs feverishly to explore. This was not a pursuit for those favoring an eight-to-five job, as often the complicated orchestration of instrumentation and preparation of living biological samples did not come together until after everyone else in the building had gone home for the evening. Once this synthesis ignited, these scientists worked in a frenzy, milking every drop of data possible from the last hours of their captive cell's life.

Initially all the attention of neuroscientists was focused on watching the calcium signals flash inside neurons, but some researchers realized that if astrocytes could in some way sense neural signals, this might be revealed by a rise in calcium in the glial cytoplasm. If there was indeed another brain, this new method, a product of mixing video games with microscopes, might reveal it.

A Droplet of Light:
Neurotransmitters or Gliotransmitters?

Stephen J. Smith and colleagues sifted out the cells filling the space between neurons, that is, the astrocytes, and grew them in a laboratory dish. They knew that unlike neurons, these brain cells have no synapses, no spiky cellular tendrils or long wire-like axons carrying impulses. Instead the astrocytes formed a thin skin over the bottom of the culture dish. Unlike neurons, these cells are electrically inert and absolutely silent, but if astrocytes could sense messages being sent through neural circuits, Smith reasoned that since calcium is the usual way signals outside cells are communicated inside, one might expect to see calcium suddenly flood into an astrocyte receiving a signal from a neuron. Knowing that neurons communicate across synapses by releasing neurotransmitters, Smith decided to apply one of the most common neurotransmitters, glutamate, directly to the astrocytes and see if the astrocytes responded.

Working in a darkened room, Smith and his associates added a fluorescent calcium-sensing chemical to these peculiar brain cells. Then they squeezed a droplet of neurotransmitter into the dish. That droplet of glutamate set off an explosion of light spreading from within the astrocytes at the point of impact, which then radiated outward in a luminescent shockwave throughout the entire population. The astrocytes had detected the neurotransmitter and triggered a cell-wide calcium alert inside their cytoplasm. But there was more.

The light signals then swirled through the entire population of cells. For tens of minutes the researchers watched the light signals pass from one astrocyte to the next, swirling throughout the dish in a fluorescent light storm. These "silent" brain cells were communicating with one another. More importantly, this communication had been sparked by the same neurotransmitter chemical that neurons use to signal to each other across synapses. The theory that astrocytes might eavesdrop on conversations between neurons had just expanded: not only could astrocytes detect neurotransmitters, they could pass the messages on among themselves using calcium ions instead of electrical signals.

Smith realized that the implications of this discovery could be revolutionary. These glial cells were communicating. Why? How? Equipped with an ability to snatch messages passed between neurons at synapses, what would astrocytes do with the information they gathered?

Firebreak Experiment

How these glial cells were detecting the neurotransmitter was soon discovered. Researchers found that, remarkably, astrocytes have on their cell membrane the identical receptor proteins to detect neurotransmitters that neurons have on their dendrites. In neurons these neurotransmitter receptors are designed to detect the neurotransmitter sent across the synapse. Why were the same receptors on astrocytes? When these receptors on the astrocytes sensed the droplet of glutamate that Smith had placed in the dish, they snapped open, allowing a spurt of calcium ions to flood inside. This in turn lighted up the fluorescent calcium tracer for the anatomists to see. It is important to remember, however, that this glial response had been triggered artificially using an admittedly crude delivery method. Whether astrocytes could indeed respond to neurotransmitter released when neurons communicate across synapses naturally was yet to be determined. Nevertheless, it was now clear that in principle astrocytes could do so. Glia were equipped with the machinery to intercept communications sent between neurons at synapses and could in turn spread the information throughout glial circuits.

Information revealed by calcium signals might be passing through the brain in an entirely different manner from the electrical signaling assumed to be the only way our brain operated. What's more, this information could flow through cells that were not neurons—namely astrocytes—cells that lacked all the special features of neurons, such as axons, dendrites, and synapses, upon which our entire concept of brain function rested. For the first time, scientists could see with their own eyes that there was another brain. But how did it work?

The intriguing question of what the astrocytes might do with the information they gained from listening in on neuronal signals lay ahead. The immediate puzzle to solve was how this new form of glial communication operated. How could glia transmit calcium messages from one glial cell to another in a communication network spanning the entire population of astrocytes in the culture dish—and by extrapolation, across an entire brain?

The most likely possibility was that astrocytes passed calcium messages between cells through protein channels joining adjacent cells. Such intercellular channels, called gap junctions, are well documented in astrocytes. Like docking points between two spacecraft, gap junctions permit the exchange of potassium ions and small molecules. In a simple but elegant experiment, Stan Kater and colleagues, pioneers in calcium imaging at Colorado State University, reasoned that if the astrocytes were physically separated from one another, a calcium wave sparked in one astrocyte could not propagate to its remote and isolated neighbors.

To test the idea, they simply scratched off a line of astrocytes from the bottom of the culture dish, forming a cellular firebreak that would block a wave of calcium if it was being passed directly from cell to cell. The result could not be more definitive: the calcium wave spread through the cells on one side of the barrier and then jumped across the cell-free "firebreak" without the slightest difficulty. This was proof that in addition to being able to pass messages between astrocytes through protein channels linking adjacent cells, astrocytes communicated with one another by broadcasting some unknown signal through the culture medium that triggered a calcium wave in distant cells in a chain reaction. Like neurons, which communicate across synapses by sending chemical messages (via neurotransmitters), astrocytes communicated with one another by sending some type of signaling molecule through the space between cells. Neurons and glia were sharing the same communication channels. However, neurons communicate through synaptic connections in linear circuits like telephones, but astrocytes communicate by broadcasting signals widely like cell phones. Nowhere in any theoretical model of brain function had this type of communication been considered.

EXPOSING THE SPY: THE GLIAL STING

The calcium response ignited in astrocytes when the neurotransmitter glutamate was introduced in cell culture was a revolutionary discovery, but the experimental conditions were highly artificial. Can astrocytes in the brain detect neurotransmitter released naturally from synapses rather than applied artificially from a pipette? Could glia do this in association with a brain function of important consequence? How can you find out?

You suspect that glia might be monitoring neuronal communication at synapses and perhaps intervening to regulate the flow of information across the synapse. How are you going to prove it? The situation is identical to the problem you would face if you suspected a spy was monitoring communications between central command and field agents. You suspect that the spy is monitoring your coded dialog and interceding in communications by modifying the messages. How can you root out the spy?

One way to crack the case would be to find the bug the spy is using to eavesdrop on the coded messages. This was the first approach scientists used to explore the possibility of glia intercepting neuronal communication. They performed a broad sweep, rounding up and testing every conceivable type of message a glial cell might intercept to monitor neuronal communication at synapses. What scientists found astonished them.

They discovered that glia had sensors that could detect a large number of neuronal signaling molecules, including *all* the various neurotransmitters neurons use for synaptic communication. Glia were also sensitive to ion fluxes that surge with the flow of electrical information through neural circuits and to a host of other potential cell-signaling receptor molecules that could in theory enable glia to monitor neuronal information processing.

The second approach adopted by scientists was to bug the suspected spy just as Smith and his colleagues had done in cell culture, by placing a calcium-sensitive fluorescent dye in the astrocytes. The challenge was to do this inside a living animal and see if a calcium-sensing molecule inside the astrocytes would glow brightly when neurons passed information across a synapse.

The first trap scientists set to expose glial synaptic snooping was at the connection between neuron and muscle. This target was chosen instead of synapses buried deep in brain tissue, which present extreme technical challenges for such experiments. Instead, Smith and graduate student Noreen Reist monitored information at the synapse controlling muscle contraction, which is easily accessible in an anesthetized frog simply by peeling back the skin over a muscle.

Muscles contract in response to electrical commands sent down motor neuron axons to a synapse on the muscle called the neuromuscular junction. When the axon fires, the neurotransmitter chemical acetylcholine is released from synaptic vesicles in the nerve ending. The acetylcholine stimulates neurotransmitter receptors on the muscle fiber, causing the muscle to contract. Communication at the neuromuscular junction is identical in every way to communication at any other synapse between two neurons in your brain. Just as a postsynaptic neuron fires an impulse upon receiving a synaptic signal, so does a muscle fiber. This electrical energy spreads over the entire muscle cell, causing it to contract.

There are no astrocytes outside the brain, but anatomists have known for 150 years that, just as at synapses inside the brain, the neuromuscular junction is surrounded by glia, specifically the terminal, or perisynaptic, Schwann cells. These cells had long been thought to be completely inert structural elements unrelated to synaptic transmission. Smith and Reist had other suspicions.

In 1992, they filled the terminal Schwann cells at a frog's neuromuscular junction with a calcium-sensitive dye, and using their video and laser scanning microscopes watched for calcium signals in the Schwann cells when they gave the nerve fiber an electric shock to make it fire impulses. After a burst of impulses stimulated contractions of the muscle, they saw the terminal Schwann cells surrounding the neuromuscular synapse begin to glow.

This was a major advance beyond the studies they had performed in cell culture. The new experiments proved that glia could detect normal information transmission across synapses. Their studies showed that terminal Schwann cells did this by sensing neurotransmitters released by motor neurons at the synapse. Now the researchers wondered, were the Schwann cells acting on the intercepted information to control neuronal events?

GLIA CAUGHT IN THE ACT

This new finding only confirms that glia are monitoring neural communications. It is far more difficult but more critical to prove that glia are acting upon that information to influence commands and alter events. How would you go about discovering this? The time-honored method in counterespionage is to plant a false message and see if the suspected spy acts on it.

Richard Robitaille, a neurobiologist at the University of Montreal, devised a plan to plant a false message in these terminal Schwann cells, and then he watched to see if this false information altered communication between the neuron and muscle. He used a very slender glass pipette to inject selected chemical agents into the Schwann cells surrounding the synapse. These chemical agents were known to be used as messengers by cells in relaying information from cell surface receptors to the inside of the cell. He tried injecting not only puffs of calcium, but also other small messenger molecules used by cells in intracellular communication. Neither the nerve fiber nor the muscle would receive injections of these chemical signals; only the terminal Schwann cells received the message.

When Robitaille and his colleagues delivered the artificial messages into the terminal Schwann cells, they saw that the electrical signal received by the muscle was altered. Some messages injected into the Schwann cells caused a bigger electrical response in the muscle when the nerve impulse excited it; other chemical messages reduced the electrical response in the muscle. This was irrefutable proof that glia not only monitor neuronal information passed at synapses, they alter it. The implications were shocking.

Scientists in other labs joined in the frenzy of research to explore how pervasive this communication between glia and synapses might be in the nervous system. Eric Newman, a neurobiologist at the University of Minnesota, began his scientific career studying infrared receptors in rattlesnakes. These unique sense organs allow rattlesnakes to detect warm-blooded prey by using a type of heat-seeking vision. Soon he left rattlesnake research and turned his attention to another sensory system, the light-sensing layer at the back of the eye: the retina. Combining electrophysiology and calcium imaging, he saw that when a calcium wave in retinal astrocytes—much like those Smith had observed in the culture dish—swept past a neuron in the retina, electrical firing increased or decreased simultaneously in visual circuits. He saw that these glial calcium waves were swirling through the retina on their own, but they could also be stimulated by shining a light on the retina, much as

Kuffler, Nicholls, and Orkand had found inside the optic nerve back in 1966. But this time, Newman could see glia not only responding to visual stimulation, but also changing neuronal firing in response to calcium signals. Astrocytes in the retina were "watching" the visual information transmitted through neurons when light struck the retina and then passing this information through glial networks in intercellular waves of calcium to regulate neuronal communication. Glia were participating in vision. Retinal astrocytes were controlling neurons.

Recall that our retina is an embryonic outgrowth of the brain. Do these new findings imply that glia might participate in mental functions inside our skull? What functions? Reflexes? Thoughts? Dreams? Emotions? Mental health? Memory?

To seek answers to these revolutionary questions, scientists needed to learn far more about glia, these nonneuronal cells that constitute the other brain. In increasing numbers, neuroscientists are beginning to expand their investigations beyond neurons to examine more closely what astrocytes, Schwann cells, oligodendrocytes, and microglia do in the nervous system in their usual role as "support cells." Consciousness is beginning to be raised: glial "support" for neurons could in principle extend into neuronal dependence on glia, or even neuronal control by the other brain. Glia were understood to operate during a pathology of the nervous system, but might those same glial functions apply in certain contexts in the normal physiology of brain function? The actions of glia in brain health and disease have enormous practical importance, but beyond that, investigating and understanding glial operations during disease may yield clues about what glia do in the healthy brain.

PART II

Glia in Health and Disease

MENDING THE MIND:
GLIA REPAIRING NERVOUS SYSTEM
DAMAGE AND DISEASE

Two types of glia, microglia and astrocytes, act together as sentinels watching for bacteria and viruses infecting the brain. When disease-causing agents are detected, these glia form the cellular army that mobilizes to fight the invading microbes. They seek out and devour pathogens and release toxic chemicals to rid the brain of disease-causing agents. This vigilant cellular combat is essential for our brain's normal function and survival, but more recent studies of microglia are turning up many unexpected roles for these odd brain cells. For example, chronic pain often lingers long after nerve injury has healed, and it is particularly difficult to treat. Now it is becoming clear that many drug therapies are ineffective in treating chronic pain because scientists failed to appreciate the role of glia in pain and in drug addiction. Neuronal painkillers address only part of the problem: they overlook the other brain.

The exciting potential for stem cells in treating a wide range of neurological illnesses, from Parkinson's to paralysis, is widely appreciated, and here too glia are at center stage. Mature neurons cannot divide, and when they become damaged by injury or disease, as a rule they cannot be replaced. By contrast, glia respond to brain injury by dividing and migrating to the site of damage. There they repair the injury, defend against disease, nurse the neurons back to health, and guide the regrowth of damaged nerve fibers to restore proper communication between neurons and between neurons and muscles. But new research is revealing that immature glial cells can act like stem cells and mature astrocytes can stimulate stem cells dormant in the adult brain to form replacement neurons and glia. The promise of embryonic stem cell research in relieving human suffering from brain disease cannot be underestimated, but ethical issues make it controversial. Lying latent throughout the brain are immature glia that have the ability to replace neurons lost to disease. This revelation holds great promise for future treatments if we can control these glial "stem cells" Nature has already provided.

Glia can also be the source of disease. Often these cells are the targets of infectious organisms, including those responsible for such alarming illnesses as HIV/AIDS and "mad cow disease." Glia are intimately involved in neurodegenerative diseases, in which neurons wither and die. In Parkinson's, amyotrophic lateral sclerosis (ALS), and Alzheimer's, glia are involved in both beneficial and detrimental ways. The very name "neurodegenerative disease" betrays a narrow focus on neurons that is conceptually myopic and biologically unsound. As recent research explores glial involvement in brain function, the implications for glial participation in disease are appearing everywhere.

Most unexpected to those unfamiliar with the other brain are recent findings implicating glia in many psychiatric illnesses, from schizophrenia and depression to such idiosyncratic disorders as pathological lying and tone deafness. Those who recognized the importance of astrocytes in regulating synaptic transmission were not surprised by the recent revelation that glia participate in psychiatric illness, but most researchers were dumbfounded when new brain imaging techniques began to reveal changes in white-matter structure in people suffering such psychiatric disorders as schizophrenia, depression, bipolar disease, autism, and attention deficit/hyperactivity disorder (ADHD). This discovery has forced a reexamination of the possible importance of myelin in information processing in the brain. Still in its infancy, this research is suggesting a new aspect to information processing in the brain and a new mechanism for learning and plasticity that has been previously overlooked.

Illnesses that cause loss of myelin—for example, multiple sclerosis—are the source of considerable dysfunction and suffering for many people. What has been less appreciated is that demyelinating diseases, which are diseases of glia, are also neurodegenerative diseases. Because of the critical dependence of axons on their glial partners to sustain and protect them—not simply to insulate them from electricity leaks—when glia die in diseases like multiple sclerosis, often their neural partners die as well.

The other brain has the potential to affect the entire spectrum of brain health and disease: it is something that touches us all in our everyday lives. Everything from drug abuse, hearing loss, critical periods for learning, aging, cancer, and many more health-related aspects of brain function involve the other brain. Glia are thus at the hub of brain health and disease.

CHAPTER 4

Brain Cancer:
Almost Nothing to Do
with Neurons

"If the glove doesn't fit, you must acquit!" This famous argument that cracked the prosecution's solid foundation of forensic evidence against O. J. Simpson in his trial for murder was forged in the mind of a brilliant defense attorney: Johnnie Cochran. Here was a mind capable of reducing a case of apparently insurmountable obstacles to a simple eight-word persuasive argument—poetry with the appeal of a rap lyric. A few years later, Johnnie Cochran learned the reason for the headaches and strange symptoms that drove him to consult a neurologist. Sadly, the diagnosis was a death sentence without reprieve. Following two years of fruitless battle, Cochran was gone, claimed by the ravaging attack of insurgent cells in his own brain—glia.

The disease is otherwise known as brain cancer. Few people realize that this dreaded disease almost never involves neurons. Mature neurons do not succumb to the disease, because cancer is a failure of the brakes that stop cellular division, leading to runaway growth of cells into tumors. Mature neurons are always "in park"; because they never undergo cell division, they are not in a position to become cancerous. Wayward glia are the killers.

People stricken with a serious disease usually react in the same way, with the inevitable question: Why? What was the cause of this ordeal of suffering? Was it something I did? Could it have been avoided? Is there anything that can be done now? True to his fiber as a defense attorney, Cochran sought to identify the responsible party and assign blame, to apply his skill and training to seek damages and restitution.

His neurosurgeon, Keith Black, was convinced that Johnnie Cochran's brain cancer was caused by his cell phone. The cancer grew on the left side of Cochran's brain; he habitually held the phone to the left side of his head. Cochran logged vastly more hours with it than the average person. Eventually, however, the suit brought against the cell phone companies for causing Cochran's death was dismissed by the courts.

Sitting at a richly polished conference table following a talk I had just delivered at a medical school in Virginia, I listened to neurosurgeons hungry for the latest advances from basic research that they could apply to help their patients. One doctor wearing a crisp white coat with his name embroidered neatly in red script above his breast pocket pleaded like an infantryman in the trenches calling for ammunition. "Six weeks after I see patients with glioblastoma multiforme, they are dead! What more can we do?"

The doctor was frustrated by the inevitable defeat awaiting him, despite all his knowledge, training, and skill. He could remove as much of the brain as he dared, poison the cancer cells with drugs, and bombard the insurgent cells with radiation, but with some particularly vicious forms of brain cancer, he knew that none of this would change the inevitable outcome. This outcome could meet him with such velocity and violence that he and his patients could see it by flipping one page of their calendar. On a daily basis, he was living the frustration of the fact that there have been no significant advances in the treatment of glioblastoma in the past twenty-five years. Imagine your own frustration if in your profession you were forced to work with tools a quarter of a century old, and the people you were caring for were suffering and dying.

Not all brain cancers are so vicious. Many are curable with modern treatments and surgery; some are among the most deadly cancers known.

Tumors can develop anywhere in the brain, and the initial symptoms they produce are as diverse as the many functions the brain performs. Most often the first symptoms are headache and fatigue, but problems with vision, speech, standing and walking or personality and psychological changes can all herald the formation of a brain tumor, depending on where in the brain it has begun to grow. So tightly packaged is the brain inside its bony skull that a tumor does not have to grow very large to compress adjoining healthy tissue and impair function. If it lies deep in the brain, the tumor cannot be attacked by surgery without risking devastating consequences. If it is not the type of tumor that forms a solid mass, but instead spreads through the brain like dry-rot fungus, it is as unreachable with a scalpel as clouds drifting through the sky.

A skilled neurosurgeon can readily diagnose precisely where the tumor is in the brain by reverse engineering from the symptoms to the regions of the brain known to control them. There are differences in brain tumors between males and females, between adults and children, and between the aged and the young because our glial cells differ in these situations. The neurosurgeon knows often before the definitive CAT scan not only where in the brain the tumor has begun, but probably what type of glial cell has become cancerous.

What is brain cancer? What causes it? How can it be defeated? Environment and genetics both have a role in causing cancer, but the problem rests in the mechanisms controlling cell division, one of the most complex and highly regulated processes in a cell. It is vital that cells divide only when appropriate: during embryonic development, during growth of a child, after injury, and to replace cells lost through natural causes. This multiplication of cells must be precisely balanced to maintain the integrity and structure of every tissue in our body. Cell division is so intricately regulated that it usually takes several failures in the control process to unleash runaway cell division; this is why there will never be the equivalent of a vaccine to cure all cancer.

Factors in combination are usually required to disrupt the cellular controls to the point of cancer, but when this occurs, the cells become so wildly aberrant that they are beyond any control. The only solution is to kill them. In my lab we have used gene microarrays to identify genes that are regulated abnormally in a nerve sheath tumor caused by cancerous Schwann cells. It looks as though a bomb were dropped on the nucleus of these cells. Hundreds of genes are wildly out of control. This is what makes a cure so difficult. This situation is the equivalent of looking at the devastation after an avalanche and trying to enumerate in detail the changes caused by the destruction. The goal is to find and prevent the one or two critical defects that set off the avalanche in the first place. The complex controls on cellular division are so highly inter-related that a single mutant monkey wrench is not enough to jam the mechanism, but when enough genetic or environmental hits manage to take out enough molecular gears in the clockwork, the entire mechanism springs apart into a hopeless cellular mess beyond repair. If you have a genetic predisposition toward a certain cancer, it would be extremely important to identify precisely which gene in the clockwork has the missing teeth, that is, which of the many genes controlling cell division has been mutated or damaged. Then your doctor could advise you to avoid certain environmental factors that could collaborate with this defect.

Radiation will cause cancer. The energetic bombardment breaks the DNA that makes up all the genes in the nuclei of our cells, shattering the DNA at random, creating an unpredictable pattern of cellular terrorism. Fortunately, our cells can repair their DNA to a remarkable extent, a prerequisite for life on this planet, bombarded from the beginning of time by natural sources of ionizing radiation from outer space and from the sun's ultraviolet rays.

Like Cochran, many people are fearful of radiation from their cell phones. This fear reveals a failure to appreciate the concept of dosage. There is a vast difference, for example, between a microwave oven and a cell phone. Just try cooking a burger with your phone. The word "radiation" sparks fear in the heart of the average person. But radiation is a normal part of our environment, cast down on us together with the warming rays of the sun. Radiation emanates from the smoke detectors in our homes and from dishes that use uranium salts in their ceramic pigments. But all are perfectly safe because the radiation levels are low.

Still, doubters argue, even though the power is low, cell phone signals can reach across the globe if the appropriately sensitive receptors are activated. Could brain cells be that sensitive? Cancer is a mix of genetic and environmental risk factors. Who is to say that someone with a genetic predisposition might not find their glial cells tipped down the slippery slope toward cancer by the seemingly innocuous radiation from a cell phone? There is no strong evidence for it, and it would be difficult to design an experiment to detect the effect of such a subtle causative agent in this rare population. There is comfort from the demographic evidence that fails to support a link between cell phone use and cancer. Although the evidence is still weak and controversial, in July 2008 Dr. Ronald Herberman, director of the University of Pittsburgh Cancer Institute, issued an advisory warning that there is sufficient data linking electromagnetic radiation from cell phones to brain cancer to warrant advising the public to limit cell phone use, especially by children and teenagers. France, Germany, India, and Canada had already issued similar recommendations. The advisory recommends not allowing children to use cell phones, except in emergencies, and to avoid carrying cell phones on the body. The use of head phones was recommended to keep the phone away from the brain.

The debate and research go on. This seems strange given the abundance of known agents and activities that *do* cause cancer but fail to strike the same fear in the hearts of most people. Alcohol, tobacco, sunburn, toxic organic chemicals in industrial and home products are all real but accepted risks, yet the cell phone and invisible radiation from power lines scare many. Looked at objectively, the reason is simply fear of the unknown. Everyone understands alcohol and sunburn; few understand radiation, and so they fear it.

TYPES OF TUMORS

If you care about brain cancer, you must learn about glia, for nearly all cancers in the brain arise from glial cells. There are relatively rare exceptions in which neurons become cancerous, especially in young children whose brains are still developing and whose neurons are still immature, but mature neurons do not divide and do not become cancerous. Other cells in the brain that can on rare occasions become cancerous are the skin-like cells lining the surface of the brain and the hollow fluid-filled cavities inside the brain and spinal cord. (Most researchers now include these cells within the category of glia.) Like glia, these meningeal and ependymal cells do divide, and they can give rise to tumors when their cell division goes out of control. The vast majority of brain cancers, however, are aberrant glial cells. Tumors of our peripheral nerves derive primarily from glia too, specifically the Schwann cells in all our nerves.

Tumors can develop anywhere in the brain, and their incidence is increased by a number of factors. There are also interesting sexual and age differences for different types of brain cancers. Trauma, genetics, hormonal status, immunological disorders, environmental influences, chemical agents, and viruses can all influence the risk of brain cancer. These factors either stimulate cell division or directly influence regulation of genes that control cell division. Trauma stimulates a rapid increase in astrocyte cell division, for example, which increases the opportunity for a cell to begin dividing out of control. Usually, the body's immune system destroys aberrant cells, including native cells gone berserk, protecting us from cancer. This explains why there is an increase in cerebral tumors in kidney transplant patients, for example, who undergo immune suppression to prevent rejection of the donor organ.

In general males tend to suffer more of the malignant types of brain tumors, whereas females experience more meningiomas, benign tumors of the covering of the brain and spinal cord, which are twenty times more common in females than in males.

Fifty-one percent of all brain cancers are glioblastomas, the category of cancer that includes the one that afflicted Johnnie Cochran and more recently Senator Edward Kennedy. These cancers grow diffusely throughout the brain, making them difficult to treat effectively surgically. Glioblastomas are relatively rare in young people, peaking in incidence between the ages of forty-five and sixty-five. There is a three to two male to female predominance for this type of cancer. Glioblastomas tend to be about the size of an egg, but they can grow across the corpus callosum to affect both sides of the brain, causing cell death (necrosis) and hemorrhages as they grow. This is a rapidly growing tumor, limiting the postoperative survival to between nine and fifteen months. Anyone diagnosed with late stage aggressive glioblastoma multiforme can count on a postoperative survival of about two to four months. Radiation treatment is a double-edged sword. It can be effective, but radiation can also change the tumor and damage surrounding neural tissue.

Astrocytomas are another form of brain cancer. Well-defined whitish tumors, chestnut to apple size, they can also grow diffusely. Astrocytomas account for 25 percent of all brain cancers. Pilocytic astrocytoma, 3.4 percent of all brain cancers, is usually encountered in the brain stem or cerebellum at the base of the brain. "Pilocytic" refers to the hair-like wavy fibrous appearance of this tumor. This cancer is the most frequent tumor of cerebellum and optic chiasm in children and adolescents, peaking in incidence at about three to seven years of age. Neuroblastomas, tumors of neuronal origin, do exist, although they are much less common, arising usually in the first three years of life and only very rarely seen in adults.

About 5 percent of brain cancers are oligodendrogliomas. These are tumors of the middle decades of life, of people thirty-five to forty years old. Males tend to predominate, and the egg-sized tumor usually develops in the frontal and temporal lobes. These tumors develop liquefied cysts in the white matter, which can then cause the surface of the cortex to grow abnormally.

EMPTY LINES

"I was very interested in your talk," a fit-looking neurobiology professor told me, directing me to a seat in his office after a lecture I had just given on oligodendrocytes as a guest speaker in his department. "I have an oligodendroglioma."

He slid out the extension leaf from his metal desk, propped open the lid of an IBM ThinkPad, and booted it up. With a couple of mouse clicks an MRI of his brain popped up on the screen.

You needed no special training to read this brain scan. From across the room anyone could see the ominous white cloud shadowing the left hemisphere. The otherwise normal brain tissue, convoluted like beautiful brain coral, was eaten away as if by acid, leaving a fist-sized hole.

I could not imagine that the brain that was talking to me at this moment was at the same time analyzing itself on the laptop screen, calculating the odds, mortally wounded, and still operating inside this body of my colleague, who showed no outward signs of illness whatsoever. Intellectually sharp, he displayed not the slightest indication that his brain was being destroyed by cancer as he spoke—a cancer that would be expected to kill him within three to eight years. (Presuming, of course, some other random act of fate did not intervene to take his life sooner.)

"If it wasn't for this picture, I would hardly know I had a cancer."

My eyes stole fleeting glances around his office, picking out pictures of his family, framed cover stories of his research published in scientific journals, his bushy-haired thirteen-year-old son.

"The doctor asked me if I wanted chemo or radiation—said it was my choice! You're asking *me*? I told him.

"Everyone I spoke to said it doesn't matter. Neither one will extend my life. Radiation will just delay the progression of symptoms, but the radiation will probably kill me in ten years anyway."

I was appalled that medicine seemed so powerless and ignorant of the biology gone wrong in this man's brain. I understood the apparent indifference of his doctors and the chilling detachment of my colleague. They were simply respecting the importance of facing reality with honesty and clear thinking. My colleague knew the score.

"I can't believe they know so little about these cells. If we knew more about the signals exchanged between neurons and glia that control glial proliferation—like the ones you were talking about in your seminar—there could be some rational specific treatment to slow down the division of these cancer cells instead of bombarding your whole head with radiation. That *damages* tissue. In fact, it *causes* tumors in the long run."

We talked at length, plumbing our minds for clues and possible new approaches. "Brain cancer is not like other cancers in the body," he observed. "They don't metastasize and spread all over the body the way other cancers do. They stay inside the brain. They should be easier to control."

We fished around but came up empty. The basic science just wasn't there. The most fundamental facts about these brain cells in their normal state were not even known. For him, looking at the problem with the cold objectivity of a scientist, it was a simple choice: length of life versus quality of life. He told me he had stopped his chemo.

"Let me know if you find out anything that might be helpful," he said earnestly as we shook hands.

I left him, feeling strangely like an imposter.

NEW HOPE

New hope for treating brain cancer is developing from research on the other brain. The recent work of Harald Sontheimer, an electrophysiologist at the University of Alabama at Birmingham, is a good example. Sontheimer, trained in Heidelberg, Germany, and later at Yale, is an energetic person with a close-cropped beard and mustache. He has always had a special interest in applying his microelectrodes to glial cells, an activity regarded by many as misguided because glial cells do not fire electrical impulses. Glia do, however, have nearly every type of ion channel in their cell membrane that neurons have. These ion channels regulate the flow of specific ions—sodium, potassium, chloride, etc.—across the glial cell membrane.

One of the delights of scientific adventure is that the exploration often takes unpredictable turns. One would not have predicted an intersection between brain cancer and studies of an electrophysiologist who was curious about why glia have ion channels and how they use them. But curiously, many studies do show that cell division in many types of cells is accompanied by changes in the cell's voltage. (Recall that all cells in the body have a voltage associated with them, because the salts inside the cell and outside it are different, and the resulting different electric charges create a biological battery.) Also, drugs that affect the flow of potassium or chloride ions through ion channels are known to slow cell division in many different types of cells, although it is not clear why or how.

A key feature of cancer cells is their unruly ability to spread. To facilitate this migration, many cancer cells exude enzymes that dissolve the extracellular matrix that binds cells together, loosening up a pathway between cells. Still, Sontheimer reasoned, the passages between cells in the brain are tight. How does a cancerous glioma cell squeeze through these narrow pathways to spread throughout the brain? Like an octopus that can expel water from its body to squeeze through an opening no bigger around than its tentacle, a glioblastoma cell must do much the same thing, he imagined. Perhaps a cancer cell squeezes water out of its cytoplasm so that it can shrink and slither through narrow spaces between brain cells. Sontheimer realized that the ion channels he was studying could be critical for this cell shrinking.

Too much salt causes water retention and bloating in cells. Cells regulate their water balance and volume by controlling the amount of salt—notably chloride—inside their cellular bodies. Drugs that block these chloride channels have been isolated, such as chlorotoxin, which is extracted from the venom of scorpions. Sontheimer's research found that when glioma cells, which are cancers derived from glia, were treated with this chloride channel blocker, they could no longer migrate through tiny pores of precise size separating different compartments of their cell culture dish. The glioma cells could not eliminate the chloride salt and thus remained too bloated to shrink their bodies down to pass through the tiny pores. Furthermore, this component of scorpion toxin can cross the blood-brain barrier, so it could be used as an anticancer drug in the brain. This application would mean that these cells would remain trapped in one place, where they could be removed by surgery or focused radiation. Sontheimer's group found that when the toxin was given to rats with glioblastoma, their brain tumors shrank.

To exploit this ion channel vulnerability in an even more clever way, Sontheimer and his colleagues attached toxic radioactive molecules to the chlorotoxin. Once injected into the bloodstream, these molecules deliver a lethal blow directly to the cancer cells by binding to the chloride channels that are present in abnormally high numbers in glioma cells. In this way, individual glioma cells can be taken out one cell at a time with strategic precision strikes, rather than via the crude surgical approach of removing chunks of brain tissue, which takes some healthy neurons and tends to leave a few cancerous cells behind to regrow and spread. This drug is currently in phase III clinical trials for treating malignant glioma in humans. Early results on these patients suggest that the powerful new ammunition that doctors need in their fight against brain cancer may soon be on its way.

Another new approach recognizes the importance of the immune system in eliminating cancerous cells and exploits the fact that microglia are the resident immune cells of the brain. From what we know of these tiny glial guardians, it seems obvious that microglia must play an important role in brain cancer. Microglia are indeed attracted to glioma and astrocytoma tumors in large numbers; in some cases microglia represent up to 70 percent of all the cells in the tumor. The ability of microglia to home in on brain tumors is being used experimentally to target gene therapy. In these experiments, microglia are used as cellular vehicles to deliver lethal genes into the tumor. In the same way, microglia treated with contrast agents that make them appear bright on an MRI brain scan are being developed to enable doctors to locate brain tumors. The microglia with the contrast agent are tracked to the site of the tumor where they accumulate.

However, a number of studies show that microglia inside brain tumors are often deficient in their immune response. This weakened immune response contributes to the growth and spread of brain cancers. Longtime glial researcher Helmut Kettenmann from the Max Delbrück Center for Molecular Medicine in Berlin finds that microglia can even spread brain cancer by releasing certain enzymes, matrix metalloproteases, that digest the fibrous network of proteins that bind cells together as they must to move through brain tissue. Kettenmann and his colleagues found that glioma cells injected into brain slices spread less if the slices were treated beforehand to reduce the population of microglia.

Thus in glia we see both the cause and cure for brain cancer.

CHAPTER 5

Brain and Spinal Cord Injury

GLIA AND PARALYSIS:
GLIA BLOCKING NERVOUS SYSTEM REPAIR

A few years after the horseback riding accident that paralyzed him, Christopher Reeve sat alone in his wheelchair in a spotlight at center stage, the focal point of several thousand neuroscientists who had come to hear him speak at the Society for Neuroscience annual meeting. Laypeople are rarely invited to speak at these scientific meetings, but Reeve had a message he wanted to deliver directly to neuroscientists. Speaking through carefully paced gasps of air dispensed mechanically from his respirator, Reeve, unable to move or feel anything below his broken neck, transformed the weakness of his body into a supreme strength of spirit.

Ascending from an object of pity, he had become the nation's most powerful and respected leader in the search for a cure for paralysis, a person of universal admiration and inspiration. I have heard many brilliant scientists, Nobel Prize winners, renowned professors, and scholars speak at these annual meetings in the last twenty-five years, but they all pale in comparison to Reeve. Displaying his professional skill as a performer, he held every person in the audience captive using nothing more than his eyes and voice, and his inspirational display of courage and reason. He showed himself to this audience of experts to be well informed about his condition and enlightened as to how neuroscience could defeat this horrible curse of paralysis with more focused research.

A wheelchair frightens us, because it is a reminder that we or our loved ones are only a car wreck or some other chance of fate away from being wheelchair bound ourselves. Parents of boys have the most to fear. Three out of four people in wheelchairs are men or boys, victims of male attraction to fast cars, motorcycles, sports, and a propensity to engage in violence. Spinal cord injury is permanent, sudden, and profoundly life-altering. It is a prolonged struggle: three out of five people in wheelchairs reached that condition before they turned thirty. There is a burning desire among many neuroscientists to do more for these injured people. For centuries the prospects looked dismal, but now, with a new understanding of glia, there is growing excitement that the goal of successfully treating paralysis may be near.

Reeve knew well that glia were the ultimate source of his misery, for glia cut off at his neck every effort of his broken body to heal. In fact, 40 percent of the recent grants for scientific research awarded by the Christopher and Dana Reeve Foundation in the search for a cure for paralysis have been for research on glia.

If Christopher Reeve had simply broken a bone in his arm and severed a nerve, he would have healed in time. But axons severed in an injury to your spinal cord or brain will not grow back. Why does an injury affect axons in your peripheral and central nervous systems so differently? Do axons in our brain and spinal cord lack a critical factor enabling them to repair themselves and grow back to their proper connections? Or is there something about the tissue environment in the brain and spinal cord that differs from the environment in our peripheral nerves and prevents central nervous system axons from regenerating? To understand the answers we must examine what happens when the spinal cord is injured.

Cellular Response to Spinal Cord Injury

Imagine you are on a sunny beach enjoying the surf, just like a friend of mine who related this story to me. As the waves roll in you bounce to keep the cold water from reaching your bare thighs. A big swell builds on the horizon and rolls toward you, inflating as it approaches. Its slope steepens and then peaks as steam rises from its summit. Boiling and churning it curls into a menacing maw, sucking up the ocean foam ahead of it and spilling discarded brine over its rolling crest. The monster races toward you and just as it is about to crash, you escape by diving under the wave for the calm beneath. Instantly you are stunned.

You cannot move a muscle or feel the cold surf tossing your flaccid body in the foam. The loud crunch and electric shock of pain in your neck tells you it is broken. You open your eyes and see swirling sand and beams of sunlight lancing through the murk, revealing nothing evil, simply a shallow sandbar hidden beneath the surface.

You cannot raise your head to gasp a breath. Your most urgent and forceful commands to your arms and legs to roll your body over so you can suck a desperate gulp of fresh air go unanswered. Time dilates as your mind races to escape the terrifying possibility that you are witnessing the last moments of your life. Stale air expands your lungs to the bursting point. It is impossible to fight the urge to breathe any longer. You know you are going to have to breathe water. Bubbles explode from your lungs as your diaphragm reflexively draws back saltwater deep inside. That's it; everything goes black.

When this happened to my friend, there chanced to be, against all odds, a nurse sunbathing at that very spot. By chance she was looking out to sea the instant he drove his head into the hidden sandbar and broke his neck. Any other person might have waited too long. Another rescuer, if there were one, might have lacerated my friend's injured spinal cord in the rescue and killed him. But this nurse knew how to support the fractured vertebrae in his neck as she pulled him from the surf and gave him CPR.

You have been saved from drowning, but the fractured bones in your neck have smashed your spinal cord though they did not sever it. Still, the damage cuts off all commands from your brain to your body and blocks all channels of sensation into your brain at your neck. Many of the neurons in your damaged spinal cord will survive, but their axons are crushed or severed and can no longer carry impulses. The blood-brain barrier is ripped open by the injury and blood seeps into the damaged tissue with a caustic effect on the cord, polluting the specialized spinal fluid bathing your central nervous system.

As the emergency medical technicians strap you onto a backboard to carry you to the ambulance, white blood cells seep out of your bloodstream and enter your damaged spinal cord. Encountering tissue that is foreign to them, they begin to attack. Microglia immediately sense the damaged tissue and the invading blood cells and mount a counterattack. They quickly transform from their resting bushy state into their amoeboid activated stage and rush from long distances toward the injury, squeezing between cells like rescue vehicles driving through a traffic jam. Both microglia and white blood cells release reactive oxygen molecules and other toxic substances designed to fight infection and kill invading cells, but these substances also damage injured and even many healthy spinal cord neurons and glia. Astrocytes in the damaged region also detect the same alarm signals and transform into their reactive stage, releasing toxic chemicals and cytokines that initiate the body's inflammatory reaction to injury. The injury signals reach the nuclei of the astrocytes, switching on an emergency program of genes to start making special proteins to help the astrocytes cope with the alert.

It has been only fifteen minutes since your accident, and you are not yet in the ambulance, but already many of the oligodendrocytes in the damaged part of your spinal cord are dead. Myelin begins to unwrap and dissolve, leaving bare axons exposed to the damaged tissue. If these axons remain bare, they will eventually die.

Oligodendrocytes and neurons will continue to die over the next six hours while you are in the emergency room and undergoing emergency surgery to stabilize your broken neck. The initial death of neurons and oligodendrocytes resulted not only from the blow to the spinal cord that sheared open cells, but also from the release of neurotransmitters, in particular glutamate, spilled from damaged neurons and raised to toxic levels. The disrupted blood flow from vascular damage kills other cells, as well. This initial cellular death will be followed by a second wave of death for a few days after the initial injury. This second wave of casualties results from the toxic conditions in the damaged region caused by microglia and astrocytes battling the injury. This later collateral damage is responsible for the death of more oligodendrocytes and neurons. Even axons near the damaged region that had been spared will lose their myelin over the next several days because of the unhealthy state of the tissue. This contributes to further paralysis as impulses are unable to pass through healthy axons beyond the frayed insulation. Their proximity to the cellular violence has made these myelinating glia casualties as innocent bystanders. Their death then makes them unwitting agents of more destruction.

At the hospital a picture of your fractured neck and smashed spinal cord forms in the mind of the neurologist as he pinpoints exactly where the cord is injured by quickly mapping the areas of remaining sensation on your now dissociated body. The CAT scan or MRI confirms what he knows. Surgery stabilizes your neck bones, and you are put to bed. Powerful drugs blunt the pain and induce sleep, but now everyone waits.

What everyone is waiting for is glial cells to come to your rescue. By the next day the damage in your spinal cord has grown substantially beyond the point of impact, like rot spreading out from a bruise in an apple. Only glia can stop this rotting destruction from spreading through your entire spinal cord.

As you rest the day after your injury, astrocytes along the perimeter of the damage are beginning to wall off the injured region. They multiply in number and assemble together to form a perimeter around the injury site to prevent the damage from spreading. They build robust cellular skeletons of the filamentous protein GFAP. Other proteins are secreted on the surface of the astrocytes to form an impenetrable shield. The glial scar helps restore the blood-brain barrier as it confines the injury.

Over the next week as you rest in the hospital, astrocytes in the glial scar continue to coat the barrier with slippery proteins called chondroitin sulfate proteoglycans. These greasy substances prevent cells from gaining a foothold on the barrier. Microglia inside the perimeter continue devouring cellular debris, becoming engorged with myelin fragments. By the end of the week, they have consumed all the cells and axons in the injury site. In another week there will be no astrocytes remaining inside the damaged region either; only a fluid-filled cyst inside a glia-bound scab.

Neurons begin to die soon after their axons are severed even though their cell bodies may be protected by being located far away from the site of injury. We know this is not natural death, but rather cellular suicide. When neurons are disconnected from their normal targets, genes in the neurons activate to cause them to self-destruct. In animals with a mutated gene that interferes with the self-destruct genes (called WldS), neurons do not die when their axons are cut. Why this mass suicide of these precious remaining neurons?

The reason for this peculiar mass death is that growth-stimulating proteins are released from the cell to which the now-severed axons were connected, for example a muscle fiber, skin cell, or the next neuron in the original circuit. During fetal life, these protein signals acted like beacons guiding the axons to their proper points of connection. Thereafter, the constant supply of these proteins taken up by the nerve ending and shipped to the cell body let the neuron know all is well. But during development, precisely the right number of neurons must be generated and matched to the proper number of cells needing connections, and every one of these neurons must send out its axon over vast distances to find its proper target. Neurons that became misrouted and failed to reach their proper point of contact could not tap into this life-sustaining growth-factor protein, and they died while your brain was forming in the womb. This mechanism is a very effective way to guide the proper wiring of our nervous system and eliminate misrouted connections. But now that the axon is crushed or cut, the protein growth factors from the proper point of synaptic contact can no longer reach the cell body. Like a rocket traveling off course, the neuron activates a self-destruct mechanism once it knows that it is no longer on the proper trajectory.

Yet even as cells are dying, other mechanisms are activating to begin the process of healing and repair. Some neurons at the injury site will seal up the end of their severed or smashed axon and trigger a genetic program to reactivate long-dormant genes that were last operational during fetal development, when the neuron first grew out its axon to wire up your body. The genes produce proteins that cause the axon to sprout, and these sprouts start to grow out in search of their proper targets.

Why didn't these damaged neurons self-destruct? One reason is that during this phase of regeneration, astrocytes and microglia release neurotrophic (literally, "neuron feeding") factors that sustain the neurons. Some of these proteins are the same growth-stimulating substances released by the target cells. By releasing neurotrophic factors at the injury site astrocytes keep the damaged neurons alive and coax the axon sprouts to grow. Astrocytes also begin to release angiogenic factors, proteins that stimulate the growth of new blood vessels to provide an infusion of vital food and oxygen to the damaged tissue.

Oligodendrocytes return to a more youthful state and begin to divide. These rejuvenated cells and other oligodendrocytes migrating into the damaged region extend their cellular tentacles to cling to as many damaged bare axons as possible. Immediately, they begin to wrap sheets of myelin around them to restore their insulation. This remyelination restores the ability of axons to conduct electrical impulses that was lost when their myelin sheath was damaged after the injury. As these glia repair the axon sheath, you begin to feel some recovery of sensation and the ability to move a bit more than before, but you are still paralyzed.

But as the surviving axons begin to sprout new fibers and seek out their original connections, they are stopped in their tracks at the site of injury. The result is that you will be paralyzed for life. If this were a damaged nerve in your arm or leg, these axons would grow out and eventually find their proper connection points on your muscles. Sensory fibers in your body too would reconnect with circuits carrying pain, touch, warmth, and pressure and the other sensations of the world to your brain. But in the damaged spine and brain, the sprouting axons fail in their valiant efforts to reconnect. Can glia explain why axons cannot regrow after spinal cord injury?

GLIA: THE CAUSE AND CURE FOR PARALYSIS

Why are axons in our brain and spinal cord incapable of repairing themselves, unlike axons in our peripheral nerves? Are central nervous system neurons more feeble? The most hopeful possibility would be that the tissue environment in the central nervous system is to blame. Rather than the hopeless scenario that CNS axons are by nature too feeble to regrow after injury, the tissue environment might be something that could be modified with the proper medications or treatments. If the CNS environment is the reason axons can't regrow, what is it about the environment that is blocking healing? Is some critical requirement for axon growth lacking in the environment, or is the environment in some way hostile to axon repair?

To answer these questions, Albert J. Aguayo and colleagues from McGill University in Montreal, Canada, conducted experiments in 1980 in which they severed the spinal cords of rats and surgically spliced in a segment of the sciatic nerve from the leg to bridge the injury in the spinal cord. Two months later they examined the splice to see if severed CNS axons could regrow through the environment of transplanted peripheral nerve. They found that spinal cord neurons grew into and through the splice without difficulty. Aguayo concluded that CNS neurons in the spinal cord could regenerate if the cellular environment in the brain and spinal cord was changed to be more like that of the peripheral nervous system.

What about the CNS environment fails to support regeneration of axons? A surprising clue came from the research of Martin Schwab at the University of Zurich.

"I wanted to test Albert's (and Ramon y Cajal's!) hypothesis that CNS long-distance regeneration fails due to a lack of neurotrophic factors," Schwab recalled, explaining the inspiration for the experiment he performed in 1988. As mentioned previously, neurotrophic factors are proteins released by cells that sustain neurons and stimulate them to grow. It was natural to assume that regeneration in the CNS failed because these growth factors were missing from the brain and spinal cord, especially as Schwann cells are known to release many neurotrophic factors, such as nerve growth factor.

Nerve growth factor (NGF) is the first of several powerful growth factors discovered in the nervous system. If lack of this factor was failing to sustain axon regeneration in the CNS, Schwab reasoned that simply adding abundant NGF would stimulate the growth of severed axons. To test the idea, he gave regenerating neurons in cell culture a choice to grow through either segments of peripheral nerve or segments of the optic nerve from the central nervous system in the presence of abundant NGF added to the culture medium. If the hypothesis was correct, the axons should grow equally well through either the CNS or PNS nerve segment.

Surprisingly, the results disproved the neurotrophic hypothesis.

"The neurons grew well through the sciatic nerves but not at all into the optic nerves. The conclusion from that was the concept that inhibitory factors may be present in CNS," Schwab wrote in an e-mail.

Although it was not known why, it was clear that CNS axons could resprout and grow through segments of peripheral nerve. This realization offered a possible treatment for spinal cord and brain injury. Over the next ten years Aguayo applied his technique of grafting nerve segments into various regions of the spinal cord and brain of experimental animals to provide a pathway for axon regeneration. In all cases, injured

CNS axons would grow long distances through a bit of peripheral nerve grafted into the damaged brain or spinal cord. Remarkably, splices of PNS nerve used to bridge severed optic nerve permitted neurons in the retina to retrace their path back to the brain. Moreover, these axons formed functional connections that responded to light.

The conclusion was abundantly clear by the late 1980s. CNS neurons can indeed regenerate their axons just as well as PNS axons; something in the environment of the CNS tissue blocks axon regeneration. This conclusion was confirmed by complementary studies showing that when neurons in parts of the body outside the spinal cord and brain were transplanted into the central nervous system, they were no longer able to extend axons. Paralysis results from unfavorable conditions for repair that exist inside the brain and spinal cord. To help people like Christopher Reeve, we must destroy the roadblocks to repair in the central nervous system that do not exist in the peripheral nervous system.

What could these roadblocks be? If it is not a deficiency in the neurons themselves, perhaps the answer must have something to do with the cells constituting 85 percent of all cells in the brain—the glia. Certainly glia in the central and peripheral nervous systems are radically different. In our peripheral nervous system we have only one type of glial cell, the Schwann cell. There are no astrocytes, microglia, or oligodendrocytes. On the other hand, there are no Schwann cells in the central nervous system. These very different glial cells provide an entirely different landscape in the brain and peripheral nerves. Moreover, since all the axons would have died and degenerated in the segments of peripheral nerve that Aguayo spliced into the CNS, they could not be a factor in promoting regeneration in the CNS. Could glia in the spinal cord and brain be the ultimate reason for paralysis?

Logically, since a severed nerve in our limbs or trunk can regrow without any of the types of glia from the central nervous system, Schwann cells must provide the necessary conditions to allow and perhaps even stimulate axons to regrow. To test this logic, several research groups performed a definitive test. Instead of splicing a piece of leg nerve into a severed optic nerve, these investigators extracted Schwann cells from rats and grew them in large numbers in cell culture. Then they filled small artificial tubes with these glial cells and used this tube to splice a severed optic nerve back together. The result was excellent outgrowth of the CNS axons through the Schwann-cell-filled tube, and these axons reconnected with the proper visual circuits in the brain. Glia restored sight—or at least sensitivity to light.

Schwann cells were found to secrete a large number of protein factors at the injury site, including NGF. These neurotrophic factors are well-known powerful medicines made by the body that rescue neurons from death and stimulate and guide the growth of the severed axons. As mentioned above, many of these growth factors are the same substances that guide the formation of the brain and nerves in embryonic development. In effect, Schwann cells sensing the injured axon reverted to an early embryonic stage to set the clock back and reinitiate the events that formed the nervous system in the first place. But, as beneficial as these Schwann-cell-derived factors are to CNS neurons, Schwab's experiment showed that the lack of neurotrophic factors was not the critical reason for the failure of CNS neurons to regenerate. Something else besides the lack of neurotrophic factors was blocking nerve regeneration in the central nervous system. What could that be?

The experiments splicing PNS nerve into the central nervous system provided valuable and hopeful information for the eventual cure of paralysis, but this technique is not a practical medical treatment. The central nervous system is far too delicate, miniaturized, and complex to repair with such a crude patching technique. The most reasonable approach would be to determine what about Schwann cells supports axon regeneration, and what about microglia, astrocytes, or oligodendrocytes thwarts it.

The Remarkable Growth Cone and Its Glial Friend and Foe

How does an axon find the correct pathway as it grows out from a severed stub to restore appropriate connections after injury? Imagine the journey an axon from a spinal cord neuron must take to find and reconnect with the muscles that curl your toes! The growing axon must have some way to sense the chemical and physical features of its local environment. It needs to know where in the damaged brain or spinal cord it is, which direction it needs to grow to seek out its appropriate target, and ultimately how to recognize the one-in-a-million spot on the right neuron (or muscle) that it must reconnect with to restore a functional circuit.

The severed tip of an axon transforms into one of the most beautiful and dynamic cellular structures in nature, called a growth cone. When seen in living form in a culture dish, it resembles a hand with nimble fingers delicately touching the surface ahead like your own hand reaching under the bed in search of a lost coin. These fingers are covered with an array of molecular sensors to detect not only the physical terrain, but also its chemical characteristics as the growth cone sniffs dissolved molecules sent as messages from other cells and its target tissues. The growth cone feels its way along, probing a pathway through tissue and deciding which way to proceed while trailing out the growing axon behind it. Glia are the stepping stones that this growth cone will follow to reconnect with the proper neuron.

In my studies using silicone tubes to bridge gaps in the sciatic leg nerve of a rat, I could see the Schwann cells enter the tube and form a bridge of cells for the growth cones of newly sprouted axons to follow. These and similar studies by other researchers showed that Schwann cells guide the axon across the defect by building a protein scaffolding, called an extracellular matrix. This matrix contains macromolecules known to promote the adhesion and migration of axons in embryonic development of the brain; now they served to guide the healing of the nervous system in recovering from injury.

The chain of Schwann cells remaining after the severed axon it ensheathes dies forms a cellular pathway all the way to the target organ that the growth cone can follow to restore its original synaptic connection. CNS axons also respond to these cell-surface macromolecules and growth factors just as well as PNS axons.

Recall the puzzle early anatomists faced when they saw axons inside nerves coated with Schwann cells but could find no such cell in the central nervous system? In the brain and spinal cord the myelin coating is formed by extensions from oligodendrocytes. These cells themselves are attached to multiple axons only through their many slender tentacles. This difference between myelinating glia in the central and peripheral nervous systems is one of the main reasons axons in damaged brain and spinal cord cannot repair themselves: without Schwann cells to guide them to their proper destination, the severed axons are lost.

When axons shrivel and die after injury in the peripheral nervous system, the series of Schwann cells that ensheathed the axon survive after the injury. These glia then act as cellular stepping stones for a newly sprouted axon to follow back to its original destination. Unfortunately, this path of cellular stepping stones does not exist after injury in the central nervous system. Unlike Schwann cells, each oligodendrocyte sends out dozens of cellular processes to myelinate as many different axons as possible. Even if these oligodendrocytes survive the injury, they withdraw their processes when the axon withers. The pathway leading to the proper destination is lost. Without glial guides, regenerating CNS axons have a baffling task to find their way back to their original connection and restore normal function.

In addition to the lack of a glial pathway to guide the growth of a severed axon, there are two more serious problems with axon regrowth in the central nervous system, and here too, glia are the culprit. As mentioned, when the spinal cord or brain is injured, microglia rush to the site and begin to clear away the debris and secrete chemicals to repair the damage. Astrocytes soon follow, backing up the microglia by becoming "reactive," that is, transformed into a state to respond to such a disease or injury. The reactive astrocytes assist in healing, but they also begin a process of cordoning off the area of damage to keep it from spreading, forming a glial scar around the site of injury.

Unfortunately, this scar also frustrates the growth cone's attempt to penetrate the damage site and work its way back to its intended point of contact. Not only is the glial scar a physical barrier to growth cones, the proteins astrocytes attach to this barrier repel the growth cones. The repellent effect of these proteins is astonishing. A touch by a single finger of a growth cone on one of these proteins (for example heparin sulfate proteoglycan) causes the entire growth cone to collapse, and the axon retracts like clenched fingers from a hot stove. Growth cones simply cannot grow where these molecules are laid down, and unfortunately, they are laid down right in the path an axon must use to reconnect. The axon sprouts stop right at the point of injury. If we could find ways to neutralize these inhibitory materials in the scar and prevent or dissolve the scar itself, this would in theory allow axons to regrow back through the point of damage.

Certain bacteria secrete powerful enzymes, called chondroitinases, which enable them to penetrate tissues and infect organisms. These enzymes readily dissolve the major constituents of the glial scar that block axon regrowth, the chondroitin sulfate proteoglycans. In 2002, Elizabeth Bradbury and colleagues at Kings College London reported experiments in which they injected this enzyme into the glial scar to dissolve it and found that axons were able to regenerate in the injured brain. Most intriguingly, the neurons they studied were those that die in Parkinson's disease.

Other groups are developing even more elegant methods of administering the enzyme to dissolve the glial scar. They have taken the gene from the bacteria that make the chondroitinase enzyme and inserted it into a virus. They then inject the virus into the damaged spinal cord. Astrocytes become infected by the harmless virus, but in the process they take up the gene and start making the enzyme themselves, just as the bacteria do. In experimental animals this approach helps improve axon growth and also improves motility and sensation after spinal cord injury. This is a much more practical medical approach to treating spinal cord injury, but unfortunately, this single enzyme cannot remove all types of inhibitory molecules that astrocytes deposit in the scar. This new approach offers a promising direction to follow in research to combat paralysis by enlisting the aid of glia. More research will be required to perfect the methods and assure that they are safe for use in humans.

PRUNING SPROUTS: MYELINATING GLIA, THE MAIN ENEMY OF SPINAL CORD REPAIR

It is easy to understand how scar tissue can frustrate repair after injury, but there is an even more insidious and unexpected block to growth cones in the central nervous system. The third major impediment to axon regeneration in the brain and spinal cord is oligodendrocytes. In performing their important function of insulating axons with myelin, oligodendrocytes unavoidably thwart regeneration.

In 1988 the Swiss scientist Martin Schwab and colleagues found that when a growth cone touched myelin extracted from the central nervous system, it instantly collapsed. This was extremely puzzling and unexpected. The devastating arrest of all progress by sprouting axons meant that very little if any regeneration would be possible after injury in the central nervous system, because myelin coats the very white matter tracts these axons must traverse to restore functional connections. Even samples of purified myelin chemically extracted and added to a cell culture instantly collapse all the growth cones in a culture dish. Who could have imagined that the myelin insulation on axons would repel the growing tips of axons?

"Yes, I was very surprised to find that CNS myelin has an inhibitory role [on axon regeneration]," Schwab informed me.

"The neurons [in their cell culture experiment] grew well through the sciatic nerves, but not at all into the optic nerves. The conclusion from that was the concept that inhibitory factors may be present in CNS myelin. Very surprisingly, we found that the culprits were oligodendrocytes and myelin."

I was privileged to have one of Dr. Schwab's colleagues, Dr. Christine Bantlow, visit my laboratory briefly in the early 1990s to collaborate on research related to this problem. She had prepared microscopic vesicles—tiny bubbles of myelin—which we added to cultures of mouse sensory neurons as we watched them grow under a microscope. As these minuscule droplets rained down on the growth cones, it reminded me of an aerial bombardment. The instant a myelin bomb hit even the smallest finger of a growth cone, the entire structure collapsed, halting it like a train bombed to pieces in its tracks. Within minutes there were no more growth cones left in the entire culture dish. The culture had become a wasteland of collapsed axon stubs.

Clearly, oligodendrocytes were joining forces with the inhibitory actions of astrocytes against the advances of regenerating axons. The consequences were painfully obvious, but they suggested a new therapeutic approach to treating spinal cord injury. If the molecules in myelin that have these devastating effects on growth cones can be identified and neutralized, we could help heal people with brain and spinal cord injuries.

Still, it seemed to make no sense for Nature to have booby-trapped myelin so that it halts all axon regrowth in the brain. Another puzzle. Could this be a clue that we are not only missing information, but also missing important concepts?

"With hindsight, these results correlated well with the restriction of developmental plasticity at the time of myelination in many parts of the brain," Schwab observes now from a new perspective on myelin that has emerged only in the last few years. He is referring to the well-known ability of a child's brain to repair itself and rewire its connections according to early childhood experience, which greatly diminishes after adolescence. Myelination of the human brain, coincidentally, is largely completed by late adolescence. We will return to this subject in part 3, where we discuss new insights into how glia may be involved in learning, but for the present we focus on how myelin blocks axon growth after paralyzing injury.

Dr. Schwab and colleagues began a vigorous search to identify the factor in myelin that collapsed growth cones so that he could try to neutralize it. By extracting and purifying myelin and then carefully dividing it into different chemical fractions, Dr. Schwab hoped to identify the fraction and eventually the specific macromolecule that was responsible. Eventually Schwab's group isolated a single protein from myelin that when added to these vesicles caused growth cones to collapse just like pure myelin. He named the protein Nogo.

Now that he and his colleagues had isolated this growth-cone collapsing protein from myelin, how could he neutralize it? Just as your body will form antibodies to bacterial proteins injected into your body during an immunization, your immune system will recognize any foreign protein and form antibodies that will recognize and destroy it. All that Schwab needed to do was inject the Nogo myelin protein into a rabbit and then collect the antibodies made by the rabbit's immune system. Once he had such an antibody he could inject it into the site of spinal cord injury and the antibodies would bind to the inhibitory protein in myelin like a wrestler pinning an opponent, rendering it inert. The growth cones should then slip past and reattach their proper connections. He tried this and it worked! In the injured spinal cords and brains of rats, injections of this antibody, called IN-1, greatly increased the ability of axons to grow past the site of injury and reestablish their normal connections. Most importantly, these new connections restored some sensation and some motility. The rats were better able to walk and sense pain.

Their recovery was far from complete, however. Only a small number of axons regrew. Even so, this was a vast improvement. While it was not enough in itself to get a paralyzed person out of his wheelchair and walking, it was a step in the right direction. Even a small improvement in sensation or movement would greatly improve the quality of life for people with paralysis.

As I write this I happen to be 39,000 feet in the air approaching the Dallas/Fort Worth airport on my way to a scientific meeting. Seated across the aisle in front of me in the bulkhead seat is a tall, athletic-looking African-American man with a close-cropped military haircut wearing a blue T-shirt. His arms and legs are paralyzed. Every fifteen minutes or so his female companion unbuckles her seatbelt, gets up out of her seat, and pushes as hard as she can against both of his knees to restore his posture. Then he slowly sags back again, unable to hold himself fully upright in his seat.

The dependence of such an athletic-looking man on such a dainty woman seems incongruous. What happened to him? His haircut, youth, and athletic body may be clues that he is a recent veteran. I do not know. Sometimes I see her get up and bend his head to his knees to release the strain on his spine. Throughout the flight I fidget constantly in my own seat to twist my own torso and flex my back and neck against the pain of confinement. On a long flight like this the discomfort is nearly unbearable. Looking at him, I can only imagine the silent and painless torture a plane ride causes his muscles and joints. Unknown to him his body is suffering pinched blood vessels and aching bones, nerves and joints compressed from the unrelenting confinement. Watching this scene, anyone could see how much relief and improvement in quality of life even a small amount of sensation or muscle tone would bring this man, and how welcome even the small independence of being able to sit and adjust his posture would be.

Nogo Is Not Alone

The disheartening reason that antibodies against Nogo failed to restore axon regeneration fully after spinal cord injury soon became apparent to researchers. Nogo was not the only villain. There is a gang of growth-cone-collapsing proteins in myelin. Another well-known protein in myelin, MAG, was found to have the same effect as Nogo. A third myelin protein, OMgp, which covers the surface of oligodendrocytes, also was also found to be a potent inhibitor of axon growth.

Curiously, MAG has opposite effects on growth cones during embryonic development. Early in development, MAG actually stimulates axon growth of immature neurons. This puzzle offers promise. It suggests that the growth-cone-collapsing factors in myelin could be neutralized by the neuron itself, since this activity occurs naturally through the course of fetal development. All that must be done to reverse the inhibitory action of MAG on adult axons is to restore the neuron to the immature state that allowed it to thrive in the presence of MAG. To do this, it would be necessary to discover how these inhibitory molecules are detected by growth cones and how this interaction causes the cones' cellular skeletons to collapse. If the membrane proteins on the growth cone that recognized Nogo, OMgp, and MAG could be identified, they might be blocked with drugs. Growth cones would then be deaf to the inhibitory signals on myelin and march on to reestablish their connections after injury.

In 2001 Stephen Strittmatter and colleagues at Yale University identified the Nogo receptor in axons. He was then able to devise a way to block this receptor. When this was done, Strittmatter and colleagues found that regeneration of a severed optic nerve was greatly improved, as was recovery after brain stroke. But axon regeneration was still much impaired.

Vigorous research into this receptor soon revealed its detailed molecular structure, and this gave important clues as to how it might work to collapse growth cones. Surprisingly, this analysis showed that the receptor, which sits on the axon membrane, has no way to send its sabotaging signals into the axon cytoplasm. Instead, the Nogo receptor hijacks other membrane molecules in the axon that can send signals controlling axon growth. This made the Nogo receptor a resourceful foe. It had many different ways to stop axon growth once the receptor was activated. Like a terrorist, it attached its lethal bombs to a variety of different molecular transporting systems carrying information from the axon membrane into the cell. If drugs were used to stop one pathway, there were still several alternative routes from the receptor into the axon cytoplasm.

Knowing that MAG switches from stimulating to inhibiting growth cones during development, Marie Filbin at Hunter College decided to block the signaling pathways inside the neuron at a key point where many different types of messages from membrane receptors converge to weaken and collapse the cellular skeleton. Using drugs that cut off signals at this key point, she restored the ability of adult axons to grow in the presence of MAG. The point in the signaling pathway she manipulated with drugs was the same one she had found changed in young neurons as they aged and became vulnerable to MAG-induced growth cone collapse.

As these discoveries demonstrate, tremendous gains are being made in understanding why CNS neurons do not regenerate after they are severed, and these gains have come from studying glia. Thus far, no single treatment will allow CNS axons to regenerate anywhere nearly as well as PNS axons can after injury. Why are there so many parallel backup ways for CNS myelin to block axon growth? Why does myelin have this effect on axon sprouts at all? Many neuroscientists suspect that all of this new information about myelin will eventually tell us something very fundamental and important about the brain.

Scientists are closing in on the molecular villains in myelin that cause irreversible paralysis. As they do, the new understanding of myelinating glia is revealing cures to many other human diseases, from mental retardation in premature babies to multiple sclerosis. We are learning that myelin is far more than insulation. Axons sprout not only to repair damage, but also to make new connections in learning. In the past scientists focused all their attention on neurons. Today myelin provides new insight into learning, psychological disturbances, information processing, consciousness, and, as will be discussed in a later chapter, the reason why you can't teach an old dog new tricks.

If Christopher Reeve were still with us today, one wonders how he would view the progress in research to cure paralysis that he so forcefully advocated. We need more research on glia and their interactions with neurons, especially the actions of glia in promoting and inhibiting axon regrowth. Many of the beneficial actions of glia might be tapped to stimulate axon regrowth. At the same time, the newly discovered inhibitory actions of glia might be blocked at appropriate times during treatment to improve recovery after CNS injury. There will be no "silver bullet" to end paralysis, because the injury and healing involve so many different biological processes. But multiple strategies might be applied at appropriate times in the recovery process, perhaps including blocking myelin proteins signaling growth cones to collapse; increasing growth stimulating factors secreted by astrocytes and microglia; limiting and controlling the damaging effects of astrocytes and microglia soon after injury to minimize the second wave of neuron and oligodendrocyte death; dissolving the glial scar once the wound is stabilized to allow axon growth past the damage site; inserting artificial bridges to replace lost or damaged spinal cord or nerve tissue to guide axons back to their proper destinations; stimulating oligodendrocytes to form myelin faster to restore electrical conduction and prevent the death of denuded axons; and replacing lost neurons and oligodendrocytes by transplanting neuron and glial stem cells (see chapter 11 for the latest research on transplanting glia into patients to treat spinal cord injury).

All of these approaches involving glia have been used individually and successfully in experimental animals. In 2009 Maiken Nedergaard and colleagues at the University of Rochester published studies showing that a chemical that suppressed the response of astrocytes and microglia to injury signals released by damaged neurons improved recovery from spinal cord damage in rats. The chemical is blue dye no. 1, used to color M&M candies and other foods. Because the compound is considered safe, studies in humans can begin quickly. Increased research into this other half of the brain offers great promise to help those who now rely on wheelchairs. But paralysis is only one type of nervous system injury. There are many other types of brain injury, and in all of these, glia are the center of the injury and healing process.

BRAIN DAMAGE AND ROCK AND ROLL:
ROCK STARS

Eric Clapton, Bono, Phil Collins, Peter Frampton, Bob Dylan, and Pete Townshend all share something in common besides rock music: they all suffer from hearing loss because of it. So do thousands of their concert fans and listeners blasting loud music through ear buds from iPods and other MP3 players. According to a recent survey by the Musician's Clinic of Canada, 10 percent of the Canadian population suffers hearing loss, but almost four times as many rock musicians are partially deaf (37 percent). The Rock and Roll Hall of Fame could be nicknamed the Hall of Partially Deaf Musicians, because 60 percent of the inductees have lost hearing due to loud music. The body-thumping rock and roll blasted from electronically amplified instruments and screeching microphone feedback are the obvious cause, but don't feel too smug if you are thinking that the degenerate rockers with their ear-splitting music have it coming to them; even more classical musicians (52 percent) suffer hearing loss from exposure to orchestra music so loud it fills an entire concert hall without electronic amplifiers. Exposure to loud sounds on the job and to horrendous noise during combat are other well recognized causes of deafness.[1]

Most of those hard of hearing understand that an intense noise or chronic exposure to loud sound damaged the hair cells in their inner ear. These microscopic cells with their tiny hairs sprouting up like a clump of prairie grass are so delicate that their quaking in the minute pressure changes of sound waves buffeting air molecules is where hearing begins. What few deaf musicians realize is that the loud noise also caused brain damage. And recent research implicates glia in this degeneration of the brain.

Dr. Kent Morest, a professor at the University of Connecticut Medical School, wondered what went on inside the brain after hearing loss caused by loud noise. In a series of experiments, he exposed chinchillas and rats to loud noise that damaged the hair cells in the inner ear, and then he trained his electron microscope on the part of the brain where the neurons carrying impulses from the hair cells make their first connection (the inferior colliculus).

An electron microscope is essential for this study, because the detailed circuitry of the brain is too miniaturized to be seen by a light microscope. Electron microscopy is a difficult and time-consuming process. The structures are so tiny, great skill is required to manipulate the samples, and exceptional spatial memory and reasoning are required to understand the images the microscope provides.

Electron beams will not pass through air the way light does. This means that the tissue must be placed inside a total vacuum so that it can be bombarded with focused beams of high-energy electrons. But tissue will not survive inside a vacuum. Any water inside tissue would instantly boil away violently, exploding the cells. This is why astronauts wear pressure suits on a space walk to protect them from the vacuum of space. To survive bombardment by an electron beam inside a vacuum, tissue to be examined in an electron microscope must be chemically preserved and all traces of water must be removed from the cells and replaced with hardened plastic resin. All this must be done without disrupting the fragile structures inside the cells. The objective is to see the finest structures in cells in their true form, not their dehydrated, mummified remains.

The process of embedding pencil-lead-size bits of tissue in plastic requires many complicated steps over the course of a week. Then unimaginably thin slices must be cut from the embedded tissue, for electrons cannot pass through plastic, just as they will not pass through air. These ultrathin sections are no thicker than the skin of a soap bubble.

In his lab, looking through a light microscope, Morest trims the plastic matrix with a razor blade into a bullet shape, tipping it with the bit of tissue he wants to examine, which is no larger than a pinhead. He then places the bullet into a precision slicing machine equipped with a miniature blade of pure diamond, sharpened into the finest edge possible. Sitting at this desk-sized instrument, he slowly turns a control wheel like an organ grinder and slices off a thin section, examining his progress through a powerful magnifying glass. Each minute section floats off the diamond blade onto the surface of a pool of alcohol in a miniature trough. Imperceptible air currents and static electricity in the air can wreak hurricane-force havoc with these tiny structures, and he holds his breath as he shaves each one—very slowly—seeing each slice peel off and glisten with the same iridescent rainbow of colors that swirl over the surface of a soap bubble. The colors tell him exactly how thin the section is. Ruby and cobalt slices are too thick; he delicately adjusts the angle of the knife, the speed of his slice, and various other controls until a golden/silver slice floats off the diamond and rafts precariously onto the pool of alcohol. Everything will depend on how perfectly he has sliced this microscopic section.

Now he must pick up the flimsy shimmering gold slice and attach it to a support that can be placed inside the electron microscope. This requires a very delicate tool, which he makes by plucking out a single eyelash and gluing it to the end of a wooden swizzle stick. This is the standard tool used by all electron microscopists. With this fine eyelash hair, he nudges the slice and sculls the fluid around it to coax it into the center of the trough as he watches intently through the magnifying glass. Touch the tiny slice too hard with the eyelash and it will wrap around the hair like a ruined kite on a tree limb. Using fine watchmaker's tweezers, he lowers a paper-thin circular copper grid, a bit smaller than the smallest watch battery, carefully against the floating slice. The golden section sticks to the copper grid like a moth on a car grille.

An electron microscope is an imposing instrument, about six feet high, housed inside a small darkened room designed exclusively for this instrument and the support equipment it requires—high voltage transformers, vacuum pumps, and a liquid nitrogen tank. A heavy steel column about ten inches in diameter rises up to the ceiling from a workstation console, encrusted with buttons, switches, knobs, flashing indicator lights, digital numerical displays, and video monitors. Sprouting out of the top of the column is a heavy, two-inch-diameter electrical cable, which delivers 100,000 volts to the electron gun at the top of the column. Liquid nitrogen billows out a white cold fog spilling down from the top of the column in the darkened room like liquid oxygen from a rocket on the launchpad at dawn.

After placing the sample in the microscope and energizing the electron gun, Morest sees shadowy patterns come into focus on a phosphorescent screen. Anyone not trained in interpreting these shadowy patterns would see nothing more than grey doodles on a glowing yellow-green background. But the electron microscopist, with his head pressed against the column to look at the glowing screen through a thick glass window, is transported inside a cell of this rat's brain. The tiny slice is now expanded into an immense new universe. Turning wheels simultaneously with both hands he moves the grid, scanning acres of cellular territory for hours, completely absorbed in the new world he is seeing. Hours later he emerges from the room with a telltale flat spot embossed into the center of his forehead from pressing it against the column.

Each time he increases the magnification the image twists violently, as if someone suddenly rotated a map on which you had just found your position. This twisting is caused by electrons deflected in a spiral path through the magnetic lenses focusing the beam. Increasing the magnification in an electron microscope is like seeing the ground spinning below as you descend swinging from a parachute, bringing greater and greater detail as you approach the ground. Morest must keep his bearings through all of these dizzying twists, maintaining a conceptual thread back through the slicing process to the way the tissue was oriented in the brain of the rat. He will have to analyze hundreds of sections to reconstruct the three-dimensional cellular structure correctly, a process requiring months or years of work.

In the part of the brain where the nerve endings carrying impulses from the hair cells terminate, Morest saw a specialized structure—a thick nest of synapses packed together tightly like a popcorn ball. An unusual thing about the nest of synapses caught Morest's attention—there were no astrocytes. He then looked at a sample from the brain of an animal taken right after exposure to a loud sound. As soon as he focused on the same brain area, he saw instantly that between the synapses were tiny fingers of cells that had the telltale marks of astrocytes. Months of research revealed the violent physical remodeling in this part of the brain in the wake of the damaging sound.

Right after the destructive sound, astrocytes begin to move into the synaptic nest, cupping their cellular hands around synapses tightly as if to shield them from the noise. In fact, Morest concluded, the astrocytes are responding to the overstimulation of these acoustic neurons and sopping up the excitatory neurotransmitter, which in excess kills neurons from overstimulation. At the same time the astrocytes are releasing neurotrophic factors in an effort to prevent the neurons from dying.

Failing against the damaging roar that causes these synapses to fire beyond normal limits, the synapses and nerve endings begin to wither and die. The animal is going deaf. The loud noise not only causes the hair cells in the rat's ear to die; the intense screams of the dying hair cells in the form of nerve impulses shooting through axons to the brain kills the brain cells, too, from overstimulation. During this process, astrocytes revert to the cleanup role they perform after any brain injury, removing bits of cellular debris. The astrocytes become "gliotic," that is, swollen in response to injury, and digest away the damaged synapses.

Morest's research shows how astrocytes are capable of moving their cellular extensions through our brain, mediating neural excitation like control rods plunging into a nuclear reaction pile. They respond physically to the activity of neurons and synapses, especially when neurons and synapses exceed the limits of normal operation. At first the astrocytes do whatever they can to restore the normal balance of activity in the brain circuits and protect synapses and neurons from the damaging byproducts of excess activity. Often these actions of astrocytes are life-saving for neurons, but if the damage is beyond repair, the astrocytes triage the injured synapses and neurons and quarantine off the area. This damaged part of the brain is now disconnected, setting the stage for rebuilding new connections, if possible.

Do astrocytes prune synapses only in remodeling the brain after damage caused by loud music, or is it simply that this experiment reveals what astrocytes do everywhere in our brain at all synapses conveying thought, emotion, or wonder? Whichever it may be, the impact of these glia on brain healing and reorganizing neural circuits after injury is still profound. Further research on both the protective and destructive actions of astrocytes may allow us to exploit these activities and assist the glia in preventing or recovering from deafness. After all, these cells are responding to the injury at exactly the right time and place; all doctors must do is give these rock-star cells a hand.

OXYGEN: MOTHS AND RUST

There are countless planets in our infinite universe, but most have hostile environments, where the temperatures are freezing or blistering and the atmosphere is toxic or even dangerously corrosive. The atmosphere is so caustic on one planet in our solar system it will turn steel to powder and ignite biological material into flames. That planet is Earth, and the corrosive gas is oxygen.

We are so habituated to living with this dangerous substance we have grown oblivious to it. There are no fires on other planets because they lack oxygen. Homes and forests would never be consumed to a dust of hot powdered carbon in a violent chemical reaction in those atmospheres. Fire is so common here that large numbers of people devote their lives to fighting it. To mitigate the corrosive effects of this gas, we wrap our food in plastic, protect our wood with coats of paint, and encase our most precious document, the Declaration of Independence, inside a sealed glass case pumped full of argon gas, to purge every violent molecule of oxygen from touching it, lest it turn yellow and crumble like old newspaper. We do these things and think nothing of them.

It was not always this way on planet Earth. Pollution is the source of oxygen that poisoned this planet's atmosphere, for oxygen is the explosive waste product of plants. This dangerous byproduct is produced when plants convert sunlight into high-energy chemical fuels, sugar, and starch, in a power-generating reaction called photosynthesis. As plants polluted the atmosphere over eons, oxygen drove off the lighter harmless gasses into outer space, and the atmosphere of the entire planet became saturated with a 20 percent concentration of the reactive waste oxygen.

Oxygen is lethal. It will kill many kinds of bacteria, because they evolved on Earth before plants polluted the atmosphere with this corrosive gas. Now bacteria seek refuge inside dead bodies and deep wounds away from the killer gas. The antiseptic hydrogen peroxide prevents infection by releasing reactive oxygen at levels that will kill bacteria but that are dilute enough for our cells to tolerate. In contrast to bacteria, humans' cellular ancestors evolved during a time on Earth when this caustic pollution had already fouled the atmosphere. They had to develop ways to resist this reactive gas.

Our cellular ancestors succeeded by evolving powerful chemical defenses to contain this violent material. In the greatest feat of biological jujitsu imaginable, our cellular ancestors even devised a way to exploit oxygen. They, and we, burn oxygen inside the chemical furnaces of our own cells. Our cells harness the same high energy that fuels a rocket into space (packing tons of liquid oxygen into the void) or melts heavy plate steel into a slurry when paired with acetylene (the mixture ignites violently when sparked and creates a searing flame of blinding intensity). Oxygen generates the energy that fuels our cells and sustains our life. Our cellular machinery is now so heavily engineered around it that we cannot survive in its absence for even a few minutes. Through ages of evolution we developed complex lungs and a vast circulatory system and blood to deliver the oxygen fuel to every single cell in our bodies. But our bodies can hold the ravages of this reactive gas at bay only within narrow limits and for only so long. Eventually, oxygen will get us all, as it slowly eats away the proteins, enzymes, and DNA of our cells, until they are weakened to the point of collapse—an end we call dying of old age.

There is a class of chemicals that combat oxidation (the chemical reaction seen in steel as rusting and in fuels as fire) called antioxidants. Oxygen ravages by stealing electrons from other atoms, violently ripping them away from the outer shell of electrons orbiting the nucleus of more stable molecules. With these electrons gone, the unbalanced and electrically charged molecule crashes into other molecules, destructing in chemical reactions that redistribute the remaining electrons in a way that restores some balance among the molecules. The electrons may be redistributed to a balanced state in this reaction, but the original molecule (protein, enzyme, or DNA) has been consumed in the process and the cell sustains some damage. This cellular corrosion accumulates over time. Antioxidants are generous. They have electrons to spare, and they donate them readily to appease the electron-plundering raids of oxygen. Antioxidants sacrifice themselves to reactive oxygen to protect other molecules vital for the life of a cell.

This is the chemical foundation for much of the health food industry. Health advisors sagely urge us to consume antioxidants—such as vitamins C, E, and A, green tea, and blueberries—to perhaps ward off cancer and old age. These natural antioxidant compounds cool the burning fires of oxidation in our bodies to safe levels. But there are many other powerful antioxidants in your body that you will not find on the shelf of your health food store. One of the most powerful antioxidants in the body is glutathione, and this lifesaving compound is packed in highest concentration inside glia, specifically astrocytes. As a consequence, astrocytes readily survive toxins and pathological conditions that can produce high enough levels of reactive oxygen to kill neurons instantly.

Neurons tested in a dish by adding hydrogen peroxide or other oxidants quickly die. But if they are grown on a layer of astrocytes, they flourish. Astrocytes lacking glutathione, however, cannot rescue neurons from death by oxidation. One of the reasons glia have such high concentrations of antioxidants is that astrocytes and microglia defend the brain by exploiting reactive oxygen as a weapon. Glia release oxidizing agents to attack invading bacteria and diseased cells, while the antioxidants shield the glia in this toxic battle. Sensitive always to the local environment and the needs of neurons, astrocytes will release antioxidants to blanket neurons under threat of oxidation. Like flame retardant from a fire extinguisher, the antioxidants released by astrocytes protect the lives of neurons in danger.

The release of antioxidants from astrocytes is one of the body's primary defenses against neurodegenerative disease, cancer, and aging. Eat fresh vegetables and sip green tea, but as you do, think about the battle for your life being waged furiously by the astrocytes in your brain. Their vigilance and selfless struggle help you survive into old age on this hostile planet.

CHAPTER 6

Infection

MAD COWS AND ENGLISHMEN: PRIONS MEET GLIA

They call themselves the Forest People (Fore). Numbering 35,000 souls, they live in several small hamlets in the mountains between 4,000 and 7,500 feet on the Purari side of the Ramu-Purari Divide of eastern New Guinea. Their 1,000-square-mile world is a dense moss-covered rain forest, cradled between the Yani River to the west and the Lamari River to the east. The summit of Mount Michael rises to 12,000 feet in the west.

The house for men is situated at the hub of each hamlet. Women are not permitted to enter except on rare ceremonial occasions. Women live in a row of small huts opposite the men's house, separated by the large open cooking area. The children live with the women.

At first the young woman in the village tried to conceal her increasing clumsiness, but soon it was obvious to everyone. She had not suffered a fever nor had she felt ill, but before long Namugoi'ia was overtaken with intense and ceaseless shivering. The trembling grew worse, reaching a point where she could no longer walk even with the aid of a stick. Her speech became slurred. Her emotions flared uncontrollably. She was prone to fits of deranged laughter.

When she became too ill to walk, the villagers dragged Namugoi'ia from the hut and left her to sit outside. She grew steadily weaker, as they knew she would. When her eyes began spinning in their sockets, she no longer had the balance even for sitting. Now they laid her alone inside a hut and left her to starve. A victim of sorcery, she could not be helped. Many like her choked to death or helplessly rolled into the house fire pit and burned to death.[1]

Upon hearing of the deadly new disease, Dr. Carleton Gajdusek (*Guy-duh-check*) was instantly overtaken with intrigue. Gajdusek was a young American pediatric specialist visiting Australia on a fellowship to study child growth in primitive cultures. Abandoning his fellowship from the National Foundation for Infantile Paralysis, he went straight into the deep jungle to meet with Dr. Vincent Zigas of the Australian Public Health Service, who had just reported a deadly neurological epidemic devastating the natives in the most remote regions of the Papua New Guinea highlands. Was it an infection? Genetic? The result of some toxin or malnourishment?

In a hastily written letter of March 15, 1957, to his former boss, Dr. Joseph Smadel at the National Institutes of Health in Bethesda, Maryland, Gajdusek wrote, "I am in one of the most remote recently opened regions of New Guinea (in the Eastern Highlands), in the center of tribal groups . . . still spearing each other as of a few days ago . . . This is a sorcery-induced disease, according to local people . . . May I send to you specimens and blood serum, etc., if it proves feasible? Most are now going to Melbourne."[2]

An anthropologist for the Australian government, Dr. Charles Julius, interviewed the dead woman's husband, Au'ia, who explained what had precipitated his wife's misery and gruesome death. He claimed that two men, his bitter enemies from a neighboring village, had obtained a fragment of Namugoi'ia's skirt. Then they must have wrapped it in the leaves of the karuku tree and placed it on a sorcerer's stone heated by the fire. They proceeded to beat the parcel violently with a stick, calling Namugoi'ia's name and saying, "I break the bones of your legs; I break the bones of your feet; I break the bones of your arms; I break the bones of your hands; and finally I make you die." As soon as Namugoi'ia started trembling, he realized that his two enemies must have obtained some personal possession from his wife and used it to cast the fatal curse.[3]

The guilty parties were promptly hunted down and killed in revenge, but this would not help Namugoi'ia. Victims of this wicked spell never got better.

"Joe, I stopped all my polio-fund salary when I left Melbourne in February, and since January will have no income until I finally get back to the States," Gajdusek wrote Smadel at the NIH. "I am now living off my small savings and have no grant at the moment, with plenty to do in a dozen fields, on my return; however, this is nothing to put off and nothing to drop. I write first to let you know about it, and second to ask whether there might be a grant you could dig up. I am in no position to make a more formal application, being devoid of mail, radio, or telephone and all else in the bush where I am working on the disease—

"All I need, however, is enough to insure my keeping alive once I get out, if this runs to several months (as it certainly should if I can afford to stay) . . . With or without help—I intend to stick this one out a bit."[4]

Together Zigas and Gajdusek trekked through the rugged mosquito-infested jungle to locate isolated tribes in a region unvisited by outsiders until recently. Along the way they provided medical treatment to the natives, who, they found, suffered from all manner of infections, parasites, festering skin ulcerations, and contagious diseases, while they gained the natives' trust and collected medical histories and samples from victims of the deadly disorder the natives called kuru.

Gajdusek passionately relished his swashbuckler role as jungle doctor and adventurer sleuthing out a mysterious disease deep in remote jungle regions. In another letter to his former boss Smadel, Gajdusek pleaded for the opportunity to pursue his jungle research on the mysterious neurological disease:

April 3, 1957
Dear Joe:
   I am writing to you just after returning from a patrol down to the southern fringe of the South Fore region, and to the Lamari River and Kukukuku and Awa linguistic areas. This is a little-visited region, uncontrolled at its extremity; and tribal wars, bow-and-arrow murders . . . have occurred here since my arrival. Takabu (a traditional native way of throttling a sorcerer with strangulation, and stone-beating to break the femur, fracture the ribs, and pound the costovertebral angles) is a frequent source of "emergency" medical problems, as are the wicked arrow-wounds. The patrol was directed to this little-known region to determine the

*southern extent of the kuru disease which has so occupied me and
is the subject again of this note.*

*Here in Moke, in a matfloor hospital which was quickly
built with native materials—in which we have a microscope,
hemocytometer, a host of lab reagents and equipment, and all
the diagnosis instruments that such a "bush" hospital would be
expected to possess, including opthalmoscope and tuning forks—
Dr. Zigas (the Kainantu physician and only M.D. in the Eastern
Highlands), who discovered the disease, and I have set up a kuru
investigation section. Thus far it is operating largely on his and my
own enthusiasm and his courage in daring to leave the European
population relatively unattended while we work far out in the
bush—that is, as long as the Administration which employs him
will tolerate his "research" diversion.[5]*

Gajdusek described with excitement forty-one cases of kuru that he
had collected and then pleaded for money to pursue his research:

*I now write to assure you that my last letter was no ruse; I have the
"real thing" in my hands. I can and will continue to support myself
and the project on my own funds and those that Zigas can wheedle
from the Port Moresby P.H.D. (now a rather complex bureaucracy,
with many petty jealousies—far away from this remote mountain
jungle post, but our lifeline). Will you let me know if you think
even a temporary salary, which I shall throw into support of this
project to pay for the axes, beads, tobacco, and other trading items
with which we purchase bodies (with autopsy permission) and food
for our patients—a normal U.S. salary will go mighty far here—
could be arranged to permit me to carry out the work for several
more months, at least long enough to watch the evolution of a
number of current cases to their fatal outcome. If any added
expense account could be arranged, all the better. For $100 worth
of flown-in supplies, we hope to have a completely new building up,
quite ample to house our 30-odd patients and another 10 or 20,
all under good conditions; this would be built of largely native
materials . . .*

Should you be able to make a grant, based on little more than

*our assurance of my integrity and medical judgment—I tell you
Joe, this is no wild-goose chase, but a really big thing; everything
in my medical training makes me confident—let me know, so
that I can plan to extend my work here for at least several further
months. On my own, I can hold out for one or two months and still
have enough to get home via Europe. . . . I stake my entire medical
reputation on this matter . . .*

*This class of illness is not my specialty, but I can handle
research on it as well as anyone . . . Should anyone in the States be
wondering of my whereabouts, let them know, since my chances of
keeping up much correspondence from here on anything but kuru
is most remote.*[6]

Australian scientists and government authorities were livid. They already had a program of research into the disease, and they regarded Gajdusek as a ruthless claim jumper. After Australian scientists had shared their research on the curious neurological disease with him he had instantly grasped the project for himself.

"Your taking over a job that had been allocated to [Dr.] Anderson was quite indefensible," Sir Macfarlane Burnet wrote Gajdusek on April 9, 1957. "Anderson has bitterly resented it and I sympathize with him." Burnet was director of the Walter and Eliza Hall Institute of Medical Research in Melbourne and Gajdusek's mentor during his current fellowship. It was from Burnet's son that Gajdusek had first learned of the new disease. (A renowned scientist, Burnet would receive the Nobel Prize in Physiology or Medicine in 1960.)[7]

But Gajdusek pressed on in his research, recording his observations in his field notebooks and sending a steady stream of letters back to his colleagues. "Kuru is becoming ever more interesting and baffling . . . We managed to collect specimens in clean enamel kidney dishes washed with catchment water out-of-doors to avoid dust and smoke from all the houses," Gajdusek wrote to Smadel.[8]

"Dr. Gajdusek has been, to say the least of it, unethical," Dr. John T. Gunther complained formally in his letter of April 9, 1957, to Burnet.[9] Gunther, director of the Papua and New Guinea Public Health Department, accused Gajdusek of entering the area under false pretenses and fraudulently introducing himself to Zigas as a representative of the institute. "Another American invasion," Dr. Roy Scragg, who succeeded Gunther as director, complained in a phone call to Gunther.[10] As a matter of national pride and scientific ethics, Gajdusek had to be stopped.

Deep in the jungle, Gajdusek was engrossed in his single-minded quest to solve the medical mystery. "He [Zigas] tells me that *all* (yes, astoundingly, all 100%) of 28 cases collected two to four months ago are now dead!" he wrote to Smadel.[11]

Many potential suspects emerged from the medical clues Gajdusek collected, but they did not converge on any prime suspect for the disease. "Toxic poisoning must be seriously considered; cases are overwhelmingly found in females—with fewer and fewer males being found, Dr. Zigas says. They are all dying he says."[12]

Scragg fired off a stern radiogram to Gajdusek, ordering him under his authority as director of the Australian Public Health Service to leave at once.

BURNET DEEPLY CONCERNED YOU TAKING OVER
INVESTIGATION PLANNED FOR ANDERSON AT REQUEST OF
AUSTRALIAN GOVERNMENT. YOUR INTEREST IN PROBLEM
FULLY APPRECIATED BUT MY INSTRUCTIONS REQUEST YOU
LOOK INTO LUFA AREA ONLY. ANDERSON NOW ADVISES
ARRIVING WITHIN ONE WEEK. ACCORDINGLY ON ETHICAL
GROUNDS REQUEST YOU CONSIDER DISCONTINUING YOUR
INVESTIGATIONS KURU. WILL ADVISE DEFINITE TIME OF
ARRIVAL ANDERSON AND SUGGEST RENDEZVOUS GORKA
TO DISCUSS PROBLEM. PLEASE ADVISE ME BY RADIO YOUR
PLANS DEPENDENT ON ABOVE.[13]

The young pediatrician at the center of this international scientific dispute radioed his reply from the jungle.

INTENSIVE INVESTIGATION UNINTERRUPTIBLE. WILL
REMAIN AT WORK WITH PATIENTS TO WHOM WE ARE
RESPONSIBLE.[14]

As they hacked their way through the jungle at a furious pace against their ever-dwindling reserves, Gajdusek and Zigas realized that the epidemic loomed larger even as the mystery of this horrible disease intensified. "New cases have now developed under our eyes; all our 60 to 70-odd study cases are slowly (some rapidly) deteriorating; and we have raised the number of cases we have ferreted out of the Fore's 8,000 to 10,000 population to 75 active cases . . . Some centers have up to 6–10% of the population now sick with active progressive disease; and many centers can attribute fully one-half of all deaths to kuru in any of the past five years. Could any more astounding and remarkable picture be found anywhere?"[15]

While Australian administrators battled over turf and ethics and sought to eject him from the jungle, Gajdusek and his party pressed on, enduring sweltering heat in the steamy jungle, bone-chilling cold in the high mountains, mosquitoes, poisonous insects and spiders, biting flies, bees, crocodiles at river crossings, and tropical disease. From Gajdusek's notebooks: "It has poured every day of our patrol and we have only made one walk thus far without being drenched to the skin by torrential rains, and that one rainless walk included several waist-deep fords of streams."[16] "Leeches infested the trails on the high ridge . . . the feet of most of the cargo-boys were bleeding from leech bites."[17]

Dr. Smadel at the NIH appealed to the National Foundation for Infantile Paralysis for funding to sustain Gajdusek and his search in the jungle, but his appeal was denied—paralysis, not kuru, was their interest. Without money or medical supplies the investigation would soon collapse, leaving Gajdusek penniless and without any means of returning home to the United States.

On May 21, 1957, Dr. Thomas Rivers, director of the National Foundation for Infantile Paralysis, wrote to Gajdusek:

> *The red tape of the United States Government is considerable, as you know and there is a certain amount of it in the National Foundation for Infantile Paralysis. I have taken the bull by the horns and have sent to Sir Macfarlane Burnet in Melbourne, Australia, a National Foundation check for $1000 made out to you.*
>
> *I have a little money in the Medical Director's fund with which I can take a flier on certain projects or for certain outstanding young men. I am told that you are such a person, and I hope that this flier on you will not be considered one of my mistakes.*[18]

"I have a sort of exasperated affection for Gajdusek and a great admiration of his drive, courage, and capacity for hard work," Burnet wrote in acknowledging receipt of the funds. "Also there is probably no one else anywhere with the combination of linguistic ability, anthropological interest, and medical training who could have tackled this problem so well . . . Kuru is one of the most fascinating disease problems I have ever come across."[19]

"I thought it might be helpful if I gave you some unofficial and informal background about him," Burnet offered in a handwritten note to Gunther, who had charged Gajdusek with "unethical behavior."

He is quite an extraordinary individual of American birth but brought up as a child in Central Europe and multilingual. There is no question about his intelligence or training in pediatrics and virology, and I found myself very interested by his enthusiasm for the pediatric and cultural study of the development of children in primitive communities. On the other hand, his personality is quite extraordinary, and is almost legendary amongst my colleagues in the U.S. [John] Enders (Boston) told me that Gajdusek was very bright but you never knew when he would leave off work for a week to study Hegel or a month to go off to work with Hopi Indians. Smadel at Washington said the only way to handle him was to kick him in the tail, hard. Somebody else told me he was fine but there just wasn't anything human about him.

My own summing up was that he had an intelligence quotient up in the 180s and the emotional immaturity of a 15-year-old. He is quite manically energetic when his enthusiasm is roused, and can inspire enthusiasm in his technical assistants. He is completely self-centred, thick-skinned, and inconsiderate, but equally won't let danger, physical difficulty, or other people's feelings interfere in the least with what he wants to do. He apparently has no interest in women but an almost obsessional interest in children, none whatever in clothes and cleanliness; and he can live cheerfully in a slum or a grass hut.

He is not a first-class scientist in any field, but I doubt whether anyone in the world has anything like his knowledge of children in primitive communities in very many parts of the world.[20]

As if by destiny, a unique combination of scientist, medical doctor, and cultural anthropologist had come on the scene with all the requisite abilities a government program would have assembled in a large staff of experts to undertake this difficult investigation. "I am rapidly learning a smattering of Fore, and Kimi and Keiagana word-lists and medical lore are already mine. Fore language is near-essential to getting the histories on which our findings of strong family-incidence is based."[21]

Finding kuru patients was arduous work, but sadly, not difficult given the proportions of the epidemic. Obtaining vital autopsy material from the superstitious natives, however, presented insurmountable difficulties. "Last week we had our first death in the Hospital of a kuru patient; and at 2 a.m. in rain and cold wind, a *docta boi* [a native medical assistant] and I did a complete autopsy in our treatment/ laboratory hut by lantern light, and then at first cockcrow got the body borne homeward with the mourning mother well rewarded with axes and salt and laplap (which she appeared to take as much interest in as in the death—which, it must be said, she had long, long ago resigned herself to). Thus, we got the 'dastardly deed' done without awakening much local curiosity or attracting too much attention to our butchery . . . The brain had to be sectioned with a carving knife and it was no easy matter, for although it was removed within one hour of death, it was mighty soft—and I fear that no pathologist will think much of my sections."[22]

"Fear and suspicion have been ticklish problems, and every [medical] procedure is suspect . . . Their cooperation depends upon respecting their unwillingness to have a near-fatal case die away from home . . . unless we get the body back home for burial."[23]

Obtaining the brains of kuru victims was the biggest difficulty, but this was the most essential material for the medical examination. "Naturally, everyone would like to get their hands on kuru brains; we were lucky to get two and may get further ones, but [the natives] do not like the idea of opening the head, although other dismemberment does not seem to perturb them. However, death away from their remote villages does!"[24]

Sorcery, obviously, was not the cause of this fatal neurological illness, but in truth, medical science was no more enlightened about the cause of this cruel death nor was modern medicine any more effective in curing it than the witchdoctors and their sorcery. As time passed, this realization among the natives began to erode the investigation.

"Local Fore resistance to hospitalization is increasing," Gajdusek warned Burnet in June 1957. "They know damn well that we do nothing for the disease but prolong its misery by supportive measures, and they are anxious to return to their technique of starvation and neglect in darkness, which ends in a speedy exodus once the illness is truly incapacitating."[25]

A cure remained a distant hope, but what the doctors *did* know was that the brains of these people had turned to sponge—they had "spongiform encephalopathy" in proper medical terminology. The brains were full of holes. The neurons had degenerated and filled with bubbles as though they had been boiled alive. There was also "astrocytic proliferation which, however, may occur as a tissue reaction to a wide range of noxious stimuli," Elizabeth Beck commented after Gajdusek's lecture to the Royal Society of Tropical Medicine and Hygiene in London on February 21, 1963. She was one of many scientists gathering from around the world at a meeting several years after Gajdusek's fieldwork to try to solve the baffling case of this horrible disease.[26]

Most in attendance favored the genetic explanation, but the evidence was not a good fit. Nearly all the cases were women, but young children of either sex could be afflicted. Possibly the disease was related to female hormonal factors, but then how would you explain the children? In the jungle, Gajdusek had tried to treat patients with testosterone injections, but they were useless. The incidence of the disease simply did not track with the ratios required by the rules of inheritance. In some hamlets the incidence of this disease among females reached such staggering frequency that it could not possibly be inherited, accounting for half of all the deaths of the Forest People. Yet the disease seemed to track along family lines.

Carleton Gajdusek considered and investigated every conceivable type of infectious agent—bacteria, viruses, ticks, venereal infection, transmission of disease through mother's milk, fungi—but nothing fit the pattern of this puzzling disease. Curiously, many individuals suddenly developed the disease years after leaving the forest and while living in towns where the disease was unknown. How then could the disease be contagious? No one outside this small forest region ever caught the disease, even though Forest People wandered widely.

Perhaps, Gajdusek proposed, this plague was some type of very slow-growing virus, much like scrapie in sheep, which was suspected at the time to be viral. Scrapie also turns the animal's brain to mush and causes very similar symptoms. To fit the evidence, the virus would have to be one that could lie dormant in the body for years, then suddenly overwhelm the victim and dissolve her brain. Yet children could get it too, so the virus could act within a few years.

"Dr. Gajdusek is clearly one of the harbingers of a new kind of virology," Dr. Gordon Smith summarized at the end of the London meeting.[27]

The biggest piece of the puzzle loomed large as a tantalizing clue: Why were men largely untouched?

After Namugoi'ia died of the shaking disease, a funeral was held. The whole village attended, and afterward her aunt on her mother's side collected her body and brought it to the sugar cane garden. There, using a bamboo knife and a stone ax, she chopped off the hands and feet, then sliced open the arms and legs and stripped the muscles off the bone. That done, she slit open the chest and belly and removed the organs, taking care not to damage the gallbladder, which would ruin the meat if it ruptured. She severed the head from the body and cracked open the skull with several blows of her stone ax to remove the brains.[28]

All the body parts were cooked with a bit of ginger and eaten by the woman's close relatives and their children. The entire body was consumed: brain, viscera, genitalia, the marrow of bones, even the bones themselves were pulverized and eaten with green vegetables. Men seldom ate the bodies of women, believing them dangerous, and rarely would men eat adult men, believing that this diminished a man's vitality and made him vulnerable to enemy arrows. Women consumed the bodies of men, women, and children. The children ate whatever their mothers provided—or more precisely, whomever. The reason for the odd pattern of this disease was not biological but cultural.

In 1966 Gajdusek and his colleagues proved the disease was infectious. After Kabuinampa, a twenty-three-year-old-woman, died of kuru in December 1963, her brain tissue was collected within the hour and frozen. In February 1964, a 0.2 milliliter suspension of her brain tissue (about ten drops) was injected into the brain of Joanne, a two-year-old chimpanzee. Dr. Gajdusek waited patiently, and eighteen months later, in August 1965, the chimp showed the first clear signs of the shivering illness. The condition rapidly worsened, displaying all the characteristic stages of kuru as the chimp succumbed to the disease.

Elizabeth Beck performed the histological analysis of the brain tissue, confirming that the chimp's brain displayed the characteristic moth-eaten damage of kuru victims, as expected. These results were replicated in several chimps, but there was a problem. An exhaustive search in tissue taken from kuru victims and the infected monkeys failed to detect any virus. There were also no chemical traces of the body's antibody reaction to a viral infection. Kuru was infectious, but frustratingly for researchers, the germ eluded all efforts to locate it.

Incidence of the disease peaked in the 1950s. Then cannibalism was outlawed by the Australian government, and kuru went extinct. Or so it was believed.

Cannibal Cows and New Kuru

Between 1996 and 2004, neurologist John Collinge from University College London and his colleagues trekked back to New Guinea to have a second look for kuru a half century after the peak outbreak of the disease. In 2006, they reported that they had found the last eleven cases of kuru. The detailed family histories Collinge recorded revealed a surprise. The disease could have an astonishingly long incubation time: up to at least fifty-six years from the last meal on human flesh.

"You are not going to eat steak, are you?" my American colleague inquired at a dinner in a posh restaurant in London after a scientific meeting on glia. Mad cow disease was prominently in the news at the time, and England was the epicenter of the emergence of this deadly disease. The effect this disease has on cattle is not unlike the effect of kuru on the Forest People, and cattle acquired it the same way: through cannibalism. In the case of cattle, however, the cannibalism was unwitting. Bits of their dead brothers and sisters had been added to their processed feed to increase its protein content. Cow brain and spinal cord in the feed carried with them the germ for mad cow disease. When people consumed beef infected with mad cow disease (properly called bovine spongiform encephalopathy or BSE), they, like the Forest People, suffered the brain-eating disease and all its devastating symptoms.

In people the disease is called variant Creutzfeldt-Jakob disease (vCJD). It strikes people just as unpredictably as kuru, and it can have a very long period of incubation. There is no cure for BSE in cattle or vCJD in people, and both are invariably fatal. Approximately two hundred thousand cases of BSE in cattle have been confirmed. The first infected cow was detected in the United Kingdom in 1986, and the numbers peaked in 1993. The practice of adding cow brain to the animals' processed feed was abolished in the 1990s, and because of this and other measures, British beef is now considered safe.

However, the extended period of incubation revealed in the recent research on kuru does little to alleviate concerns of anyone who may have consumed tainted beef up to that point. Even today, these people are still at risk.

Sporadically there are reports of cattle infected with BSE in the United States. Because the disease has such a long incubation time, an infected cow may not have developed symptoms before being butchered. The meat could be tested for the germ, but it would be impractical to test every animal in the food supply.

In 2001, ninety-two cases of human vCJD had been detected; eighty-eight of them from the United Kingdom, three in France, and one in Ireland. Should I risk the deadly brain-dissolving disease or forgo the British beef for some safe Italian pasta? I was hungry. I found it indeed a savory, juicy steak, cooked to perfection: medium rare. I still think of it from time to time.

Before BSE became a concern for human health, people fell victim to Creutzfeldt-Jakob disease, which had nothing to do with cattle. Ten to fifteen percent of these CJD cases are inherited, but a significant fraction are transmitted by a modern sort of cannibalism: transplantation of body tissues, such as cornea, as well as by injection of human growth hormone and other contaminated biological materials extracted from humans. The neurological symptoms and brain degeneration of kuru, vCJD, CJD, and BSE are, as mentioned, most similar to scrapie, which affects sheep.

What sort of virus can survive the manufacturing process of producing beef byproducts for enriching bovine feed? In 1967, Tikvah Alper and colleagues at the MRC Experimental Radiopathology Research Unit in Hammersmith Hospital in London reported that the infectious agent in scrapie was extremely resistant to the most powerful known mechanisms for sterilization. The infectious slow-growing virus in scrapie could not be killed with radiation or intense ultraviolet light. Such radiation chops DNA and RNA to fragments, killing all other forms of life, viruses among them. The only conclusion Alper could draw was that the germ that caused scrapie had to lack genetic material (DNA or RNA); thus it was not a virus or any other known microbe. What kind of new germ was this, and how could it reproduce itself without genes?

In 1967, biophysicist John Stanley Griffith of Bedford College, London, proposed three ways that, in theory, a single protein could indeed self-replicate and cause the disease. Stanley B. Prusiner and his colleagues succeeded in isolating this infectious protein in the 1990s and called it prion protein (PrP). Prusiner accepted and greatly advanced the idea that a protein could self-replicate without any genetic material and act as an infectious agent.

Curiously, his research revealed that perfectly healthy animals had the gene to make this prion protein. Researchers soon realized that disease must result when normal prion protein changed into a new form that is toxic to cells. Indeed the scrapie prion protein had been changed from the normal one: enzymes that digest proteins had no problem munching up normal PrP, but the infectious PrP from scrapie resisted destruction by the powerful enzymes.

We now know that prion proteins become resistant to attack much like an armadillo, by folding up into a new, impervious shape. The refolded protein has the ability to cause a normal PrP to change its own shape permanently into that of the disease-causing form. Thus, like the legendary bite on the neck from a vampire, scrapie PrP propagates itself by contacting a normal PrP molecule and causing it to change its shape to the deadly form. The number of scrapie prions increases explosively, building their ranks in ever-increasing numbers by converting innocent proteins into the ghoulish form in a molecular pyramid scheme destined to overtake all the normal prion protein in the cell. Prions, not a slow-growing virus, are the cause of all the spongiform encephalopathies. The reason some prion diseases are inherited is simply that genetic mutations in the normal PrP cause the protein to kink and change its shape more readily into the toxic and self-propagating configuration.

Looking back on Gajdusek's achievements, one wonders if the baffling disease of kuru could have been cracked so quickly by any other individual or government-sponsored program of research. Without any planning or funding, indifferent to protocol, scientific ethics, governmental regulations, and at great risk of contracting the hopelessly fatal and hideous disease, this singular pediatric specialist persevered. The demands of the investigation were outside the bounds of nearly any established scientist to endure. Reaching the correct answer amid so many perplexing possibilities as he did was a remarkable achievement.

The expedition required an adventurer with physical stamina to match the jungle.

[We] have completed some 1000 to 1500 miles of mountain-climbing and walking (often with bush-cutters to clear the trail) trying to track down kuru epidemiology since March 14th.[29]

It required a medical doctor with the selflessness of a Mother Teresa content to live among impoverished and diseased people.

We have often crowded into these tight huts through the small, low doorway to sit among the fleas and lice . . . Their ectoparasites are causing us no end of suffering. We shower and completely change clothes—only to pick children up into our arms shortly thereafter, and soon find ourselves itching violently as a result.[30]

It required an anthropologist who delighted in experiencing a primitive culture and was willing to embrace alien ways that would clash with his own culture's standards of behavior.

They [the Kukukuku] tend to be light-fingered, and will pick up anything we do not keep an eye on. They do not hesitate to push the keys, move the carriage, and interfere—in any way that intrigues them—with the typewriter while they crowd around me as I type.[31]

It was facilitated by an individual with peculiar persuasions that enabled him to be accepted by the local people in ways that would have been repellent to most.

I returned to the men's house in the village and again found myself besieged by those wanting to palpate and examine my genitalia— and more insistent youths who wished to engage [in homosexual activity]. Again, all ages appear keenly interested in the possibility of talking their visitor into such practices; and there was much ribald gesticulation and suggestion . . . I should be most interested in fathoming the role of homosexuality and pederasty in this culture.[32]

In his field notes of the following day:

Everyone in our party had some Kukukuku embracing him for at least part of the way up to Anji; and again, some suggestions of [homosexual activity] were the only form of aggression besides genital-handling. This was all done in a most hospitable, friendly, ribald, and exuberant fashion; and on the high tide of Kukukuku enthusiasm, we were carried into their stockaded village. We spent some three or four hours in their settlement before starting back for Aurooga, and during this time had a good chance of seeing much of our Kukukuku hosts.[33]

Carleton Gajdusek received the Nobel Prize in Physiology or Medicine for his work on kuru and CJD in 1976. In 1997 Stanley Prusiner received the Nobel Prize in Physiology or Medicine for his research on prions. In 1996 Carleton Gajdusek was arrested for sexually molesting a seventeen-year-old boy in Bethesda, Maryland, one of fifty-six boys he had brought back to the United States from the South Pacific since the 1960s. After openly admitting to the sexual offense (which he did not view as harmful), Gajdusek was convicted of child molestation and sentenced to prison for nineteen months. Upon his release in 1998 he returned to Europe, where he continued a vigorous career in research into prion diseases, traveling as a welcomed collaborator to different research groups in several different countries. He died in December 2008.

PRION DISEASE: A SEARCH BEYOND NEURONS

In the mid 1990s, the mechanism by which prions, this misfolded protein, caused damage to brain tissue was unknown. Many researchers recognized that the accumulations of this protein inside cells and in plaques surrounding cells resembled other neurodegenerative diseases such as Alzheimer's and Parkinson's, but in those diseases the protein accumulations were different molecules from PrP.

"No one knows exactly how propagation of scrapie PrP damages cells," Stanley Prusiner wrote in a 1995 *Scientific American* article.[34] His search was focused on neurons, and his work showed that they were capable of hosting the replicating prions. "In cell cultures, the conversion of normal PrP to the scrapie form occurs inside neurons, after which scrapie PrP accumulates in intracellular vesicles known as lysosomes," he reported in the article. Astrocytes were another matter.

"Astrocytic gliosis is found upon microscopic analysis of the CNS," Prusiner stated in his Nobel lecture on December 8, 1997, referring to the first figure in his presentation that showed astrocytes stained to reveal abundant GFAP, a protein found in high levels in diseased or stressed astrocytes, in the brain of a mouse infected with scrapie prion. The involvement of astrocytes in prion disease was obvious; indeed, the accumulation of numerous reactive astrocytes in the infected brains were hallmarks of kuru, CJD, BSE, and other prion diseases. The question was whether the astrocytes were responding to the cellular damage or contributing to the disease. In his Nobel lecture, Prusiner concluded that astrocytes were only responding to the injury. This was the prevailing view at the time.

"In scrapie, GFAP mRNA and protein rise as the disease progresses, but the accumulation of GFAP is neither specific nor necessary for either the transmission or pathogenesis of disease," he said.[35]

Slowly scientists became more open to the possibility that astrocytes could be contributing to the disease, particularly by releasing cytokines and other agents that are toxic to neurons. Gajdusek, for example, published several papers on the involvement of all types of CNS glia—microglia, astrocytes, and oligodendrocytes—in prion diseases beginning about the time of his prison sentence and continuing after his release.

The evidence from the experiment presented in Prusiner's Nobel lecture that astrocytes are not involved in the neuron destruction in prion disease was not the final word. In 2004, for example, researchers in France reported studies showing that prions can infect cultured neurons or astrocytes. Moreover, both types of cells can sustain the replication of PrP. Prion protein was replicating in the other brain, unseen by neuroscientists fixated on neurons.

Astrocytes can contribute to prion disease by manufacturing and spreading the prion to infect neurons, but do infected astrocytes themselves in some way kill neurons? Lisa Kercher and colleagues at the NIH Laboratory of Persistent Viral Diseases in Montana reported the results of studies in 2004 in which they switched on the PrP gene selectively in either the astrocytes or neurons of laboratory mice and then infected the animals with scrapie. The results showed that in either case, whether astrocytes or neurons contained the PrP, brain degeneration resulted. Astrocytes infected with scrapie PrP can kill neurons even if the neurons lack the prion protein.

It is now clear that prion disease is a disease of glia as much as it is a disease of neurons. How much further might we have advanced in the search for the causes and cure of prion disease had researchers not so easily dismissed the other brain? Now, decades later, we are playing catch-up.

Astrocytes respond to brain injury in prion disease and they contribute to the death of neurons. Astrocytes replicate the prion protein and they contribute to the formation of PrP amyloid plaques in prion disease (just as we will see in a later chapter they contribute to the formation of the amyloid plaques of Alzheimer's disease). Astrocytes infected with the diseased prion release cytokines and neurotoxic agents, and their ability to maintain normal levels of glutamate around neurons is impaired. As a result neurons die. Finally, astrocytes also interact with oligodendrocytes. When oligodendrocytes become diseased, the myelin sheath insulating axons suffers damage. What more might be found to help solve the prion puzzle if we broadened the search even further?

A study published in 2006 by researcher Grazyna Szpak and colleagues at the Institute of Psychiatry and Neurology in Warsaw examined the brains of forty CJD patients at autopsy. They reported that microglial activation and microglial immune inflammatory responses were characteristic features in CJD. Many of the microglial cells themselves were also filled with bubbles (vacuoles), just like neurons in prion disease.

Other autopsy studies of people who have died of CJD have revealed that microglial cells also harbor the damaging form of PrP. Studies in cell culture show that microglia can accumulate and replicate the disease-causing forms of prion protein. The researchers conclude that microglia contribute to the spread of the infection, because unlike neurons, microglia move rapidly and over long distances through the brain, leaving infectious PrP in their wake.

In response to prion infection, microglia produce toxic molecules (cytokines, reactive oxygen, proteases, and complement proteins) that kill neurons. Microglia activated by the abnormal form of the prion protein also release substances that stimulate astrocyte injury responses and boost astrocyte cell division. Thus, microglia may play a major role in initiating the pathological changes in prion disease.

Microglia seek out and selectively kill neurons infected by scrapie prions. Thus, microglia directly cause the death of neurons in prion disease. This suggests that drugs to limit activation of microglial cells may be beneficial in treating patients suffering from prion diseases by decreasing the neuronal damage caused by the neurotoxic molecules microglia release to fight the infection.

This view is controversial, however. Others argue that the neuroprotective actions of microglia, so important in other brain injuries, as well as the critical immune functions of microglia in the brain play a beneficial role in prion disease. Microglia fighting the disease and eliminating infectious PrP from the brain may be one reason the disorder seems to lie dormant for so many years before symptoms appear. Once the microglia lose the battle and brain injury accelerates, however, patients slip rapidly into an agonizing death as bits of their brain are destroyed, the

disease assaulting one by one the brain's vital functions. Some researchers suggest that augmenting these beneficial actions of microglia would
be the best therapy. Microglia do indeed target and kill neurons that are
infected with the disease-causing prions and thus they do damage neural circuitry, but in doing so they may protect the brain by limiting the
spread of the diseased prion. Microglia also engulf the deposits of PrP
outside neurons, thus eliminating or slowing the accumulation of PrP
plaques. Once infected by prion, however, the ability of microglia to
consume particles of PrP becomes impaired. Dysfunctional microglia
might even contribute to the disease, making some people more susceptible to prion infection than others.

In prion disease microglia even assume some of the vital functions
of injured astrocytes, which become impaired by prion disease. Normally astrocytes take up neurotransmitter (such as glutamate) from
synapses. This is essential for synaptic transmission and to prevent glutamate from rising to toxic levels. In CJD, microglia transform to take
over this vital function as astrocytes become infected and die. The transporter molecule in the cell membrane that absorbs the neurotransmitter
glutamate into astrocytes starts to be synthesized in microglia during
prion infection. Now equipped with the glutamate transporter of astrocytes, microglia step in for their fallen glial comrades to lower the toxic
levels of this neurotransmitter in damaged brain tissue. This protects
neurons from death due to overstimulation by the excess glutamate.

Finally, microglia could be helpful in diagnosing prion disease.
Microglia develop distinct cellular changes in response to prion infection, and these alterations can be detected with appropriate diagnostic
techniques. Much as monitoring changes in blood cell count informs
doctors of the type and severity of infections in the body, one can imagine that careful monitoring of changes in microglia could provide critical insight into infection in the brain.

Interestingly, current evidence suggests that oligodendrocytes are not
capable of supporting replication of the infectious prion protein. This
resistance sets them apart from both neurons and astrocytes. However,
oligodendrocytes and myelin suffer damage in prion disease. Other
studies indicate that oligodendrocytes are killed by oxidative injury accompanying prion infection.

Recall that PrP is a normal protein in neurons that becomes mutated and infectious in prion disease. The biological role of the normal PrP in cells is still mysterious. In 2005 it was reported that normal PrP is found in purified myelin and in oligodendrocytes. In 2007, Frank Baumann and colleagues from the University Hospital of Zurich, Switzerland, reported that a mutation in a particular part of PrP caused myelin breakdown in both the central and peripheral nervous systems of mice. This study suggests that myelin integrity must be maintained by some unknown action of the normal PrP in myelin. By studying the role of PrP in myelin, we may learn more about the normal function of this protein in cells.

Finally, a most unexpected finding for a disease of the central nervous system: recent experiments show that Schwann cells (cells of the peripheral nervous system) can replicate sheep scrapie prion in culture, suggesting that prion infection might be spread through peripheral tissues. If so, even the finicky eaters in New Guinea were not above the risk of kuru. What might that mean for consumers of beef?

Epilogue

Earlier this year I attended a scientific meeting where an authority on prion disease presented his latest research on the involvement of glia. When I spoke with him after the talk he shared his personal opinion with me that he and others were not convinced that the evidence rules out a slow-growing virus as the basis for prion diseases. "What about the radiation experiments?" I asked.

"Some viruses are very resistant to radiation," he said. "I would say about ten percent of scientists in the field believe that prion disease could involve a slow-acting virus, but it is difficult to get the papers published." He cited additional points supporting the viral cause as opposed to the generally accepted prion theory but then stopped when his comments betrayed frustration in trying to publish controversial findings. I finished his sentence for him: "There *is* that Nobel Prize . . ."

Advancing alternative thinking in any human endeavor is always a challenge, but in science truth is not determined by majority vote. Beyond the exciting information and new insights, this enduring reality is the larger lesson in the story of the other brain.

OUR BLACK PLAGUE: HIV AND GLIA

November 5, 1991, Los Angeles—Los Angeles Lakers basketball star Ervin Johnson, known since the age of fifteen as "Magic" because of his mastery of the game, announces to the world that he is infected with the HIV virus.

"I will have to retire from the Lakers."

The packed pressroom falls silent. Seasoned sports reporters react with stunned disbelief.

"We sometimes think only gay people can get it, that it's not going to happen to me," Johnson says. "And here I am saying that it can happen to anybody, even me, Magic Johnson."[36]

Without self-pity, Magic courageously and calmly explains the sad situation to an audience of journalists, many struggling unsuccessfully to hold back tears.

For many people, the illusion that AIDS was a distant medical curiosity was shattered that day. The disease had become personal. A smug sense of security had been violated. The emaciated HIV-infected homosexual and IV drug abusers with flesh melted from bone sockets and hollow eyes staring out helplessly had been replaced by the familiar smiling eyes of the star athlete and hero whose body was the epitome of strength and vigor, his career struck down at its peak.

We are now living through a modern Black Plague. The Black Plague of the Middle Ages claimed the lives of an estimated 25 million people in Europe. But according to the 2006 United Nations global AIDS report, 22 million souls have been taken by AIDS in Africa alone. Fifty million people in Africa have been infected by the virus, each one slowly enduring their incurable infection. In our own country, a census by the Centers for Disease Control and Prevention in April 2005 reported HIV infection among sexually active homosexual men in the United States ranged from 18 percent in Miami to 40 percent in Baltimore. Forty-eight percent of the men infected were unaware of their own infection before the census testing revealed it to them.[37] Despite our advanced scientific understanding of infectious disease and our impressive arsenal of medical treatments, the world is exploding with an epidemic of a deadly viral disease. The disease is spreading around the globe, with the fastest rates of infection now growing among the billion people in China.

AIDS attacks the brain. In fact, the brain is one of the earliest targets of the deadly HIV virus. The infection leads to dementia, characterized by movement disorders and cognitive impairment. Patients become forgetful, unable to concentrate, and their mental function slows markedly. HIV infection of the brain causes psychiatric abnormalities, including mania, apathy, and emotional instability. At autopsy, brains of HIV patients with dementia show loss of cerebral mass, loss of dendrites, dead neurons, holes (vacuoles) in the brain tissue, and white matter defects. Prior to today's highly effective antiretroviral therapy, one-fourth of HIV patients suffered neurological impairments. This condition is now reduced in the early stages of the disease, but as drugs allow people infected with HIV to live longer, more people ultimately become afflicted by the associated brain injury.

Many types of viruses attack the nervous system. Two famous examples are polio, which causes paralysis, and herpes, which causes cold sores on the lips and in the genital area. These two examples illustrate a prominent characteristic of all viruses: by virtue of the way they gain entrance into a cell, viruses are very selective. This explains why most viruses do not pass readily between different species of animals. We do not catch feline leukemia virus from our pets, for example, and generally they do not catch our colds. Thus far, bird flu virus does not readily infect human cells, but scientists are concerned that a simple mutation could allow it to recognize human cells, thus igniting a devastating epidemic of human disease.

The polio virus infects motor neurons in the spinal cord selectively, killing them and paralyzing its victims. Thought and reasoning remain clear, but nervous commands from the brain do not reach the muscles, and with the lines of communication cut, the muscles waste away. If the motor neurons to the legs are attacked by the polio virus, patients are confined to wheelchairs, but if motor neurons to the lungs are attacked, the victims must be confined to iron lungs or use other means of constant artificial respiration.

Herpes virus on the other hand infects sensory neurons. Like polio, herpes infection is incurable, and the virus resides permanently inside sensory neurons to provide an inexhaustible source of virus for periodic outbreaks of infection throughout the life of the patient. Herpes simplex virus type 2 acts below the waist, and herpes simplex virus type 1 causes the common cold sores on lips.

What type of neuron is infected in the brain of HIV patients? How does the HIV virus gain entry into the neuron? Does the virus sneak up the axon from nerve endings to the nerve cell body as herpes virus does, or does it perhaps attack proteins only on the surface of dendrites or the cell body?

Autopsies reveal that the HIV virus devastates the brain, riddling it with pustules and leaving a wasteland of neurons. But researchers soon discovered that the HIV virus does not infect neurons at all. HIV infects glia.

No more dramatic proof of the importance of glia in brain function is needed than the devastation of the brain seen in AIDS. Like all viruses, HIV can enter only cells with appropriate types of proteins on their cell membranes. The HIV virus attaches to these specific proteins and pierces through the cell membrane. Once inside the cell, the virus hijacks control of DNA synthesis machinery in the cell's nucleus, causing the hostage cell to become a factory for producing swarms of new HIV virus. These new viruses burst from the cell to infect other cells in a chain-reaction explosion throughout the body. But neurons do not have the type of protein receptor on their cell membrane that the HIV virus needs to attach to, and so neurons cannot be infected by the HIV virus. The protein magnet for attaching HIV viruses to cell membranes is called CD4.

HIV attaches to the CD4 membrane protein present on a particular type of white blood cell that fights infection: the t-helper lymphocytes called "T cells." HIV patients track the progression of their disease by counting the steady depletion of CD4 positive T cells in their blood as the virus targets these cells specifically for infection and kills them. As these cells are depleted from the blood, HIV patients lose the ability to fight infection, and they become vulnerable to attack by all manner of microbes.

The HIV virus enters the brain from white blood cells in the bloodstream that become infected and penetrate the blood-brain barrier, which has been weakened by the disease. In the brain microglia, distant cousins of the white blood cells, share the CD4 membrane receptors that allow HIV virus to infect white blood cells. Microglia become infected by HIV; they then move through the brain, spreading the HIV virus widely through the brain ventricles, scattering virus particles behind them all along the white matter tracts of myelinated axons that link distant regions of the brain. Oligodendrocytes do not become infected, but as will be discussed later, they are attacked and injured by the microglia gone haywire from HIV infection. Microglia not only disperse the HIV virus throughout the brain; they also become permanent reservoirs of the virus. In this way, microglia maintain HIV infection against the efforts of the immune system and medicine to clear it from the brain and body.

Astrocytes can also become infected by the HIV virus. In comparison with microglia, astrocytes are less well equipped to replicate the virus once they become infected, but the infection causes cellular changes in astrocytes that turn them into assassins. Together astrocytes and microglia cause the destruction of the brain in HIV infection, breaking neuronal circuitry, killing neurons, and causing dementia.

Peripheral nerve damage is also an early sign of HIV infection, but nerve axons are not touched by the virus. A sixty-five-year-old woman in Bangalore, India, suddenly began to experience weakness in her arms and legs. As in polio, the weakness progressed rapidly, and she sought treatment from a neurologist. After two months, she had become anemic, malnourished, and showed signs of dementia. Her doctor took a biopsy from a nerve in her leg, which revealed devastating loss of the myelin insulation, inflammation, and destroyed axons. The muscle fibers appeared normal. Looking more closely with the electron microscope, Dr. Mahadevan and colleagues reported in 2001 the startling discovery: the woman's Schwann cells were filled with HIV virus.

The doctors determined that the patient had received several blood transfusions for anemia, which they suspected had infected her with HIV. Tests of three other patients suffering nerve damage revealed the same finding of HIV virus inside Schwann cells. In all cases, the nerve damage from infected Schwann cells was the *first* sign of HIV infection. The damaging loss of axons after HIV infection of Schwann cells illustrates dramatically the critical dependence of nerve cells on these glial cells.

The Black Plague of modern times is very much a disease of glia. The killer virus is smuggled into the brain inside glia, which harbor and reproduce the lethal invaders. The brain succumbs to the loss of essential support from its other brain when attacked by HIV infection.

In every infectious disease of the nervous system, glia are at the center of both disease and healing. This is in keeping with our understanding of the role of glia in protecting and serving neurons. But to do this task properly, glia must be sensitive to the changes in functional activity resulting from the disease. This ability to sense neural activity and distress, and the ability to initiate profound beneficial or detrimental changes in neurons, suggest that glia could, in principle, be modifying brain function all the time, not simply during disease. If so, could some forms of mental illness spring from glia gone bad? Could the other brain move the mind to madness? Could it help us in remedying mental illness?

CHAPTER 7

Mental Health: Glia, Silent Partners in Mental Illness

Someone suffering mental illness is frequently described as "unbalanced." This is precisely how glia may support mental health, because glia seem to be particularly important in maintaining balance and setting the general tone of excitability in the brain.

MADNESS

Mental illness has been a mystery for centuries. The invisible biological roots of mental illness force the afflicted to suffer not only from their disease, but also from society's inability to comprehend and cope with it. Supernatural causes, moral weakness, parental failings have all been blamed for mental illness. Historically, psychiatric disorders were ignorantly lumped together with neurological disease. From this muddle of ignorance and confusion, physical neurological problems such as seizure have been mixed historically with mental illness.

In investigating whether the mysterious seeds of madness can be found in the other brain, let's start with the story of epilepsy. The earliest medical records of epilepsy come from Egyptian and Babylonian writings dating from 1050 BC.[1] The belief that epilepsy arose from supernatural causes prevailed for centuries. During an epileptic seizure, a person's soul and body were believed to have been violently seized by a supernatural power. In the Gospel of Mark 9:17–27,

> "Master, I have brought unto thee my son, which hath a dumb spirit; and wheresoever he taketh him, he teareth him; and he foameth, and gnasheth with his teeth, and pineth away: and I spake to thy disciples that they should cast him out; and they could not."
>
> They brought the man's son to Jesus, whereupon he suffered a fit.
>
> And when he saw him [Jesus] straightway the spirit tare him; and he fell on the ground, and wallowed foaming. And he [Jesus] asked his father, "How long is it ago since this came unto him?" And he said, "Of a child." . . .
>
> He [Jesus] rebuked the foul spirit, saying unto him, "Thou dumb and deaf spirit, I charge thee, come out of him, and enter no more into him." And the spirit cried, and rent him sore, and came out of him: and he was as one dead; insomuch that many said, "He is dead." But Jesus took him by the hand, and lifted him up; and he arose.

Prior to the nineteenth century, ridding the body of demonic possession through religious exorcism, whipping, or ingesting toxic substances were the methods used to try to cure epileptics. Branded as witches, the mentally ill and epileptics suffered torture, stocks and whipping posts; death by hanging, drowning, stoning, or fire; or they were locked away in prisons or asylums in fear of contagion and crime. Over the centuries, remedies for epilepsy included drinking cups of blood from recently dead humans; ingesting powdered human skull, mistletoe, digitalis, silver nitrate, zinc oxide, or vulture liver; or purging the evil spirits by bloodletting, vomiting, pressing a hot metal iron against the head, or trephining (boring holes in the skull). Even the moon was thought to have an influence on epilepsy and insanity, explaining the origin of the word "lunacy."[2]

Yet the clear-thinking Greek physician Hippocrates knew better. In 400 BC he wrote that epilepsy "is not in my opinion any more divine or more sacred than other diseases, but has a natural cause, and its supposed divine origin is due to men's inexperience, and to their wonder at the peculiar character."

Legally sanctioned (often gruesome) punishment for witchcraft ended in the mid-eighteenth century, and the idea that mental illness had a treatable biological cause gradually began to supplant the belief in the supernatural. In the preface to his book *On the Curability of Certain forms of Insanity, Epilepsy, Catalepsy, and Hysteria in Females* published in 1866, Dr. Baker Brown, who was the senior surgeon at the London Surgical Home and president of the Medical Society of London, advocated medical treatment and surgery for mental disorders, which he believed had a biological basis and could be cured:

> Does not common charity lead us to think that cases treated by friends and spiritual advisors, as controllable at the will of the individual, may be in reality simply cases of physical illness amenable to medical and surgical treatment? Is it not better to look the matter steadily in the face, and instead of banishing the unhappy sufferers from their home and from society, endeavour to check their otherwise hopeless career towards some of the latter stages of the disease, to restore their mental power, and make them happy and useful members of the community?[3]

In his book, Brown documents scores of epileptic and insane patients whom he treated successfully through surgery:

Case XLVII. ACUTE HYSTERICAL MANIA—FOUR
MONTHS' DURATION—OPERATION—CURE.

Miss——, aet. 23, was sent to me by Mr. Radcliffe, stating that she had been brought over from Ireland as an insane patient, and that everything had been settled for her admission to some asylum, when he was induced to consult me on the last day before her entering one. He stated that the paroxysms [epileptic seizures] always came on at half-past five or six every evening: I replied, if the attacks depended on peripheral irritation, that an operation would at once prevent recurrence of the attacks. She was accordingly admitted into the London Surgical Home Feb. 6, 1864.

When admitted, said she had taken no food for three days, and asked for a cup of tea, which was given her. Enema was also administered.

3.45 p.m. Was seized with a fit, throwing her arms up over her head, and then appearing as if comatose. In about twenty minutes revived: the lips began to quiver, and she gradually became conscious, saying, "I want a knife—I want blood!" She asked for the matron's hand, that she might bite it off . . .

5 p.m. Mr. Baker Brown saw her ; as soon as he came near her, she seized his shoulders with great violence ; was wild, and would not answer questions ; but gradually became soothed, and allowed an examination . . .

Operation, 5.30 p.m. Was very violent under the first attempts to administer chloroform. She was long in being brought under its influence, but when once thoroughly anesthetized, bore it exceedingly well.

The clitoris was excised, the elongated nymphae [*labia minora*] removed, and the fissure of the rectum divided. The wounds were dressed in the usual manner, and the patient having had two grains of opium administered, was ordered to be constantly watched.

In twenty minutes awoke from the chloroform. Was calm, and slept at intervals during the night.

Feb. 7, 10 a.m. Visited by Mr. B. Brown. Present—Mr. I. B. Brown, junior, House-Surgeon, and Matron. Pulse quick but steady ; tongue brown and furred ; breath offensive ; gums spongy ; pupil natural ; countenance rather flushed ; skin moist and warm . . .

Feb. 8. Lint removed from rectum, and wounds dressed. Is calm and rational ; passed a quiet day.

Feb. 10. Very restless ; obliged to restrain hands and legs.

Feb. 11. Better ; says her head feels heavy ; countenance cheerful ; manner quiet and rational.

Feb. 12. Very excited and irritable ; constantly managing to free her hands ; will allow no one near her.—2 p.m.: Is quite maniacal ; has managed to irritate the wounds, and also the mammae. To have one grain of opium in pill, and ten grains of bromide of ammonium three times a day.

Feb. 13, 6 a.m. Hands again free ; repeat opium. Slept afterwards till 4 p.m., when she awoke calm and rational.—9 p.m. Slept again.

Feb. 14. Very restless, and at times violent. Bandages removed and jacket substituted.

Feb. 15. Much better ; rational, and conversing cheerfully.

Feb. 16. Improving.

Feb. 17. At her urgent request, hands were freed, but shortly after she became excited.

March 1. Much improved ; has written to her sister, and amused herself knitting and reading during the day.

March 2. Allowed to dress ; seemed to enjoy the change, and is very cheerful.

March 4. Visited by her sister ; has been quietly cheerful all day. Is certainly improving wonderfully.

March 20. Took a walk, and enjoyed it.

March 25. Spent the day away from the Home with her sister ; returned looking quite well and all the better for the change.

April 2. Discharged quite cured.[4]

As recorded in his book, Dr. Brown applied his surgical scissors to forty-eight female patients with epilepsy and other psychiatric illnesses. All of the women and girls he treated by the same procedure were "cured."

One hundred years later, in 1966, three states still had laws against marriage by people with epilepsy (West Virginia, North Carolina, and Virginia), and thirteen states had eugenic sterilization laws against people with epilepsy.[5]

The melding of madness with epilepsy is inevitable not only from a historical perspective, but also from a biological viewpoint. Since marked and varied psychiatric and behavioral changes accompany seizures, epilepsy offers insight into the pathology of brain dysfunction underpinning mental illness. Behavioral changes experienced in epilepsy include aggression, altered sexual interest, elation, euphoria, emotional instability, guilt, hypergraphia (keeping extensive diaries and detailed notes), irritability, obsessional behavior, paranoia, jealousy, religious experiences, sadness, and a sense of personal destiny. Interestingly, many religious leaders are alleged to have had epilepsy, including the Buddha, Mohammad, George Fox (founder of the Quakers), the apostle Saint Paul, Saint Cecilia, and others.[6]

We now will consider briefly some psychiatric illnesses so that the intersection between them, brain seizure, and glia can be explored.

SCHIZOPHRENIA AND DEPRESSION: A NEW UNDERSTANDING

Before I became a neuroscientist, I used to live in New York City in a run-down apartment in a crummy part of town. The nicest thing about it was the bird's-eye view it offered of the street below. One day as I peered through the dingy window stuck shut by layers of peeling paint, I saw a red pickup truck drive past very slowly. This was the prearranged signal warning me that they were coming to pick me up. We had other ways to communicate. Flipping on the TV, I scanned the channels for the local news. John Newman, the reporter on channel 4, looked me in the eye and winked; then he said the secret words that only he and I understood. There was no time to waste.

The subway station was only two blocks away. I walked briskly to the entrance, trying to blend anonymously into the crowd, and then raced down the concrete steps into the dark underground station. As my eyes adjusted to the darkness I saw a man next to the turnstile holding an umbrella. There was no rain. I diverted my path and ratcheted through the other set of turnstiles leading to the train platform. My heart pounded as I waited anxiously for the train. The crowd around me thickened. Then the man with the umbrella walked up and stood shoulder-to-shoulder with me. I couldn't move. Finally the jarring vibration and rattle of metal announced the approaching train. The man touched his nose and then very deliberately adjusted his glasses. I understood his warning. Nose = knows. Glasses = eyes = see. He had spotted me. A whoosh of cool air billowed out of the dark tunnel as the train's head-lights approached with a deafening screech and rattle of steel wheels against rails. I slipped behind the man and pushed him onto the tracks.

The scene above is entirely fiction, but the incidents in it were experienced by patients with schizophrenia—including, tragically, the murder of an innocent subway commuter. It illustrates an essential question Dr. Michael Eleff, associate professor of psychiatry at the University of Manitoba, posed in explaining this disorder to me. "How do you know things?" At some point in the narrative above you no doubt began to question the reality of what you were reading. Schizophrenia is a disease that affects perception and undermines the part of our brain that knows what is real and what is not.

One in one hundred adults will have schizophrenia during their lifetime. How can it be that so many suffer the disorder, with its debilitating hallucinations, delusions, and deranged thoughts, and yet it seems rare and unfamiliar?

First, it is essential to correct the common misconception that people with schizophrenia are dangerous to those around them. Statistically, people with schizophrenia are no more dangerous than anyone else. The rarity of killings such as the one described above and the freakish thinking that motivates them amplify the impact of a death at the hands of someone suffering schizophrenic delusions or hallucinations. Far more homicides are committed for other motives (robbery, rage, retribution), but these commonplace tragedies fail to generate the same level of fear. Other mental illnesses—substance abuse, for example—are responsible for many more deaths and homicides of innocent people than schizophrenia is, yet we accept the risk of dying on the highway from a drunk driver. The horrific loss of life is no less tragic to the family and friends of innocent victims on the road, but incapacitation by drunkenness is comprehensible. The deranged thinking of a person suffering schizophrenia is not. People with schizophrenia are, however, dangerous to themselves. Of people with this devastating mental illness, 10 to 13 percent kill themselves. Roughly half the population with schizophrenia attempt suicide at least once.

Muscular dystrophy, by contrast, is classified as a rare disease by the National Institutes of Health, but no one is unfamiliar with it, and most people are sympathetic to those afflicted by the illness. Only 1 out of 540,000 people in the United States is diagnosed with muscular dystrophy each year, or about 500 people in the entire country. Yet schizophrenia, which presents a lifetime risk of 1 in 100, attracts little public attention. Why?

"This tends to be an illness that pulls people out of society and makes them relatively invisible," Eleff explained. At the same time, there is a shameful stigma associated with this mental illness that is not attached to other diseases. Family members of those afflicted with schizophrenia do not discuss it openly as they might if their loved one was diagnosed with cancer or diabetes. "The illness deprives people of a social voice," Eleff says. Most people with schizophrenia suffer their lifelong illness in isolation.

> I'm 52 and I have had schizophrenia for the last 36 years. That includes about 8 years of prodromal symptoms, 12 years of un-treated psychosis, and 16 years of treatment with antipsychotics. Between 1980 and 1990, I was experiencing a very disabling psy-chosis, alone and very poor. I eventually got in trouble with the law in 1988 and received three years probation with the condition that I see a psychiatrist for those three years. I've been to jail, been actively alcoholic, attempted suicide, and was homeless for six months in 1980. My story illustrates an interplay of biology and sociology that can make schizophrenia a devastating illness. There is the illness itself, and the way we, as a society, treat people who develop it.
>
> The public is developing an interest in schizophrenia and movies like "A Beautiful Mind", are quite popular . . . Psychosis is like a heart attack. It is a serious health event. You should only have one episode in your life.
>
> —FROM IAN CHOVIL'S HOMEPAGE (WWW.CHOVIL.COM)

A mind unhinged is a mind unreachable. It is natural to avoid such people. The homeless schizophrenic, the criminal, the drug addict, the neighbor, the family member with schizophrenia are isolated, intention-ally and unintentionally. All illnesses are disturbing, but schizophrenia is in a class of its own.

The causes of schizophrenia are poorly understood because this disease undermines the foundation of our mind, eroding the perception of reality. Just as a person with schizophrenia has no basis to distinguish real from fabricated voices he hears in his own mind, science has little knowledge to cure such patients. We do have medications and other treatments for schizophrenia that help control the symptoms, but a cure is elusive because perception, consciousness, and internal visions of reality in our mind are beyond current scientific understanding.

Neuroscientists recently have learned much about this mental illness. It has a strong hereditary component. If one identical twin has schizophrenia, the chances are fifty-fifty that the other twin will as well. This has encouraged researchers to look to genetics for the roots of the disease, and this research is yielding many fruitful leads. Most of the genetic abnormalities suggest that this unbalanced mental state is the result of faulty signals and connections in the mind. These problems can include deficits in certain neurotransmitters in the brain that process perception, fear, and memory, most notably dopamine and glutamate. The problems also include imbalanced connections in brain circuits, which can be seen with functional brain imaging and by examining schizophrenic brains at autopsy. Both of these lines of research have recently revealed an imbalance of another kind that has been previously overlooked: an imbalance in glia.

The schizophrenic brain is physically different from the normal brain. Whether this difference results from developmental defects, scarring from deranged patterns of mental activity, or the years of drugs required to restore some balance to the mind is not certain, but most likely all three contribute to the physical changes in the schizophrenic brain.

The brains of people with schizophrenia often show decreased mass in certain areas and enlargement in the fluid-filled cavities at the core of the brain. Some of this tissue loss is neuronal, but much of it is glial. Is this brain tissue loss the result or the cause of mental illness?

An answer is suggested by the recent finding that there are many abnormal genes detected in people with schizophrenia. This surprise finding was revealed by modern gene chip analysis, a technology developed in the effort to sequence the human genome. This new method surveys thousands of genes at once. Previously a researcher would have to guess which gene might be contributing to a disease in order to test it specifically, but now researchers can test thousands of genes at once in large populations of people and can then sift the data for unusual genetic defects that the patients share. In the case of schizophrenia and depression, this unbiased search revealed some big surprises.

Some of the genetic abnormalities found by this wide-ranging search made good sense, because they involved genes controlling neurotransmitter function, but other genes turned up that were completely unexpected. One of the largest categories of genes found to be abnormal in schizophrenia and major depression are those involved in regulating the development of oligodendrocytes and the formation of myelin.

Brain imaging and autopsy studies confirm that schizophrenic brains lose myelin and oligodendrocytes in regions involved in processing sensory information, fear, and memory. This might be explained as a consequence of abnormal function of certain neural circuits in the schizophrenic brain, in turn causing changes in myelin or loss of oligodendrocytes, but finding abnormal genes in myelinating glia as risk factors for schizophrenia suggests that these glial cells are involved in the mechanism that causes schizophrenia, major depression, and bipolar disorder.

At the same time, drug treatments for psychiatric illnesses could affect white-matter integrity either through direct effects on oligodendrocytes or by altering the electrical activity of neural circuits in ways that subsequently affect myelin. Recent research shows that many of the antipsychotic drugs, such as Quetiapine, used to treat schizophrenia can affect (positively and negatively) development of oligodendrocytes and the formation of myelin.[7] This should not come as a surprise, because oligodendrocytes have receptors for several neurotransmitters, including serotonin, glutamate, and dopamine. Quetiapine acts on the dopamine and serotonin receptors. Excess serotonin has long been known to cause demyelination. In 1977, for example, B. A. Saakov and colleagues published an electron microscopic study documenting severe breakdown of myelin in dogs injected with serotonin. Although their conclusion could not have been stated more unambiguously—"It is concluded that serotonin has the property of injuring myelin and glia"—a possible implied connection between white matter and psychiatric illnesses such as schizophrenia and depression escaped the notice of most neuroscientists.[8]

As scientists begin to explore the other half of the brain, our understanding of how the brain operates normally is expanding; at the same time, completely new insights into how the brain fails in psychiatric illness are coming into focus. Interestingly, these insights represent not a new discovery, but rather an amazing rediscovery.

Hallucination to Lobotomy and Back

The Portuguese neurologist Egas Moniz introduced prefrontal leukot-omy in 1936, later called lobotomy, as a treatment for schizophrenia. The original name derives from two Greek words *leukos* (white) and *tomos* (to slice). The procedure did not remove neurons; it destroyed connections between the prefrontal region and the rest of the brain by severing the white matter (myelinated) fiber tracts to the forebrain. These cables appear white because of the thick layer of fatty myelin insulation deposited by oligodendrocytes. This treatment was based on the hypothesis that serious psychiatric disorders can result from imbalanced activity in different parts of the cerebral cortex. Severing the connections to the forebrain could restore some balance of information flow through the higher brain.

This procedure was highly effective in calming patients and releasing them from schizophrenia. Moniz received the Nobel Prize in 1949 in recognition of the revolutionary benefits possible from this psychosurgery, but later the procedure became stigmatized because it could sacrifice too much of a person's identity in the exchange. Although intelligence was unaffected, patients lost higher executive functions, which are carried out by the prefrontal region of the brain. They became docile and found it difficult to plan and sustain mental focus. Moreover, the procedure was abused in the 1950s as a treatment to control difficult patients.

It is not necessary to sever the axon cables to perform a prefrontal lobotomy; in fact, the first method developed by Moniz was to inject alcohol into the white matter tracts connecting to the prefrontal lobes. The myelin insulation around these axons is essential for electrical impulses to flow through them, and once alcohol dissolved this insulation communication was effectively blocked. (The alcohol injection would also have damaged the nerve fibers, but it need not do so to be effective.) Recent functional MRI imaging shows that prefrontal cortex and the emotional center of the brain, the amygdala, have fewer myelinating glia in patients with schizophrenia.

Another treatment for depression and schizophrenia is electroconvulsive shock therapy. To understand the medical basis for electroshock therapy, and possible involvement of glia in the therapeutic response, we need to trace the origins of the method and explore the discovery of human brain waves. The search will reveal glia as the inspiration for using powerful electrical currents to shock the brain back into balance.

Unbalanced Glia in an Unbalanced Mind

Medically induced seizure was the first and remains one of the most effective therapies for depression, but its use has always been controversial and stigmatized because no one really understands why it works. Medical historian and psychiatrist Max Fink describes the serendipitous roots of shock therapy, which grew from seeds of suspicion that glia were the cause of mental illness. In the mid 1930s a Hungarian neuroscientist, Ladislas von Meduna, noticed a strange coincidence—epileptics almost never suffered from schizophrenia. What is more, he observed that schizophrenics who experienced an epileptic episode were often cured. From the beginning Dr. Meduna suspected glia as the vital connection; now he had to find a way to test his theory.[9]

After some preliminary animal experimentation, Dr. Meduna decided to treat a schizophrenic patient who was hopelessly ill by deliberately triggering a brain seizure. The patient was catatonic, completely detached from the world and unresponsive. After experimenting on guinea pigs with various drugs to trigger seizures, Dr. Meduna settled on injecting camphor oil intravenously. The injections were repeated at three- to four-day intervals, and by the fifth injection, the schizophrenic patient was no longer psychotic. The man became coherent, alert, and talkative. In 1935 Meduna published his findings in an article titled "An attempt to influence the course of schizophrenia by biological means." The very concept of considering mental illness as a biological process in the brain was revolutionary. Soon, electrical shock replaced camphor injection as the preferred method to induce seizure, but insulin shock was also widely used.[10]

Glia were the source of inspiration for this revolutionary mending of the mind. Dr. Meduna's earlier studies as a neuropathologist interested in cellular changes in the brain after head trauma had led him to write papers describing reactions of microglia to injury. He had found clues left in the brains of people with schizophrenia and epilepsy who had died, not infrequently of suicide. These same clues continue to accumulate today, and they strongly implicate glia in schizophrenia, epilepsy, and other mental illnesses and diseases.[11]

Scientists have long known that the cellular structure in the brains of epileptics and the mentally ill is altered. But is this the result of a sort of brain scarring from the mental disturbance, or is it an underlying cause? In the region of the human brain where a seizure begins—an epileptic focus—the astrocytes are changed and their numbers increase substantially. These astrocytes are larger (hypertrophied), and they exhibit a different molecular composition from other astrocytes. In particular, the GFAP protein that forms their cellular skeleton is unusually abundant. This makes it possible for a neuropathologist to spot the part of the brain where the seizure began, simply by treating the brain tissue on a microscope slide with a stain to reveal concentrations of GFAP, much as a forensic pathologist uses chemicals to uncover a latent fingerprint at a crime scene.

Scientists also noticed that the brains of people suffering from chronic depression, schizophrenia, and certain other mental illnesses lose mass. Intriguingly, much of the tissue lost in the brains of schizophrenics is caused by loss of glia, in particular oligodendrocytes and astrocytes. Meduna wondered, could an imbalance of glia be the cause of such mental illness as depression and schizophrenia on one hand and epileptic seizure on the other?

When you consider that many mental illnesses are now understood to arise from defects in synaptic transmission, the significance of abnormal glia is not difficult to deduce. Modern treatments for schizophrenia, bipolar disorder, and depression are based on drugs that restore the balance of neurotransmission in particular brain circuits. In particular, the neurotransmitters serotonin, glutamate, and dopamine are deficient in circuits involved in cognition, perception, and emotion in these patients. Drugs known as neurotransmitter reuptake inhibitors are often the best treatment. These drugs block the cellular process of removing neurotransmitter from the synaptic cleft after it is released by the nerve terminal. In effect, clearing neurotransmitter from the synaptic cleft erases the chalkboard so another message can be conveyed. But if reuptake is slowed with drugs, the chalkboard will not be erased so quickly. The amount of neurotransmitter in the synaptic cleft will be sustained a bit longer, increasing the time for the message to be read and thus strengthening the weakened connections in the minds of the mentally ill.

This is exactly what astrocytes do at a synapse. Astrocytes are the cells that absorb the excitatory neurotransmitter glutamate that is released into the synaptic cleft. Once denigrated as a lowly housekeeping servant for the regal neuron, the glial cell, it is now clear, is in a position of ultimate control over the synapse in sickness and in health. Glia can seize control over the flow of information between neurons, sometimes with detrimental results.

Consider the range of other psychiatric disorders and drug addictions where imbalances of neurotransmitter at synapses result in an imbalanced mind. Obsessive-compulsive disorder is treated with serotonin reuptake inhibitors. Amphetamine-derived drugs, such as ecstasy (MDMA), disrupt serotonin synapses. Methamphetamine (meth) affects synapses that use the neurotransmitter dopamine, as does Parkinson's, a degenerative disease. Marijuana, alcohol, cocaine, amphetamine, caffeine, benzodiazepines (such as Valium), nicotine, heroin, PCP (angel dust), and tranquilizers all affect synaptic transmission in the brain. The role of glia in mental illness is scarcely explored, yet glia are the cells primarily responsible for clearing neurotransmitter from synapses. If astrocytes fail to do their job, would the consequences for synaptic function and cognition be any different from the effects of drugs that alter neurotransmitter levels?

Could better treatments for drug addiction and mental illness develop from research into drugs that act on glia at synapses? Almost certainly some of the drugs used now to treat mental disorders—including ADHD, mania, depression, anxiety, and schizophrenia—act in part through their effects on glia. Might we find new and better drugs to treat mental disorders by designing glial drugs to add to our arsenal of neural drugs? The only thing separating us from this promise for better treatments is the will to break down intellectual barriers and study glia with at least as much energy and money as has been applied to the study of neurons.

Without benefit of this modern insight, Meduna simply presumed the obvious imbalance: schizophrenic brains had too few astrocytes, epileptic brains had too many. The reason these two conditions appeared to offset each other could simply be that they restored the normal critical balance of astrocytes in the brain. Remarkably, 95 percent of Meduna's patients treated by induced brain seizure after acute schizophrenia got better; 80 percent treated with induced seizure within the first year of their illness were cured. For depression, high rates of success were typical in Meduna's studies, and the treatment remains about 80 percent effective today."[12]

But there are several intriguing loose ends in this story. Most notably, why are there more astrocytes at the focus of an epileptic event? Is this evidence of glia as culprit, innocent bystander, or good Samaritan coming to the aid of neurons in trouble? These possibilities are not mutually exclusive, and glia may be there for all these reasons. Their presence does, however, strongly hint at connection between glia and electrical activity in the brain, at least in the extreme cases where normal brain wave patterns spin wildly out of control and the entire brain and body seize.

BRAIN WAVES AND MADNESS: GLIA IN BRAIN SEIZURE AND ELECTROSHOCK THERAPY

Phenol, that prickly antiseptic smell startling your senses as you enter a hospital, is the scent of alarm, marking a place where life and death change sides. But to Dr. Hans Berger it must have been a comforting familiar fragrance of sanitation at the Psychiatric Clinic at the University of Jena, Germany, where he was the director and chairman of psychiatry and neurology in the 1920s. Meticulous in manner, punctual and rigid in his routine, Berger could be irritating to coworkers and was resented by subordinates who did not share his precision of habit. His scientific research was not widely appreciated, and ultimately, his biographers conclude, he lost his position as director for political reasons.[13]

In the 1920s, when Berger was searching for a connection between physical and mental energy in the brain, fascination with the occult and supernatural raged. Séances to communicate with dead spirits flourished in the United States and Europe. Revolutions in science were unfolding as the world pivoted into a new century, and interest in clairvoyance and telepathy resonated with a deep human desire to comprehend the unfathomable mystery of life. Berger was a believer, and he had scientific proof.

Unlocking the door, Berger entered the room in a separate small building on the grounds of the psychiatry hospital, revealing a world resembling the set of a Frankenstein movie: stainless steel instruments, electronic dials, and machines with electrodes for a person's scalp. Searching for a physical, scientific basis for elusive mental function, Berger finally found it in the rapidly expanding science of electricity. An artificial beam of light danced upon photographic paper, tracing patterns in jagged spikes and waves: the wake left by patterns of human thought. The spot of light moved as though by telekinesis, like a medium revealing answers (from beyond an impenetrable barrier) through involuntary motions on a Ouija board. Berger was conducting important research on brain function, but he also believed he had found the scientific mechanism for mental telepathy. He guarded his scientific results fastidiously. He alone had unlocked the mystery of human brain waves and mental telepathy through systematic scientific investigations, working alone and at night at the hospital in absolute secrecy. Many of the experiments he performed on mental patients, but many were conducted on his own teenage son.

Berger had discovered that the human brain emits waves of electromagnetic radiation, much like those of a radio transmitter. The waves, energized by our thoughts, change with sensory stimulation and with mental focus and attention. Without showing any outward sign, these electromagnetic brain waves broadcast the private workings of the human mind. Berger conducted these experiments for five years before announcing his findings to the world. The possibility that glia could have any relevance to brain waves could not have occurred to anyone at the time, and would not until the twenty-first century.

As the head of a psychiatry clinic, Berger was swept up by the political turmoil of Germany in the 1930s. The mentally ill were the first targets of the growing National Socialist (Nazi) movement, ensnarling psychiatrists at the center of political events. Euthanasia was first perfected in psychiatry clinics for the purpose of eliminating the mentally ill, thus relieving society from the burden of supporting useless, defective beings and eliminating a potential source of pollution of the Aryan gene pool. The process was referred to by the antiseptic term "social hygiene."

By August 1941, seventy thousand patients from German mental hospitals had been euthanized, a prelude to the subsequent mass elimination of Jews, homosexuals, communists, and gypsies to follow. So efficient was the program that Professor O. Wuth, chief physician for the army, worried, "Who will wish to study psychiatry when it becomes so small a field?"[14]

Medical doctors and scientists applied their skills with cold logic. The gas chamber was the product of careful scientific research into how to mass-produce killing as efficiently as possible. In 1941, the gas chambers at psychiatric hospitals were dismantled and shipped to Auschwitz and Treblinka. The same doctors, technicians, and nurses often followed the equipment.[15]

The second front in the eugenics program was forced sterilization, and this too originated in the psychiatry hospitals. Genetic purification through forced sterilization would terminate the "defective" and undesirable on the threshold to the next generation. Psychiatrists reviewed the case files and provided professional recommendations about which patients should be sterilized. Physicians systematically applied X-rays to the testicles and ovaries of "patients" and afterward surgically removed the organs to determine scientifically the most effective time and dose to achieve sterilization.[16]

Many scientists and physicians fled the perversion in their country: Sigmund Freud and Albert Einstein are notable examples. Others were coerced to comply as a matter of personal and professional survival. Still others were swept up in the war effort and cooperated out of patriotism and the imperative to prevail in the war.

It is said that Berger was not favored by the Nazis.[17] At universities throughout Germany, faculties were purged of undesirables as Nazis took control, and top administrative positions were assigned to Nazi sympathizers. Records uncovered after the fall of the Berlin Wall have recently exposed complicity in Nazi war crimes of top administrators and faculty at the University of Jena, where Berger was head of the hospital and conducting his research on brain waves. The documents reveal that the university was a center of the Nazi eugenics program, in part because of its immediate proximity to the concentration camp Weimar-

Buchenwald and in part because there was strong support for the National Socialists in this region of Germany. An analysis published in 2005 by scientific historian Dr. Susanne Zimmermann, anatomist Dr. Chris Redies, and their colleagues reveals that at least two hundred human specimens in the current collection on public display at the famous Anatomical Institute at the University of Jena are the remains of mental patients euthanized during the Nazi regime. Two hundred more specimens were traced to corpses obtained from Nazi executions of other undesirables, primarily Jews and petty criminals. (These anatomical specimens have since been removed from display and a plaque installed in honor of those who were killed.)[18]

By nature, Hans Berger had difficulty getting along, and many of his colleagues questioned the scientific value of his research on the interface between psychic and physical energy as revealed by brain waves. In 1938 he was summarily forced to resign by the Nazis. The real madness of that time reverberated sympathetically with his own tendencies to a sort of personal madness, the despair of melancholy that afflicted him periodically.[19] On June 1, 1941, Hans Berger walked into the hospital and went directly to a preplanned spot. There he drew his last deep breath of antiseptically tinged air and hanged himself. What he did was irrational; where he did it was not.

"I would not recommend him to my children as a hero," German neuroscientist and anatomist Dr. Chris Redies summarized in relaying answers from Dr. Susanne Zimmermann, a medical historian at the University of Jena, whom I contacted in an effort to illuminate Berger's past. Contradicting the often repeated accounts in his biographies, Berger's retirement was not forced, Zimmermann concludes. In fact, she says, he served on the selection committee for his successor, and the chosen individual was fired immediately after the war because of his Nazi activities. According to Zimmermann, Berger served as judge on the *Erbgesundheitsgericht*, the court that imposed sterilizations on people who did not fit the Nazi scheme of social hygiene, and his diaries contain anti-Semitic comments. Berger's suicide, she says, was the result of major clinical depression while he was hospitalized in a vain attempt to find a physical explanation for his mental illness. His suicide was not a political statement. It was the result of mental illness.

Zimmermann's research has uncovered many unwelcome revelations relating to the euthanasia program in Jena during the Nazi period. Her work has not met with unanimous approval because her recent findings have discredited honored doctors in Jena, some of whom are still living.

Historical truth is difficult to reconstruct from shadows of events long past, which shift with one's perspective. In the spring of 2006, I visited Zimmermann and Redies in Jena in an effort to determine for myself the truth of her findings. Flipping through the photocopied pages of official records from the 1930s neatly indexed in a five-inch-thick black binder, Zimmermann stopped as she came to a series of documents recording the proceedings of a court of appeals to consider the forced sterilization of mental patients. The cases included mentally retarded children, schizophrenics, epileptics, and a sixty-one-year-old alcoholic man. As she read the record of one case, I imagined the anguished and imposing scene as a husband pleaded with the appeals court not to subject his young wife to forced sterilization. Every appeal on record was denied. Balancing the black binder on her knees, Zimmermann pointed to the signature at the bottom of each decree. I read the signature written in an unmistakable, precisely penned hand and was sickened: *Hans Berger.*

Berger's discovery of human brain waves may be the most important discovery in electrophysiology of the twentieth century. Today EEG recording is a fundamental investigative tool for scientific research on the brain, and it is an essential medical diagnostic tool. Its scope expands beyond science and medicine to encompass social and legal realms; in our modern world, brain waves have become the ultimate definition of death.

The great irony in Berger's tragic story is that the brain waves he discovered are now appreciated as a key to relieving clinical depression, from which he suffered, as well as many other mental illnesses. As mentioned, one of the most effective treatments for depression is electroshock therapy, which ignites a firestorm of electrical activity in the brain. From the ashes the mind arises calm and at peace.

But how does resetting brain waves with electroshock treatment release mental patients from the grip of depression or schizophrenia?

Glia may be a clue to the puzzling connection between brain seizure and therapy for psychiatric disorders that include depression, mania, and schizophrenia.

BRAIN WAVES: GLIA IN EPILEPSY AND DEPRESSION

Hans Berger was the first person to record the violent brain waves in an epileptic brain during seizure. He found that the brain waves of epileptic patients were much larger during an epileptic attack but nearly flat afterward, and he learned that patients with Alzheimer's disease and multiple sclerosis also showed altered brain waves. Multiple sclerosis is a disease affecting glia (oligodendrocytes), suggesting a possible involvement of glia in brain waves. In studying changes in brain waves in children, Berger found what was later understood as a further clue that glia affected brain waves. The waves could be recorded with his instruments in children only after two months of age, a time when glial cells had coated nerve fibers with insulating layers of myelin over extensive regions of the brain.

Brain waves arise from the combined action of electrical activity in large numbers of neurons; they are like the roar in a baseball stadium resulting from the combined conversations among thousands of individuals in the stands. Most of the time these conversations are uncoordinated, creating a steady background noise, but certain stimuli, like the crack of a bat, coordinate their activity at particular moments to create surges above the background noise. Opening and closing the eyes, arousal, and sleep all profoundly affect the coordinated activity of thousands of neurons in our cerebral cortex, and this activity is reflected in the brain waves measured from a person's scalp.

How glia, which do not communicate using electrical signals, could possibly affect the electromagnetic radiation from a person's head is puzzling. To understand this we must examine more closely the source of electrical power in neurons that energizes communication between nerve cells.

When an electrical current flows through a single neuron, return currents pass through the fluid surrounding the nerve cell. This happens because electricity must always flow in a circuit. Break the circuit and the current stops. This current flow around the neuron creates an electric field around it that looks much like iron particles aligning between the two poles of a magnet. The electrical currents streaming through the fluid space between all cells in the brain simply flow like water, according to the path of least resistance, and the various currents of individual neurons contribute collectively to the ocean of brain current recorded by the EEG. The electrical currents flowing inside our head can be detected as electrical signals by using electronic amplifiers attached to the scalp, just as a stereo receiver detects and amplifies minute electromagnetic waves in the air.

In our brain, this current flow is not constant; it surges in tides and waves according to the combined action of all the neurons at work in our cerebral cortex. This is why when we are relaxed, brain waves oscillate in a slow wave pattern, like waves on the shoreline rolling in at a regular pace. If there is a torrent of activity in the brain, the waves collide and no longer roll into shore regularly, but instead crash with increasing frequency and in a more haphazard pattern. Simply opening our eyes will cause a sea change in brain wave pattern, as Hans Berger first observed when he asked his son to open his eyes while he recorded brain waves through his scalp.

Recall that glial cells also have a voltage. Astrocytes cannot generate electrical impulses; theirs is a steady battery-like voltage, in contrast to the pulsed electric discharges of neurons. Many pathological and normal brain functions—sleep, hypoxia, hypoglycemia, and ischemia (stroke), for example—are associated with slow changes in glial voltage contributing to the EEG. In the retina, for example, a flash of light stimulates visual neurons to fire, increasing potassium outside the retinal neurons. The light flash makes retinal glial cells (called Müller cells) positively charged by 8 millivolts as they mop up the excess positively charged potassium ions dumped by the retinal neurons. The same process of charging and discharging glia as they absorb and move potassium ions through glial networks probably occurs throughout the brain, and these glial voltage fluctuations register on the EEG.

Thus, not only do astrocytes regulate the electrical power of neurons and thereby affect neuronal activity and the resulting brain waves, they also contribute directly to the slower waves of electrical currents in the brain by moving positively charged potassium ions through glial networks.

Glia contribute to brain waves in yet another way, because in addition to the potassium extruded from neurons and absorbed by astrocytes, neurons also release dopamine, glutamate, and other neurotransmitters at their synapses. These neurotransmitters are also charged molecules, and as they are taken up by astrocytes this too creates an electric current; transporting these substances into the astrocytes changes the balance of charged molecules passing into the cytoplasm of the astrocyte.

All of these glial responses to neuronal firing become accentuated during intense brain activity, and the contribution of glial currents to brain wave activity increases proportionately. During a brain seizure, for example, glia accumulate so much potassium that they depolarize substantially, losing about 35 millivolts from their normal voltage of −100 millivolts. A spreading depression in voltage through the brain can be measured by EEG, like a brownout through a failed power grid, often induced by lack of oxygen (hypoxia) or lack of blood flow, as occurs during a brain stroke. Glia in the region of a spreading voltage depression can lose all of their voltage, and these areas of the brain can become the epicenter for an epileptic event. This explains how glia contribute to brain waves and seizure, and how they may participate in schizophrenia, depression, and other psychiatric disorders.

Glia as Neuronal Brake and Gas Pedal

Glutamate is of particular interest in epilepsy, mania, and drug abuse. Glutamate is the principal neurotransmitter at synapses in the higher brain (cortex and hippocampus) that send messages of stimulation between neurons. Other types of synapses, called inhibitory synapses, use different neurotransmitters to calm neuronal firing. One of the most common neurotransmitters at inhibitory synapses is GABA (gammaaminobutyric acid). This is the neurotransmitter that activates synapses that are the target of the calming (anxiolytic) drug Valium. Glutamate and GABA are the chemical accelerator and brakes, respectively of our states of arousal. Astrocytes have receptors for both GABA and glutamate. Thus, astrocytes in a sense "know" our moods. Moreover, astrocytes can release or absorb glutamate. In this way, astrocytes can excite or depress neurons.

Neurotransmitter spilling from the neuronal synapse activates glial glutamate and GABA receptors, stimulating a rise in calcium in the astrocyte. The increase in calcium triggers the intercellular communication between astrocytes by initiating a wave of calcium passing from cell to cell. As discussed in chapter 3, sending signals through waves of calcium is the main mode of information flow between glia, and this transmission operates outside the lines of communication between circuits of neurons. This calcium signaling through astrocytic networks disperses and integrates the local activity among groups of synapses or neurons over larger areas of the brain. This gives astrocytes a more global influence on general levels of excitation in the neuronal brain, where, in contrast, neurons can only communicate through discrete points of contact via synapses. For this reason, astrocytes acting in a global manner are poised to influence large-scale changes in excitability in the cortex, which is the fundamental dysfunction in epilepsy.

The rise in calcium causes astrocytes to release many substances, including neurotransmitters and other factors that directly affect neuronal excitability and survival. In regulating neuronal excitability in this way, astrocytes contribute to and regulate seizure during epilepsy, but they also regulate states of arousal, such as sleep (chapter 13), and influence the death of neurons from overstimulation. Many researchers suspect that the stabilizing influence of astrocytes on the neuronal brain could become impaired in psychiatric illnesses, producing deranged thinking and hallucination.

Too much glutamate and the brain not only becomes hyperactive, but neurons, like engines revved beyond redline, are damaged or killed. Excessive levels of glutamate can originate from hyperactive neuronal circuits themselves, but astrocytes can also release glutamate, adding fuel to the fire in epilepsy, for example, and contributing to neuronal death from hyperexcitation.

However, astrocytes also clear excess glutamate from synapses, a critical function of these brain cells that has long been appreciated. It is reasonable to interpret the increased number of glia in epilepsy as a glial response to restore normal glutamate levels caused by hyperactivity in seizure. Now it is also known that patients suffering from bipolar disorder have fewer glia in the regions of the brain handling mood.

Thus, one can imagine how deficient numbers of glia in these brain regions controlling thought and mood could contribute to glutamate imbalance in the hyperactivated or depressed brains of bipolar patients.

Recent research has found that astrocytes markedly change their calcium signaling in animals in which seizures have been induced experimentally. Normally the amount of calcium signaling between astrocytes in brain cortex is relatively moderate, but after a seizure these astrocytes typically show large oscillations in calcium signals. They sweep through the cortex in strong waves, presumably releasing more glutamate and tipping the brain toward seizure. The evidence suggests that these changes in astrocyte calcium signaling are permanent changes following repeated seizures, rather than echoes in the wake of increased calcium signaling in astrocytes induced during the seizure itself. Possibly this change in astrocyte calcium signaling after seizure could be beneficial, but early research suggests that damping the excessive astrocyte signaling with drugs improves the outcome and limits the death of neurons in animal models of epilepsy.

As mentioned previously, interesting new research reveals that patients with bipolar disorder and schizophrenia also have fewer oligodendrocytes, the myelinating cell of the brain. These glial cells wrap axons with myelin, but they are not thought to be involved in glutamate regulation. Nevertheless, too much glutamate can be just as toxic to myelinating glia as it is for neurons. This could be another way glia participate in mental illness, because when the myelin insulation becomes frayed, so does mental function.

Excess glutamate in the brain may be one of the main causes of death of myelinating glia in the brains of people suffering overproduction of this neurotransmitter. The effects of glutamate on myelinating glia might influence cognitive function even though these cells are not associated with synapses. When one considers that prefrontal lobotomy works by severing the connections to the forebrain, and seeing how the process completely changes a person's personality, it is not difficult to imagine how a pathological loss of myelinating glial cells in these forebrain tracts could lead to psychiatric disorders such as schizophrenia and other mental impairments. Breaking the insulation on critical communication cables in the brain will disrupt communication as effectively as severing the cable. This may account for how a prefrontal lobotomy relieves a psychotic patient's symptoms and stabilizes debilitating mood swings. The surgical procedure simply severs connections to the frontal lobes, but a breakdown in communication with the forebrain resulting from disrupted insulation on the nerve axons could accomplish the same thing.

If this approach of altering transmission through communication lines in the brain could be exploited more selectively, it might possibly provide an effective treatment for some mental illnesses. Despite its infamous reputation, some doctors and patients argue that prefrontal lobotomy may be unjustly maligned largely because the techniques that were used were too crude. Far better than a lifetime of treatment with mind-altering drugs, precision surgery to restore a balance in communication lines disrupted in the psychotic brain might be the ultimate treatment for severe mental illnesses such as schizophrenia. What better way to selectively regulate impulse flow through brain circuits than to manipulate the insulating glia that control the flow of impulses through nerve axons?

Electroconvulsive shock has a therapeutic effect on clinical depression and schizophrenia by resetting brain waves, but electroconvulsive shock therapy may also activate a beneficial injury response in the brain. Since astrocytes and microglia are the first line of defense in any brain injury, it is obvious why altered glia would be seen in regions of the brain giving rise to seizures. This injury response of glia to brain seizure may also be one of the cellular mechanisms that explains the changed brain function induced by electroshock therapy. Both microglia and astrocytes release growth factors in response to brain stress and injury. These growth factors sustain neurons under neurotoxic conditions that would normally kill them, and in the healthy brain these glial-derived growth factors promote neuronal growth and health. Both microglia and astrocytes release many different natural inflammatory agents to aid in the healing process, and all of these glial responses probably contribute to the therapeutic effect of electroshock treatment.

There is one other powerful way glia could participate in the therapeutic effects of electroshock therapy. Many types of glia can act as stem cells, lying latent in the brain and waiting to be stimulated to generate new neurons to replace those lost to injury or disease. Surprisingly, it has recently been discovered that all major antidepressant drugs stimulate the birth of new neurons in the hippocampus, the part of the brain that is essential for memory. It is now known that astrocytes can control whether neural stem cells develop into neurons or astrocytes. Other types of stem cells in the brain that are more primitive and plastic (changeable) than either neurons or astrocytes can give rise to cells that are glial and then subsequently transform into either oligodendrocytes or neurons.

These various actions of glia suggest multiple mechanisms for the therapeutic benefit of electroshock therapy and they also place glia at the root of brain seizure and mental disorders of various types. These glial activities also account for the changes in astrocytes seen in the brains of patients after brain seizure or electroshock treatment.

From this perspective, it should not be surprising to find differences in glia associated with various mental illnesses. By releasing glutamate and other substances (for example ATP), astrocytes intensify the excitability of neurons during an epileptic event. Like pouring fuel on a fire, the neurotransmitters released by glia stimulate the synapses connecting neurons into circuits. Other substances released by astrocytes during seizure (adenosine, for example) dampen neuronal excitation.

Glial Drugs for Mental Illness

By reasoning from the premise that schizophrenia results from abnormal communication and processing of sensory and internal information in the brain, one can reasonably infer that the cells controlling conduction of impulses through axons could have a critical influence on the disease. If defects in the myelin insulation around axons prevent impulses from reaching their proper connections at the proper time, the internal processing of information in the mind will be impaired. Blocking the flow of information through such "frayed wires" would have the same effect as cutting the cables. Remember that the first prefrontal lobotomies, which were performed to treat schizophrenia, did not involve a knife, but rather an injection of alcohol into the white matter fiber bundles to the forebrain that would have destroyed the myelin insulation. It probably damaged axons too.

One of the genes that is abnormally expressed in schizophrenics codes for a growth factor, neuregulin, which glial biologists have long known regulates the development of myelinating glia in our nerves (Schwann cells) and in our brain (oligodendrocytes). There is a new branch of research to find drugs to compensate for the loss of this growth factor and stimulate the formation of oligodendrocytes or prevent their death in the brains of schizophrenics. At least a dozen other myelin-associated genes have recently been identified as abnormal in schizophrenics, offering dozens of new branches of investigation into the roots of this illness and new therapies to treat it.

Other genes that are abnormal in schizophrenics affect the development and migration of neurons and glia in the brain. Recognizing the important function of glia in directing migration of neurons in the fetal brain and the outgrowth of neuronal connections (see chapter 11), many neuroscientists are now exploring the possible involvement of glia in subtle developmental defects that could result in schizophrenia.

As mentioned above, a large class of genes implicated in schizophrenia includes genes related to neurotransmitters. Many hallucinogenic drugs, such as PCP and ketamine, produce hallucinations similar to those experienced in schizophrenia. These drugs act on excitatory synapses in the brain that use glutamate as a neurotransmitter. One important class of glutamate receptors are called *NMDA receptors*, which have some very unusual properties. Much like a two-key lock on a vault, NMDA receptors will not open up and operate unless the neurotransmitter glutamate and another substance, D-serine, bind to it simultaneously. The reason for such high security is that NMDA receptors are the critical glutamate receptors triggering storage of new memories.

Glia, it turns out, hold the second key required to open these critical receptors. Serine is one of the common amino acids in the body, but D-serine is different from the others. Amino acids are organic (carbon-based) molecules with a three-dimensional shape, which provides them with left-right symmetry. Just as all gloves have four fingers and a thumb but are left-handed or right-handed, amino acids differ, too, in being left-handed or right-handed. Scientists refer to the left and right mirror images of the same molecule by L (Latin *levo*, "left") and D (Latin *dextro*, "right"). Proteins are made by snapping different kinds of amino acids together into long chains. We are all familiar with the difficulty presented by the necessity of having two different gloves (left and right). When one is misplaced, an opposite glove cannot be substituted. Just as in clasping hands, the correctly "handed" amino acids must be joined to make chains. Nature decided to eliminate this problem by using only one form: the left-hand version. All natural amino acids forming the proteins of all life on earth are built from only the left-hand form of amino acids. Nature has no right-handed gloves. This is one way scientists examining amino acids extracted from meteorites are able to determine if the organic molecules were synthesized in extraterrestrial organic chemical reactions or are contaminants from earthly biological sources. Extraterrestrial chemical reactions might produce left and right forms of amino acids indiscriminately, but amino acids produced by life on earth will be *only* the left-handed form.

There is an exception. The right-handed form of the amino acid serine (called D-serine) can be detected in the brain. This right-handed form is useless for building proteins, but its odd character suits it as a messenger molecule to carry chemical messages between cells. D-serine is the unique second key required to open NMDA receptors. Only when NMDA receptors open are the connections strengthened to form memories.

The peculiar thing about this is that when the brain is examined for the enzyme that converts the natural L-serine into the unique D-serine, it is found inside the astrocytes that tightly surround synapses using glutamate for a neurotransmitter. Glia thus control the neurotransmitter system in the brain that is essential for forming memory. Glia synthesize and release D-serine into the synaptic cleft to allow NMDA receptors to activate when glutamate is released from the nerve terminal.

Recent genetic analysis has found that some people with schizophrenia have defects in the D-serine synthesizing gene, and this has stimulated intense research to develop drugs resembling D-serine to treat schizophrenics. One wonders how many other drugs targeted to glia at synapses might offer new treatments for this and other debilitating mental disorders.

The extent to which glia participate in psychiatric disorders is only beginning to be explored, but the fundamental importance of glia in a wide range of neurological disorders has long been recognized. Surprising new research shows glia participating in many neurodegenerative diseases, providing strong evidence for the importance of the other brain in normal brain function and mental illness. Neurodegenerative disorders such as Parkinson's disease, Alzheimer's, ALS, and Huntington's are caused by the death of neurons. Glial cells are now understood as both friend and foe.

Beyond illuminating causes and possible cures for these life-altering neurological diseases, collecting information on glia at work in mending the mind may provide insight into the vexing question of why glial cells along axons would respond to electrical impulses, as we had seen in our Schwann cell experiments. The answer remains elusive, because unlike astrocytes, myelinating Schwann cells have nothing to do with synapses. Yet, as implied by mental disorders associated with defects in myelinating glia, information flow in the brain and thus our thoughts—normal or abnormal—appear to be regulated by processes operating beyond the synapse, processes in which myelinating glia are operating.

CHAPTER 8

Neurodegenerative Disorders

Astrocytes play a powerful role as silent partners in mental illness. They regulate the electrical power source for neurons (potassium ions), clear and release neurotransmitters from synapses, and respond to neuronal distress by releasing growth factors and stimulating the birth of new neurons. These same functions give astrocytes a life or death influence on the survival of neurons in neurodegenerative diseases such as Alzheimer's, Parkinson's, and others, and in assisting in recovery from brain injury.

ALS (LOU GEHRIG'S DISEASE): TRAPPED INSIDE

As if a biblical miracle—he who cannot walk shall fly—Stephen Hawking, world renowned astrophysicist, soared in zero-gravity weightlessness on April 26, 2007, inside the belly of a Boeing 727 in free fall toward the Atlantic. His trademark grin, twisted by paralysis, beamed with radiant elation. Hawking, confined to a wheelchair since his early twenties from a progressive motor neuron disease, is respected as a brilliant scientist and author of the book *A Brief History of Time*. Defying his disease, he pecked out the text word by word using a computer program that responds to feeble twitches of his head. Selecting each word from a list, nudging it as if it were a puzzle piece, Hawking assembled each sentence until he had written his book. It is a book that no one else could have written no matter how able-bodied. Father of three, Hawking is above all an inspiration.

"Shortly after I came out of hospital, I dreamt that I was going to be executed. I suddenly realised that there were a lot of worthwhile things I could do if I were reprieved," he wrote.[1] That reprieve enriches us all.

Hawking's disease, amyotrophic lateral sclerosis (ALS), is also known as Lou Gehrig's disease because it toppled the star athlete from the pinnacle of health and fitness. Gehrig is remembered most fondly for his grace in accepting a fate that would rob him of everything, including his life.

> Fans, for the past two weeks you have been reading about the bad break I got. Yet today I consider myself the luckiest man on the face of this earth.
>
> I have been in ballparks for seventeen years and have never received anything but kindness and encouragement from you fans.
>
> Look at these grand men. Which of you wouldn't consider it the highlight of his career just to associate with them for even one day? Sure, I'm lucky . . . I may have had a tough break, but I have an awful lot to live for.
>
> —LOU GEHRIG'S FAREWELL SPEECH, JULY 4, 1939,
> YANKEE STADIUM, NEW YORK[2]

Within two years of his farewell speech, Gehrig was dead. ALS causes paralysis by killing motor neurons, the nerve cells in the spine that issue commands to muscles. Before it strikes, ALS issues no warning, typically unleashing its sneak attack suddenly in adulthood. Scientists can clearly see the devastation in the wake of the crippling attack: the motor neurons in people afflicted by ALS die. At present, there is no cure for ALS.

The odd thing is that in ALS, only motor neurons are assassinated, with impressive precision. Every other variety of neuron in the spine and brain remains perfectly healthy. How this disease executes such a surgical strike on these particular neurons is still a mystery.

In 1993 a genetic clue to ALS was found in the form of a resemblance in the genes of several people suffering from the disease. Biologist Daniel Rosen and colleagues at the Massachusetts General Hospital discovered that one gene on chromosome 21 had a small mutation in its genetic code. The gene was known to make an antioxidant enzyme called superoxide dismutase 1 (SOD1). The research team found eleven different genetic mutations among thirteen families with the disease, and in every case, the mutations were in the gene that coded for SOD1.

Not everyone who has ALS has inherited the disease. The genetic clues implicating the SOD1 enzyme in ALS lead to the obvious conclusion that anything that might damage this molecule would result in ALS. What does this odd enzyme, containing copper and zinc atoms at its protein core, do?

SOD1 is an enzyme that is found throughout the cytoplasm of cells, where it searches for toxic forms of oxygen molecules called free radicals. As explained in chapter 5, free radicals are oxygen atoms that have stolen electrons from other atoms. The excess of negatively charged electrons they hoard makes oxygen free radicals extremely chemically reactive and corrosive to proteins and body tissue. SOD1 strips off the extra electrons from oxygen free radicals and attaches them to water molecules. Quenching the excess electrons in this way creates hydrogen peroxide and stable oxygen. Without this antioxidant protein on patrol in our cytoplasm, free radicals would build to toxic levels and ravage the proteins from which all cells are built. The loss of normal functioning SOD1 in motor neurons could easily lead to their death. This slow corrosion through oxidation seen in SOD1 dysfunction would also be consistent with the adult onset of ALS. As in aging, the slow accumulation of cellular damage eventually results in sudden failure and death of neurons.

Proof of this theory came in experiments on mice with the gene causing ALS. When researchers prevented the mutated SOD1 gene from being made in the mice's motor neurons, the mice were cured: they did not develop the ALS motor neuron disease. Further proof came in experiments where the mutant SOD1 gene was inserted into normal mouse motor neurons. These motor neurons died, proving that having the defective gene in motor neurons causes ALS.

But the control experiments in these studies yielded a surprise, as control experiments so often do. In the control experiment on mice with the gene causing ALS, researchers blocked the mutation in cells surrounding the motor neurons, leaving the neurons stuck with the unhealthy SOD1 genes. They expected, naturally, that the motor neurons would die, because they did not have the functional SOD1 gene, and the mice would suffer the effects of ALS. Instead, the animals remained healthy. The researchers concluded that the cells surrounding motor neurons in some way prevented the neurodegenerative disease through a mechanism involving SOD1. In 2007 two groups independently identified the critical lifesaving cells as astrocytes.

The researchers grew motor neurons in cell culture together with normal astrocytes or together with astrocytes that had mutations in the gene that codes for SOD1. The motor neurons died when they were grown on astrocytes with the defective SOD1 gene. The results of this experiment suggest that astrocytes are the motor neuron assassins.

Researchers also found that astrocytes have a coconspirator. SOD1 activity in microglia was also involved in the motor neuron death. These glial cells did not appear to be a trigger for the disease, but defects in SOD1 in microglia contribute to the rapid progression of disease in its later stages.

Researchers found that even the culture medium taken from mutant astrocytes killed motor neurons, suggesting that the mutant astrocytes must be releasing something that is toxic to the neurons. The relation between astrocytes and motor neurons is special, because disrupting the gene that codes for SOD1 in other types of cells grown together with motor neurons has no effect. This special neuron-astrocyte bond also explains the selective death of only motor neurons in ALS.

Astrocytes with mutant SOD1 become reactive (gliotic). Astrogliosis has always been a well known hallmark of ALS. But what toxic substance is being released by these astrocytes with the mutant SOD1 gene? This question is still being investigated, and recent evidence strongly suggests that an impaired ability to remove the neurotransmitter glutamate released from synapses could allow glutamate to rise to neurotoxic levels. In theory any number of consequences of SOD1 defects in astrocytes could cause them to release a variety of toxic agents that would kill their motor neuron neighbors. Several candidate molecules have been investigated, but as of this point, the lethal toxin being released from astrocytes and crippling ALS patients has not yet been identified. Nevertheless, we now know where the poison is coming from.

This discovery was made simply by expanding the scope of search beyond the universe of neurons to include the neglected cells in the other brain. As Hawking graciously conceded when one of his scientific predictions was disproved in 2004, sometimes information can escape from a black hole.

MULTIPLE SCLEROSIS: COLLATERAL DAMAGE IN GLIAL WARS

You reach for the cup but it slips from your hand, spilling hot coffee over the table. You snatch it up, but the handle slips from your grasp again and the cup crashes to the table. By reflex you grab the cup a third time in a flash of anger, squeezing it firmly and with deliberation as you lift it high above the table. What's wrong with your table skills this morning? You are just sleepy. After all, for weeks now you have felt exhausted, somehow never quite able to sleep enough to recharge your batteries. Maybe you have chronic fatigue syndrome. Your doctor might be able to do a blood test.

A few weeks later you awake, but when you open your eyes, there is a hole in your visual field that no amount of eye rubbing will clear away. Everywhere you look it appears as if there were a permanent smudge on a camera lens, a debilitating and annoying hole in your vision. Even outside the blemish the bright colors the world once offered are washed away like faded laundry.

This is often the way multiple sclerosis first reveals itself. The problem is that wires in your brain are shorting out. Impulses from your senses never reach your brain, and commands from your brain fail to make it past the points of broken insulation on the axons that travel to your muscles. As with a short circuit in any electronic device, the malfunctions of MS can be wide ranging. Your vision or balance can be disrupted, as can normal bladder, bowel, and sexual function. MS can impair coordination and cause tremor, weakness, fatigue, numbness, and pain. You will have difficulty doing the things you once took for granted, such as reading, driving, talking, and walking. Speech can become slurred, making you appear perpetually drunk. One-third of people with MS experience cognitive and psychological disturbances. Your memory may fail, and your vital abilities for foresight and planning can diminish. Sometimes there are personality changes and emotional instability. It all depends on which of the brain's circuits are damaged.

The disease is rarely fatal, but the cruelest aspect of this progressive disorder is that it is characterized by multiple remissions. For a time things get better, and it seems as though your suffering was all a bad dream. The smudge on your retina goes away and colors flood back into the world. You feel a bit foolish for the self-pity and dread that you allowed yourself to feel when you were tested by the disability. But then suddenly there is another attack, and it eats away a bit more of your brain, body, ability, and connection to the world.

This is an attack on the nervous system, but it is not an attack on neurons. MS targets oligodendrocytes. Your glial cells never give up the fight against these deadly attacks. They win a battle on one front, reclaiming lost territory and function, but they lose the war eventually from the relentless attacks coming from all sides. Over years your attachment to the world and the things you once loved slips away. There is nothing known to science yet that will stop it. You are in good company; one in seven hundred people suffer from MS. The severity of the disease varies from minimal or transient symptoms to a severe progressive disease.

Current thinking about MS is that it is the result of friendly fire from your own immune system. As discussed in chapter 2, your brain has its own immune system, the microglia, which act as sentinels against disease. The white blood cells protecting the rest of the body do not normally reside in the brain, but they can enter the brain if the blood-brain barrier is breached by disease or damage. When this occurs, there is serious trouble. The T cells of the immune system that have been activated by an infectious agent will squeeze across the blood-brain barrier. In this state they scout every part of the body for the invaders, but after a quick survey they usually leave the central nervous system. In MS, some T cells become activated and stay in the brain where they don't belong. This is when the trouble starts. Microglia detect the T cells and attack, causing an inflammatory response in brain tissue. This battle gets out of control and ultimately leads to the destruction of myelin. If the disease continues, the myelin-producing cells, the oligodendrocytes, are killed. These regions of destruction appear in the brain as plaques, wastelands of cellular debris—no myelin, dead oligodendrocytes, and armies of microglia and some infiltrating white blood cells that have crossed the damaged blood-brain barrier.

Although MS attacks only myelinating glia, the intimate interdependence between axon and glia widens the damage. For reasons that are not understood in detail, axons begin to shrivel and die in patients with chronic MS. In advanced stages of MS, neurons in affected regions of the brain are lost. This suggests that the myelinating glial cell does more for the axon than simply provide electrical insulation.

The whole situation is aggravated by the trespassing T cells, and given the important role of astrocytes lining blood vessels in the brain and contributing to the blood-brain barrier, many experts suspect that astrocytes are participants or even instigators in allowing the T cells to pass. Once the battle gets out of hand, it is not difficult to understand why there is so much death and destruction. The cytokines released by microglia alter the adhesion between cells, including those forming the blood-brain barrier, and the chemokines, small signaling proteins, activate leukocytes (white blood cells) and recruit them to the site of inflammation. Astrocytes, after they become activated, also secrete these same substances. Once the leukocytes enter the brain, they release their own inflammatory agents, as they would when battling an invading organism in the blood, and these agents provoke the microglia and astrocytes to a life-or-death battle, in which microglia release toxic substances that poison cells or poke holes in cell membranes with protein lances called complement.

But the defenders are also helpful in repairing the damage and in stimulating new oligodendrocytes to divide to replace ones killed in battle. Newly minted oligodendrocytes mature into cells that will remyelinate the axon to repair the damaged insulation. This is why MS patients temporarily improve.

There is hope for treating MS because scientists believe they know why it starts. It is a problem of the immune defenses of the brain going out of control. With more research, it seems possible that drugs directed at these three types of glia (astrocytes, oligodendrocytes, and microglia) could treat MS patients. A better understanding of how oligodendro-cytes develop and decide to myelinate an axon or not could suggest effective ways to stimulate the repair of damaged myelin. Experimental treatments in animals in which Schwann cells have been transplanted into damaged regions of the central nervous system have shown that these myelinating glia of peripheral nerves will myelinate CNS axons quite readily. This proves that the problem in multiple sclerosis is not with the nerve axons, but rather with the oligodendrocytes, and suggests that it may be practical to transplant Schwann cells from peripheral nerves into the damaged regions of the brain in MS to replace lost oligodendrocytes and remyelinate brain axons. Finally, it seems possible that scientists will soon have the knowledge to transform stem cells into myelinating glia (see chapter 11), which could be delivered to patients to replace the lost oligodendrocytes. This replacement technique has already been done quite successfully in experimental studies on rats. Stem cells have been taken from many different sources for these experimental studies, including bone marrow and highly plastic Schwann-cell-like cells in the olfactory region (the brain region associated with the sense of smell), which avoids the ethical and practical difficulties of obtaining stem cells from human fetal tissue. Much more work needs to be done before these transplantation treatments could be applied to people. Doctors must be certain that these surrogate myelinating cells do not themselves run amok and create worse problems. This is not a trivial concern, especially with stem cells, which by definition divide readily and transform into a wide variety of cells. They might easily form tumors in the brain. The answers, and the ultimate treatments, await a better understanding of microglia and myelinating glia.

Could MS, which shows how dependent the axon is on its myelinating glia, offer a clue to why electrical activity in axons can be sensed by Schwann cells and oligodendrocytes? The dependence of axons on glia ensheathing them raises the question of whether glia might need to monitor impulse activity in axons to respond appropriately. It is reasonable to postulate that the vital process of making myelin insulation to allow efficient transmission of electrical signals between neurons might benefit if the process were influenced by impulse activity. Certainly, physical therapy is helpful in patients with multiple sclerosis. There are many reasons for improvements from physical therapy, but stimulating myelination by impulse activity in axons appears to be one of the ways to benefit patients with demyelinating disease.

HEART ATTACK AND STROKE: POOR PLUMBING

The wondrously elegant machine that is the human body often succumbs to the most mundane failure: poor plumbing. The cardiovascular system pipes blood to every cell in the body to supply it with oxygen and nutrients and to remove waste products. The complexity of a system that can plumb a pipe to feed every cell in the body is unfathomable. But when it comes to plumbing, complexity is not a good thing.

The cardiovascular system of the human body is subject to all the catastrophes of any complex plumbing system: sudden leaks and ruptures, clogs, and pump failure. As in an old home, the aged plumbing of the human body often fails suddenly and with a catastrophic cascade of consequences. Senator Tim Johnson of South Dakota was stricken in mid-sentence while speaking on the telephone to reporters in 2006. A silent and painless rupture of blood vessels starved neurons in a part of his brain controlling speech. Drained of their vital fluids, the neurons quickly died during the telephone call, leaving the senator fully aware of the catastrophe as he experienced the failure but unable to form the words for help. This type of plumbing failure, a rupture of blood vessels, is one of two types of brain stroke.

One and a half million Americans suffer heart attacks every year, and heart attack is the second-leading cause of death in the United States.[3] Sometimes there are early indications that the heart, the main pump, is failing, but many times, it just stops. The brain is the most vulnerable of all organs to heart attack. If the heart cannot be restarted quickly, brain cells will die. You have no more than a couple of minutes for someone around you to figure out that your heart has failed and somehow get it restarted.

In aged cardiovascular systems, as in old plumbing, clogs become a frequent source of problems. The remedy is the same: run an auger through the pipes to clear the clog, which in medical terms is called angioplasty. Blocked arteries in the heart starve cardiac muscles, bringing its incessant beating to a grinding halt. Blocked arteries in the brain starve brain cells, killing the section of the brain supplied by the pipeline, and with it whatever function this part of the brain performed. The rest of the brain remains aware and helpless as it witnesses the death of part of the mind from stroke. Again, like a clogged drain, the system works fine until the day it becomes blocked, and then the sudden brain stroke rips your life away from its familiar routine onto a new course that you are powerless to change. Stroke is the third-leading cause of death in the United States, according to the American Heart Association and the CDC. Every forty-five seconds someone in America suffers a stroke; 275,000 of these people will die every year, but thousands of others will live the rest of their lives missing part of their brain function. Six and a half million Americans are doing so at this moment.[4]

Recent research has revealed that the waves of calcium flowing through astrocytes greatly increase in frequency and intensity after a stroke. As we've seen, similar increases in calcium waves through astrocytes have also been found after epilepsy. Since a rise in calcium can cause astrocytes to release the neurotransmitter glutamate (which is neurotoxic in high concentrations), this glial agitation following stroke could cause the death of additional neurons that is known to occur long after the initial stroke. Drugs that dampen these calcium waves in astrocytes are being tested in experimental animals with promising results in saving neurons from the second wave of cell death following stroke.

From the standpoint of the brain, the real magic of the cardiovascular system is the specialized cellular interface between the bloodstream and brain cells. This cellular interface is called the neurovascular unit, and in comparison to the straightforward plumbing system that supplies it, this microscopic machine is a marvel. All the exchange of molecules between blood and brain occurs through this neurovascular interface. Like any critical exchange, this process must be highly regulated and dynamic to meet the changing demands for supplies. The amount of oxygen delivered to brain cells must match their changing demands on a minute-by-minute basis in the specific population of brain cells that are being exerted. The waste products generated by brain function must be removed rapidly, regardless of how quickly they build up under demanding circumstances. Nutrients, drugs, and hormones must traverse the boundary between blood and brain under proper circumstances, but the unique cellular fluids bathing the brain must be maintained pristine and separate from common body fluids.

Devising a system to monitor, regulate, and adjust the transfer of nutrients, wastes, and oxygen between the blood and brain requires a sophisticated and complex array of sensors, coupled to processors and exchangers that far exceeds the greatest aspirations of human engineering. The system that accomplishes this is a microscopic partnership of cells in the walls of cerebral blood vessels and cells that monitor and respond to the changing demands of neurons. These cells are called perivascular astrocytes.

The Great Wall: Blood-Brain Barrier

The brain operates on a knife edge. A glass of beer, a cup of coffee, a minute without oxygen will tip our mental function off balance. To maintain the brain in an absolutely constant environment, the brain exists in a highly privileged shelter, separated almost entirely from the rest of the body. Brain tissue is sealed inside something called the blood-brain barrier.

Although the brain is riddled with blood vessels and capillaries, their walls are sealed off from the brain by this special barrier, which admits only a select number of materials. As with all barriers, this one has its practical drawbacks. Most drugs that enter the bloodstream cannot penetrate the barrier; thus drugs that could be helpful in many brain disorders cannot be administered to the brain through the blood. Whether the drug is ingested or injected, brain tissue will never encounter it unless the drug can be made to pass across the blood-brain barrier. Without this barrier, however, the brain would be wildly destabilized as materials—ions, water, nutrients, antibodies, and all the other substances in the cerebrospinal fluid and bloodstream—fluctuated throughout the day. What makes this barrier between blood and brain?

The cells of the capillaries in the brain are unique by being very tightly sealed to one another. Glial cells ensheathing these capillaries support the blood-brain barrier and regulate it. These particular astrocytes are specialized for this function. They regulate the permeability of the blood-brain barrier in health and especially in disease, and they exchange nutrients and ions between blood and brain. Despite their demanding requirement for oxygen and glucose, neurons are not in direct physical contact with cerebral capillaries: they rely on astrocytes to keep them alive.

The famed neurosurgeon Wilder Penfield probed the brains of patients with his electrodes in an effort to discover where thoughts and memories are stored and how information moves through the brain. He noted in 1933 that he could see an epileptic event beginning in the brain of a patient on the operating table by observing the rapid change in color of the cerebral cortex, which Penfield had exposed by removing the skull. The blood flow at the spot in the cerebral cortex where the epileptic event began suddenly changed, transforming the tofu-colored brain tissue into a bloom that spread like a ruddy flush.

Somehow brain cells control the flow of blood on a very fine scale to feed the particular neurons increasing their firing and demands for oxygen and nutrients. The brain and blood were somehow communicating. Now we know they communicate using astrocytes as interpreters.

What Penfield could not have imagined was that this interaction between blood and brain would one day permit doctors and research scientists to peer harmlessly through the skulls of people using powerful machines to monitor microscale changes in blood flow. In so doing, scientists can literally see thoughts flowing through the human brain.

What Is a Thought?

What would it look like if you could see with X-ray vision through a person's skull and watch the brain conjuring a thought from the noisy traffic of electrical impulses inside? What if you could see the thought emerging like a picture from noisy TV static?

Lying flat on my back with my head at the entrance of the tunnel of a massive white machine resembling an industrial clothes dryer, I suddenly felt the technician rudely grab my belt buckle and jerk me toward the opening. Startled, I looked down and saw no one; it was the invisible hand of an incredibly strong magnet drawing me by my metal belt buckle into the machine's maw. I was about to be examined by functional MRI (magnetic resonance imaging), which works by detecting local changes in blood flow in the brain that accompany increased cerebral activity. This new technique has provided a window into brain function that reveals fundamental new information on how the brain processes thoughts (both conscious and unconscious), while also providing a valuable new tool for medical researchers.

Functional MRI takes advantage of the local increase in blood flow in the brain accompanying neural activity. Hemoglobin is the iron-rich protein that gives red blood cells their color and allows them to capture oxygen to deliver to our cells. The signal inside the brain that the MRI machine detects derives from the hemoglobin that has delivered its oxygen selectively to the individual or small groups of neurons that need it. Hemoglobin that has lost its oxygen (deoxyhemoglobin) is magnetic; its magnetic field alters the magnetic signal resonating in response to the probing strong magnetic field generated by the machine.

Typically about thirty images of the brain are acquired in a minute and a half, while the subject is engaged in a mental activity. The computer-generated pictures of the brain taken before and after the mental activity are used as controls, and their signals are subtracted from the brain image acquired during the mental task. What is left are the parts of the brain where neural activity changed during the mental task.

For example, a subject may listen to a Bach fugue during images ten through twenty. Those images are averaged and subtracted from the images collected during the control periods before and after the Bach fugue. The result is an image showing areas of the brain with difference in blood flow while listening to Bach. Now researchers can see which parts of the brain are engaged in the mental process of listening to music. Much more creative uses of functional MRI are applied to understand how our conscious and unconscious minds work. For example, researchers can see how the flow of information differs through the brain of a dyslexic while reading compared with the flow of information through the brain of an efficient reader.

This window into brain function has revealed the neurophysiological basis for cognitive and perceptual events and for pain; it can expose the location of hidden brain tumors and pinpoint where neurological disorders reside inside the brain. Until recently, no one imagined that functional brain imaging was revealing not increased electrical activity in neurons directly, but rather activity in glia responding to the demands of neurons for oxygen. After all, what controls the local blood flow to brain cells? The answer again is astrocytes.

In the last few years it has been discovered that astrocytes detect neural activity in neurons near them and then release molecules that cause the fine blood vessels in the brain to expand or contract. Recently scientists have succeeded in seeing and studying this neuron-astrocyte-blood-vessel communication in action in the brains of living mice and rats. After filling astrocytes in a rat brain with calcium-sensitive dyes, the scientists place the anesthetized rat on the stage of a powerful laser scanning confocal microscope. Through a small opening in the rat's skull they can watch the waves of calcium flowing through astrocytes in the brain of the living rat. When activity in particular groups of neurons in the rat's brain increase, scientists can see the astrocytes surrounding them light up as the calcium levels rise inside them and then spread through populations of other astrocytes in the rat's brain. Then the tiny blood vessels in that local brain region expand or contract to regulate the flow of blood and oxygen to the hungry neurons.

For instance, tickling the whiskers of a rat sleeping on the stage of a microscope causes astrocytes to light up in the part of its cerebral cortex where neurons analyze whisker movement. The blood vessels in this region are influenced by the astrocytes and dilate in response. An MRI machine would detect the increased neural activity in this spot by the drop in oxygenated hemoglobin in response to astrocyte control of blood flow delivering oxygen to the active neurons. By adding drugs that block particular signaling molecules or that dampen the rise in calcium in astrocytes, scientists are determining exactly how astrocytes detect the increased neural activity and what molecules astrocytes release to constrict or dilate local blood capillaries.

As significant as this information is to basic neuroscience, this neuron-glial interaction has many important implications for medicine. The brain responses in stroke and many neurodegenerative diseases involve local changes in blood flow, and now scientists know that the cell that acts as the critical regulator of this process is an astrocyte. Recent research by Maiken Nedergaard and colleagues finds that in mice suffering from a form of Alzheimer's disease, the astrocytes are sluggish in responding to whisker stimulation and act feebly in regulating the local flow of blood to neurons.

Migraine

This neurovascular unit, as the astrocyte-capillary-neuron group is now called, is also intimately involved in another form of pain and disability suffered by millions of people. Migraines are debilitating vascular headaches originating within the brain. Electrical impulses spreading to other regions of the brain change nerve cell activity and together with the resulting disturbances in local blood flow cause symptoms that can include visual disturbance, numbness, tingling, and dizziness. This spreading depression of brain waves was described previously in association with epilepsy. The same phenomenon occurs in migraine, and here too, astrocytes have a role in the process.

These vascular headaches are caused by blood vessels in the brain dilating excessively and triggering pain and inflammation in the surrounding regions. The inflammation triggers the trigeminal nerve, resulting in a severe throbbing headache originating in the meninges, the skin-like cells that cover the brain. The senses become hypersensitive, so that normal sound and light cause excruciating pain. Nausea, vomiting, loss of appetite, and mood disturbances can accompany migraine attacks, and patients often experience auras, visual halos and hallucinations caused by abnormal nerve cell activity in blood-starved cortical regions where vision is processed.

Migraines occur on a periodic basis and they disproportionately affect women. Hormonal effects are thus suspected to have a role in migraine, but it is also in part an inherited disorder. Once we know more about the molecular mechanisms astrocytes use to sense neural activity and control brain waves and local blood flow, new treatments may be devised to control the problem at its source rather than simply trying to blunt the painful aftereffects when the intricate cooperation between neuron, glia, and blood vessel goes awry.

PARKINSON'S DISEASE: TURNED TO STONE

If only he hadn't gotten impatient, but drug addiction and patience rarely cohabit. He couldn't wait, and so glia in his brain paralyzed him permanently. Manuel (not his real name) lay in a hospital bed in 1982 in

Santa Clara County Medical Center, drooling, unable to speak or walk. The doctors were stymied.

A week later a young woman, suffering from palsy and rigid as a mannequin, was admitted to the same hospital. When the doctor learned that Isabel (not her real name) was Manuel's girlfriend, his concern grew to alarm. Could it be a virus? Immediately, Dr. William Langston issued public warnings, and quickly five more patients were identified suffering the same symptoms: rigidity, tremors, inability to speak.

The doctor knew these as the symptoms of Parkinson's disease, but this illness usually affects people later in life—and Parkinson's is a gradual neurodegenerative disease. Manuel and Isabel had turned to stone overnight. The cause of Parkinson's is well established: dark-colored neurons die in a spot in the brain called the substantia nigra (meaning "black substance"). Voluntary movement is controlled by a region in the center of the brain called the basal ganglia. Here fine muscle control is regulated by two types of neurons that control the brakes and the accelerator on these movement circuits. The gas pedal neurons stimulate movement by using acetylcholine as the neurotransmitter, and the brake pedal neurons inhibit movement, using dopamine as the neurotransmitter. In Parkinson's disease, this balance becomes upset and there is not enough inhibition provided by the substantia nigra neurons. The "go" neurons become overactive, causing the body to clench up rigidly. Movement becomes stalled or halting just like a car with a standard transmission operated by someone who cannot coordinate the clutch and the gas and brake pedals smoothly. Former United States attorney general Janet Reno, the late Pope John Paul II, champion boxer Muhammad Ali, and actor Michael J. Fox are well-known Parkinson's sufferers.

The transformation of Parkinson's patients from a frozen to a fluid physical state after treatment with L-dopa is dramatic. For them, L-dopa is a miracle drug. The drug boosts dopamine, thus rapidly restoring the balance of excitation and inhibition in circuits controlling movement, but over time its power gradually lessens. L-dopa is the raw material used by neurons to synthesize the inhibitory neurotransmitter dopamine. Providing patients with more L-dopa increases the amount of dopamine in their brain. The chemical can be administered readily as a drug, because it crosses the blood-brain barrier and then transforms into the neurotransmitter inside the neurons that need it.

On a hunch, the doctor administered L-dopa to Manuel and the others. Suddenly their symptoms improved: their frozen bodies thawed and returned to life. This improvement confirmed that all the patients had suffered loss of function or possibly death of their substantia nigra neurons. But how?

Interviews soon revealed a common thread; all of these patients were drug abusers. It had happened again. Several years before, in 1976, a twenty-three-year-old graduate student who had been experimenting with drugs for nine years synthesized his own supply of the designer drug MPPP, a synthetic opiate. While cooking up a batch of the drug, he decided to speed the process along by turning up the heat. The shortcut altered the chemical reaction and produced a contaminant as a by-product.

After using this batch of tainted synthetic heroin, he began to grow rigid. He suffered tremors, and soon he could not even speak. He was admitted to a hospital and diagnosed as suffering from catatonic schizophrenia. But after it was discovered that L-dopa improved his condition, the diagnosis was changed. He was believed to be suffering from Parkinson's syndrome. The contaminated illicit drug must have attacked his substantia nigra neurons.

Two years later the patient committed suicide by an overdose of cocaine, and this provided the opportunity to examine his brain. Autopsy confirmed what the doctor had suspected: the substantia nigra neurons in his brain had been destroyed. The tragic contaminant in the home-brewed drug yielded a boon to science, however, because now there was a way to induce Parkinson's disease in monkeys for medical research.

Suddenly, the story took a surprising twist. When the contaminant in the synthetic heroin, MPTP, was isolated and injected into the substantia nigra of experimental animals, the dopamine neurons were killed, but oddly, when the compound was added to dopamine neurons in cell culture, it was not toxic to the neurons at all. Researchers were baffled. The puzzle was solved when an additional critical ingredient was discovered: astrocytes. If the neurons were grown together with astrocytes in cell culture, the neurons were killed by the drug. Furthermore, if astrocytes were first killed in the substantia nigra of the brains of experimental animals and MPTP was then injected, the neurons were spared. Astrocytes and the drug contaminant were the lethal combination for substantia nigra neurons. Researchers finally determined that astrocytes take up the compound and convert it into a different substance, called MPP+. This toxin then accumulates in the substantia nigra and kills dopamine neurons.

Beyond drug abuse, this discovery raises some interesting questions. Could astrocytes have a naturally occurring role in Parkinson's disease? Could it be that over a lifetime, astrocytes generate neurotoxins as a byproduct of breaking down other brain chemicals and these toxins eventually damage the vital neurons? Possibly other environmental factors acting on astrocytes could in a similar way contribute to this disease. A lifetime's exposure to damaging chemicals might prompt astrocytes to produce specific neurotoxins, killing the cells whose death results in Parkinson's when age finally catches up with us. This could account for why the incidence of Parkinson's increases with age.

ASTROCYTES IN CURING PARKINSON'S DISEASE

Rather than treating the symptoms of Parkinson's disease with L-dopa to compensate for the death of dopamine neurons, it would be far better if we could attack the cause of the disease. Often overlooked by neurologists, astrocytes may not only be involved in causing Parkinson's, but they may also provide new cures for the disease.

The substantia nigra neurons die in Parkinson's disease because they become clogged with abnormal proteins that should have been destroyed by the neuron's natural cleansing process. Many other neurodegenerative diseases show similar abnormal accumulations of junk proteins, including Alzheimer's, Lou Gehrig's disease, and Lewy body disease. Until recently, scientists, fixated on the protein inclusions inside patients' neurons, overlooked the fact that the nearby astrocytes are filled with the same blobs of abnormal protein. On an anatomical level, astrocytes are a part of these "neurodegenerative" diseases, and it is quite reasonable, given the new awareness of neuron-glia interactions, that these abnormal astrocytes may be a root cause of these diseases.

Other research points to new treatments for Parkinson's disease by exploiting the natural neuroprotective actions of astrocytes. As mentioned in the chapter on spinal cord injury, neurotrophic factors are proteins that are released from cells that are important for the survival of neurons. In a study published in 1993, Francis Collins and colleagues announced that they had purified a neurotrophic factor that increased survival of the dopamine neurons of the type that die in Parkinson's disease. When added to dopamine neurons in cell culture, the substance increased survival of the neurons, stimulated their development, and improved their biological activity in taking up dopamine from the culture medium. These actions were quite specific to dopamine neurons: the new neurotrophic factor did not affect several other types of neurons that were tested. The researchers named this new growth factor glial-derived neurotrophic factor (GDNF), reflecting its cellular origin in astrocytes.

At the present time, there are more than two thousand scientific papers on GDNF in the PubMed database of scientific medical studies. These studies show that the glial-derived factor sustains neurons in the central nervous system and also some neurons in the peripheral nervous system in association with many medical conditions. This remarkable growth factor has been found useful in protecting neurons from degeneration, treating patients for pain, stimulating recovery from nervous system injury, stimulating stem cells to form neurons, treating alcoholism, modulating the effects of morphine, treating mood disorders and multiple sclerosis, encouraging myelination by Schwann cells, and protecting against MPTP toxicity and hypoxia. Hundreds of papers address the role of GDNF in Parkinson's disease. Steven Gill and colleagues at the University of Bristol in England successfully treated Parkinson's patients by using a catheter to deliver GDNF into the striatal region of their brain. Larger-scale studies by Amgen Corporation were eventually halted, unfortunately, because of safety concerns. The first word in GDNF says it all: neurons depend on glia for survival. If the side effects can be understood and overcome, new treatments for Parkinson's disease may be found. Future research on growth factors supplied to neurons by astrocytes is one of the most promising avenues for new treatments for Parkinson's and other neurodegenerative diseases. Perhaps such research will uncover the underlying causes for the death of these neurons and many other "neurodegenerative" diseases.

Another well-appreciated approach to treating Parkinson's disease involves implanting cells that release dopamine into the brain region deprived of dopamine-releasing neurons. This technique has been used successfully, but the approach is hampered by ethical and political concerns over the use of neural stem cells derived from aborted human fetuses for implantation. Other cells that produce dopamine, such as retinal epithelial calls, have been used with positive effects. Astrocytes can also be engineered to produce dopamine, and as we learn more about the relation between stem cells and astrocytes, these labile glial cells may be induced to transform into dopamine neurons in cell culture for transplantation into the brain of Parkinson's patients without resorting to the use of fetal stem cells.

Another recent and surprising new treatment for Parkinson's disease is deep-brain stimulation. In the mid-1990s, surgeons implanted thin platinum electrical wires into the basal ganglia of the brain where the movement disorder in Parkinson's disease arises. The electrical pulses are remarkably effective in reducing muscle tremors and rigidity in Parkinson's disease, allowing patients to walk smoothly soon after the brain region is stimulated. The beneficial effects of the artificial stimulation can be much longer lasting than drug treatments, since tolerance to drugs develops and the treatment eventually fails.

How deep-brain stimulation relieves the incapacitating symptoms of Parkinson's disease is not understood. Many researchers are now speculating that the treatment stimulates the other brain. Appreciating the importance of waves of calcium flowing through astrocytes in coordinating the waves of neurons and regulating excitability of brain circuits during brain activity, seizure, and sleep, several researchers are starting to investigate the possibility that astrocytic calcium waves may help coordinate neural activity in brain circuits gone out of whack from neurodegenerative disorders. Deep-brain stimulation may change this important neuron-glial interaction by altering the astrocytic calcium waves that coordinate activity of large populations of neurons controlling movement.

Neurodegenerative diseases are invariably accompanied by glial responses, but the prevailing view of glia as cellular servants to neurons blinded most scientists to the possibility that glia could be the root cause of neuronal death rather than a response to it. Yet as recent studies of ALS and Parkinson's disease illustrate, neurodegenerative diseases can often result directly from glial dysfunction. The strict intellectual segregation of neuron and glia also prevented most scientists from appreciating the converse—that known "glial" diseases, such as multiple sclerosis, could result in the death of neurons. The recent realization that this could be so did not spring from late-breaking clues: the evidence was there in plain sight all along, but prejudicial thinking fueled the ignorance. The alternative view was regarded as controversial because it ran counter to current thinking.

Still, it is surprising that the long-held and well-accepted role of glia as first responders to neuronal injury did not stimulate more vigorous research into exploiting glia for new investigative methods and treatments for brain disease and injury. Neuroscientists and the scientific establishment (research funding agencies, editors at scientific journals, and even biomedical companies) were slow to move in this promising direction. Few who marvel at the wondrous new imaging of functional activity in the human brain appreciate that they are seeing the power of glia at work in the brain, both in health and in sickness, as they promote information processing and sustain neuronal function.

Now, however, neuroscience is turning the corner. As a result of this new direction in research we are beginning to reach a new basic understanding of nearly all neurodegenerative diseases, from Alzheimer's to stroke, arising from an enlightened attitude toward glia. Along with exciting new treatments, this path of discovery is yielding some surprising findings. As will be discussed in the next chapter, for example, until recently no one imagined that glia could have any role in pain. The body's neural pain circuits had been thoroughly traced out, and not a single glial cell could be found in the neurological wiring. Now we know that something was missing.

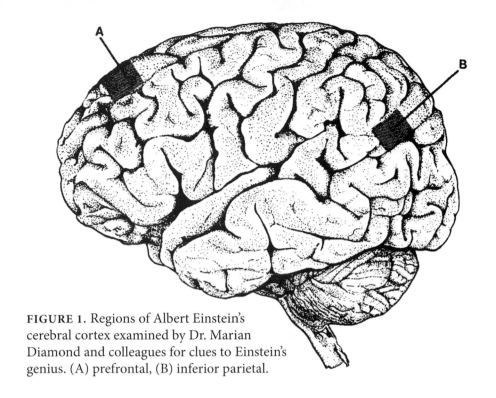

FIGURE 1. Regions of Albert Einstein's cerebral cortex examined by Dr. Marian Diamond and colleagues for clues to Einstein's genius. (A) prefrontal, (B) inferior parietal.

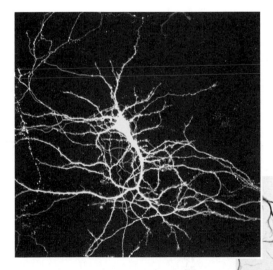

FIGURE 2. A typical neuron from the cerebral cortex.

FIGURE 3. An astrocyte, one of four main types of glial cells. Researchers found more glia than average in the sample of Einstein's brain.

FIGURE 4. A self-portrait of Santiago Ramón y Cajal, Spanish neuroanatomist who conceived the neuron doctrine, shown in his kitchen converted into a makeshift laboratory in 1885.

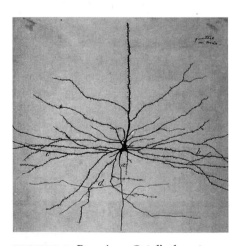

FIGURE 5. Ramón y Cajal's drawing of a neuron stained by the Golgi method reveals nerve cell structure in fine detail.

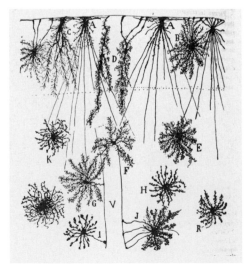

FIGURE 6. Ramón y Cajal recognized glia as distinct from neurons, but their function was not clear.

C. Golgi

FIGURE 7. Italian neuroanatomist Camillo Golgi, who developed the silver impregnation method of staining neurons.

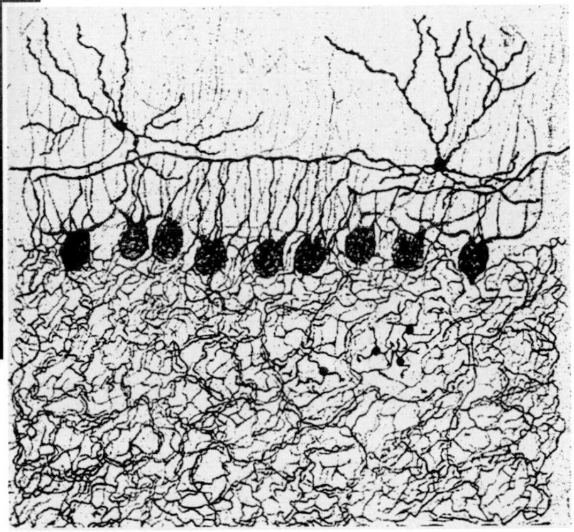

FIGURE 8. Drawing of neurons by Camillo Golgi, showing what he believed to be a highly interconnected network that allowed information to be transmitted in any direction.

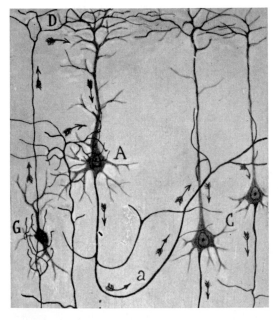

FIGURE 9. Golgi-stained neurons drawn by Ramón y Cajal, who perceived that neurons were not fused together. Information, he surmised, flows through neurons in one direction, passing to the next neuron across synaptic junctions.

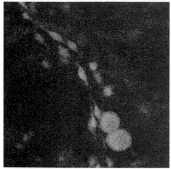

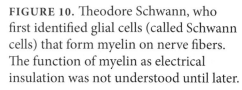

FIGURE 10. Theodore Schwann, who first identified glial cells (called Schwann cells) that form myelin on nerve fibers. The function of myelin as electrical insulation was not understood until later.

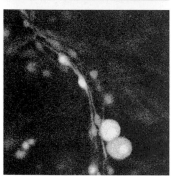

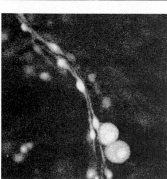

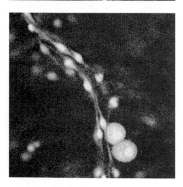

FIGURE 11. A time sequence (top to bottom) showing Schwann cells responding to electrical impulses in axons. The cells are filled with a dye that emits light when stimulation causes calcium ions to enter the cell. The large round cell bodies of two neurons can be seen responding to electrical stimulation. The spindle-shaped Schwann cells along the axons somehow sense the electrical activity in the nerve fibers.

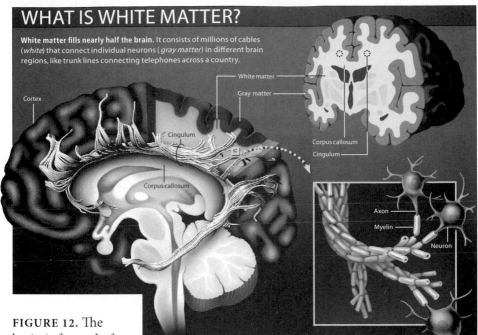

WHAT IS WHITE MATTER?

White matter fills nearly half the brain. It consists of millions of cables (*white*) that connect individual neurons (*gray matter*) in different brain regions, like trunk lines connecting telephones across a country.

Cortex

White matter

Gray matter

Cingulum

Corpus callosum

Corpus callosum

Cingulum

Axon

Myelin

Neuron

FIGURE 12. The brain is formed of two tissues: grey matter (containing neurons) on the surface layers, and white matter at the core of the brain. The white color is due to electrical insulation (myelin) coating the nerve axons.

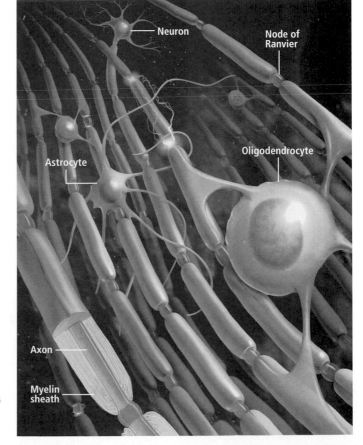

Neuron

Node of Ranvier

Astrocyte

Oligodendrocyte

Axon

Myelin sheath

FIGURE 13. Oligodendrocytes form the myelin sheath that insulates axons.

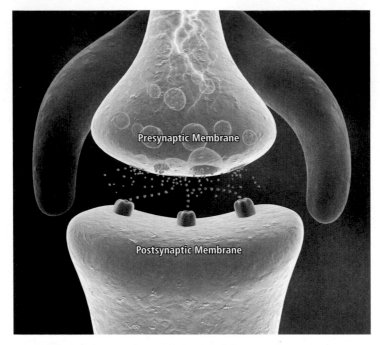

FIGURE 14. Electrical impulses travel down axons to release neurotransmitter at synapses, stimulating the dendrite of the next neuron in the circuit.

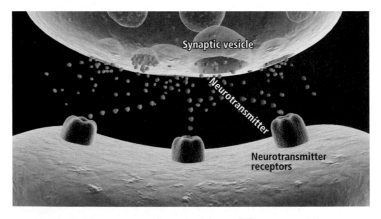

FIGURE 15. Neurotransmitter is released from synaptic vesicles and then diffuses across the synaptic cleft to activate neurotransmitter receptors on the dendrite.

FIGURE 16. Fridtjof Nansen, explorer of the nervous system and the Arctic, in 1895. Nansen speculated that glia may relate to higher intellectual ability.

FIGURE 17. A specially designed ship, the *Fram*, was captained by Nansen and frozen in the polar ice cap to study Arctic ice currents.

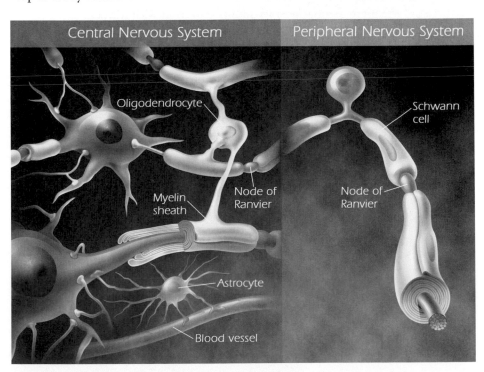

FIGURE 18. Myelin is formed by different cells in the central and peripheral nervous system. Differences between the sausage-shaped Schwann cells and octopus-shaped oligodendrocytes explain why paralysis after spinal cord injury is permanent.

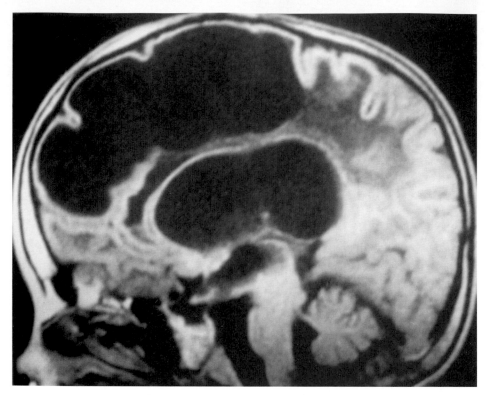

FIGURE 19. The brain of a twenty-two-month-old child with Alexander disease. The enlarged head and degeneration of much of the forebrain (black space) is caused by a genetic defect in astrocytes.

FIGURE 20. The filamentous protein in astrocytes, GFAP, is defective in Alexander disease. The same protein increases in astrocytes after brain injury, epilepsy, and most neurodegenerative diseases, such as Alzheimer's and mad cow disease.

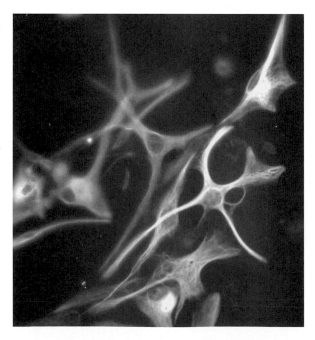

FIGURE 21. Microglia in their multi-branched resting state, ready to defend the brain from infection.

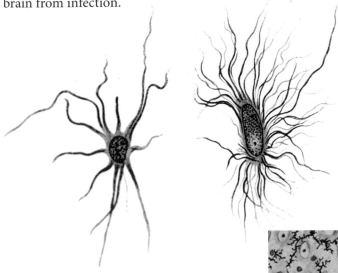

FIGURE 22. A drawing by Spanish neuroanatomist Pío del Río-Hortega, who first described microglia. The cellular branches of microglia embrace neurons protectively. Dotted structures are blood vessels.

FIGURE 23. Neurons and glia working together. Astrocytes associate with synapses and blood vessels. Oligodendrocytes insulate axons.

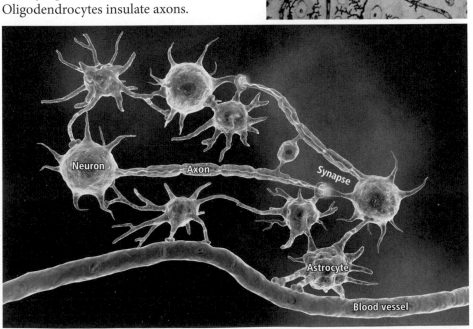

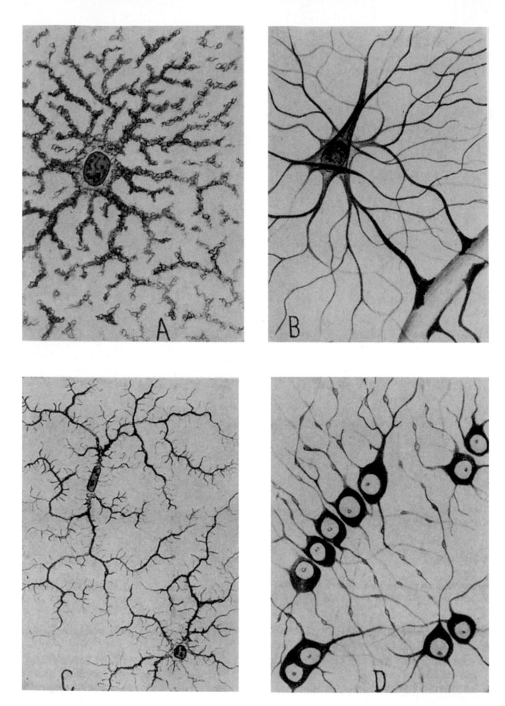

FIGURE 24. Drawings by Río-Hortega in 1920 showing the three main types of glia in the brain and spinal cord. (A) Astrocyte in grey matter; (B) astrocyte in white matter, with endfeet clinging to blood vessels where astrocytes siphon nutrients (glucose) to feed neurons and dump excess potassium ions collected from neurons into the bloodstream; (C) microglia; (D) oligodendrocytes.

FIGURE 25. The missing partner in synaptic transmission—glia. Astrocytes surround the synapse and regulate transmission by taking up and releasing neurotransmitter. Astrocytes also communicate with each other by using chemical messages.

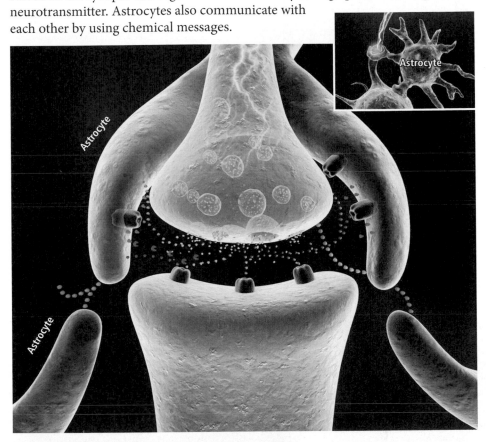

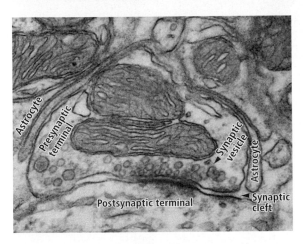

FIGURE 26. A synapse seen through an electron microscope. A synaptic cleft separates the axon and dendrite, and astrocytes surround the synapse to take up and release neurotransmitter to regulate synaptic transmission.

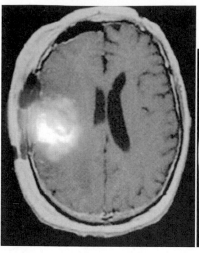
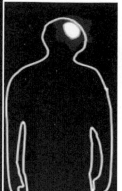

FIGURE 27. Glia, the cause of and cure for brain cancer. A radioactive experimental cancer drug seeks out and kills cancerous glia. The radioactive drug is seen here in this whole-body scan and brain scan of a brain cancer patient.

FIGURE 28. Spinal cord injury, as in this patient with a broken neck (below), causes permanent paralysis.

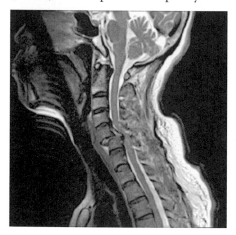

FIGURE 29. Like a hand with probing fingers, the growing tips of axons, called growth cones, reestablish connections after injury, but they collapse upon contact with CNS myelin, preventing repair of damaged axons.

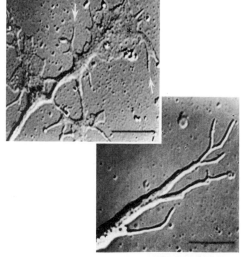

FIGURE 30. Glial stem cells for transplantation to treat paralysis caused by spinal cord injury.

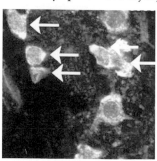

FIGURE 31. Carleton Gajdusek in white T-shirt examines tissue from patients with the deadly neurological disease kuru in a field research station in Papua New Guinea, 1957. Light was provided by kerosene lamp, the dinner table was used for desk, patient treatment, and autopsies. A human brain can be seen in the wash basin on the table. Colleagues Vincent Zigas and Jack Baker are seated left to right.

FIGURE 32. A mother dying of kuru, surrounded by her five children, husband, and sister outside the house for women.

FIGURE 33. Brain tissue from a mouse with prion disease, showing holes (above) and reactive astrocytes (right).

FIGURE 34. Stanley Prusiner in Papua New Guinea, 1980.

FIGURE 35. Hans Berger, the first person to record human brain waves.

FIGURE 36. Psychiatric hospital in Jena, Germany, headed by Hans Berger from 1919 to 1938.

FIGURE 37. Equipment used by Berger to record human brain waves.

FIGURE 38. An early EEG recording by Berger.

FIGURE 39. Stephen Hawking, renowned astrophysicist who suffers from ALS disease, floating in zero gravity inside a NASA 727 aircraft in 2007. The disease has recently been linked to glia.

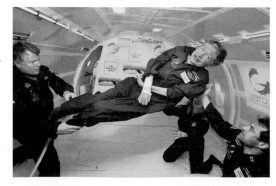

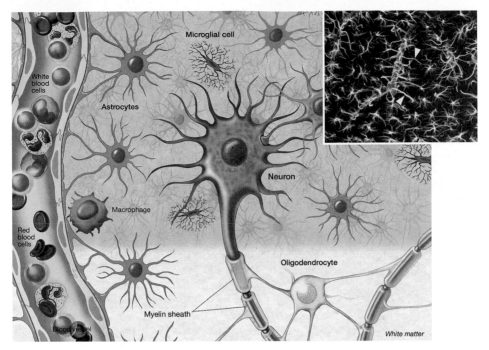

FIGURE 40. The blood-brain barrier controls the exchange of water, ions, and materials between the blood and brain. Astrocytes regulate blood vessel diameter in accordance with neuronal demand. Inset shows microscope view of astrocyte endfeet gripping vessels.

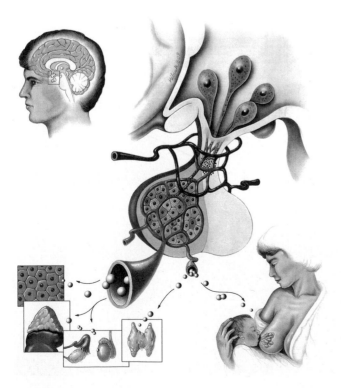

FIGURE 41. The hypothalamus controls many automatic functions in the body, including milk production, through actions on the pituitary gland. Astrocytes move into and out of these synapses to regulate the release of hormones into the blood.

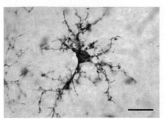
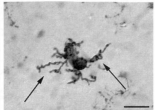

FIGURE 42. Microglia withering with age in a 68-year-old brain (right) contribute to age-related neurodegenerative diseases including Alzheimer's. Normal microglia is shown on left.

FIGURE 43. Neuroscientists seeing the true 3-D structure of astrocytes for the first time.

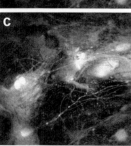

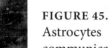

FIGURE 45. Astrocytes communicating using calcium signaling. Video sequence runs from top to bottom. The "lightning bolts" in (b) are nerve fibers that have started to fire impulses, which the astrocytes detect.

FIGURE 44. The true structure of an astrocyte revealed when filled with dye, shown against astrocytes appearing star-shaped because only the cellular skeleton is revealed by staining the cells.

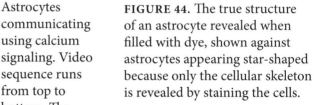

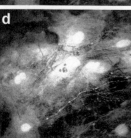

FIGURE 46. German pathologist Rudolf Virchow, who in 1856 described nonneuronal cells and named them "neuroglia."

CHAPTER 9

Glia and Pain:
A Blessing and a Curse

Pain rules our life. Nothing approaches the power of pain in motivating behavior. It drives us to toil in the fields and kill and consume animals to soothe the pain of hunger. It drives us to seek shelter from cold, heat, and storms. It confines the elderly and infirm, whose arthritic joints and aching bones defy, with protesting pain, the human desire to roam. It drives those in agony to implore someone to rip out a painful tooth or cut out a diseased part of their body to end the excruciation. It forces submission when applied in torture. In ancient times it drove people to trephination and bloodletting—permitting someone to bore holes in their skull or open a vein to relieve it. Today, it pushes the helplessly pain-ridden to acupuncturists, chiropractors, quacks, and mystics in the hope that a nontraditional intervention will subdue it. It drives many to drug addiction in the quest for a soothing balm.

Yet pain does not touch everyone with equal intensity, and at times, under the most severe assault, pain oddly loses all potency. The bones of a woman's pelvis are forced to separate and tissues tear in childbirth. A soldier fights through a frenzied battle only afterward realizing that he is shot, fatally wounded, and will die in moments. The whiff of vapor from a common solvent like chloroform or ether mysteriously defeats pain. And even the strong belief that a sugar pill will provide refuge from pain subdues it. Pain is mysterious.

Pain can also be good. More than good, it can be lifesaving. Five-year-old Ashlyn Blocker from Patterson, Georgia, can feel no pain. She is not paralyzed nor has she lost the sense of touch; she simply feels no pain. Her loving mother fears releasing Ashlyn from her guarded shelter at home to enter school, where she can play and learn with other children. Already the five-year-old is severely scarred from countless broken bones, burns, amputated fingers, and a mangled tongue bitten through many times insensibly during meals. One of her eyes is blinded.[1]

People with Ashlyn's congenital condition, called CIPA (congenital insensitivity to pain with anhidrosis), usually die before the age of twenty-five. Injury or infection will take them. They have no gag reflex, never feel the tickle to sneeze or the scratchy throat to cough. Never—no matter how severe her injury or what the cause—can Ashlyn tell the doctor where it hurts. Such children are insensible to an inflamed appendix or an ear infection, so the infections rage undetected. As babies they never cry out in pain or develop the alarm and panic reflex at the sight of their own blood. To these babies blood is merely a curiosity. Sadly, grotesquely, but understandably, her own blood was a gruesome plaything for baby Ashlyn. These people, normal in every other respect, suffer painlessly and die prematurely because of a genetic defect that weakens and kills their pain neurons before birth.

In contrast to this puzzle of injury without pain, many people suffer the converse: pain without injury. Chronic pain is not the warning slap of acute pain that saves us from further injury. That pain subsides on its own. Strangely, chronic pain often develops after an injury has healed. It intensifies when there are no longer any noxious or injurious signals to excite pain neurons, yet these neurons scream out in gut-wrenching pain nevertheless. The ceaseless intense pain controls the lives of such chronic sufferers. Pain robs them of sleep and blots out all pleasure from their lives by imposing constant misery. A normally pleasant touch or sensation ignites raging flames of pain. Putting on a pair of socks may be unbearable.

Chronic pain is mysterious and frustrating for doctors and patients alike, because it has no identifiable cause or satisfactory cure. Powerful pain medications, such as morphine, provide temporary respite, but the relief fades. Eventually dosages must be increased as tolerance develops to these narcotic drugs when they are given for long periods of time. Patients become addicts. Adding to the groggy, debilitating side effects of powerful painkillers, the patient now suffers both chronic pain and narcotic dependence. Some tortured by chronic pain resort to having their pain nerves severed by surgeons to silence the ceaseless agony. Oddly, this treatment does not always work. Pain may originate in our nerve endings, but it is processed and channeled through our spinal cord, where it is beyond the reach of a surgeon's knife. An injured nerve, even after it has healed, can cause changes in the internal pain circuits of our spinal cord that are the source of chronic pain for millions.

Some patients are forced to accumulate and consume such large quantities of painkillers that the police become suspicious. Doctors who treat patients suffering chronic pain can attract scrutiny from the authorities for the high volume of narcotic drugs they must dispense to relieve these patients' agonies—doses of narcotic drugs that could be fatal for a person without tolerance. At the risk of drawing unwanted attention from police and health insurance companies, many doctors are compelled to limit the dosage of pain medications to levels they know are no longer effective for the patient grown tolerant to them.

Richard Paey, a forty-seven-year-old father of three, saw his life shatter after a car crash damaged his spine. His body healed, but the pain did not go away—instead it intensified into an unrelenting agony. "I felt like my legs were being dipped into a furnace," Paey told a reporter for the CBS News show *60 Minutes.* "They were burning, and I couldn't move them. It's an intense pain that, over time, will literally drive you to suicide." Twice he did attempt suicide: "For me, death would have been a form of relief." This once successful lawyer, husband, and father found himself sitting in a Florida prison serving a twenty-five-year sentence for forging prescriptions to obtain the massive quantities of narcotic medications he needs to subdue his chronic pain. As an inmate, he had his narcotic medication provided in ample quantity by the state.[2] (After four years in prison Richard Paey was pardoned by Governor Crist in October 2007.)

Doctors and scientists have looked long and hard for the source of chronic pain without much success. The answer has eluded them until recently, simply because they were looking in the wrong place. "When you have eliminated the impossible, whatever remains, *however improb-*

able, must be the truth," says Sherlock Holmes in Arthur Conan Doyle's "The Sign of the Four." Improbable as it seems, raw nerves are not the sole offender in the puzzling case of chronic pain.

With an increasing appreciation of glia in nervous system function, some pain researchers have begun examining the most improbable of suspects, and they have found the culprit. These scientific sleuths were awakened to the improbable realization that the source of chronic pain is not in pain neurons themselves, but rather in glia. This insight is not only leading to new treatments for chronic pain; it is cracking the case of drug addiction to heroin and other narcotics.

CIRCUITS OF PAIN: STARTING AND STOPPING SENSATION

Before considering chronic pain and the role of glia, we need to understand some fundamentals of pain circuitry. It may come as a surprise to learn that your pain neurons are not located in your brain. They are not inside your spinal cord either, where you will find the motor neurons that issue commands to your muscles. Pain neurons are squeezed like an afterthought between each vertebra of the bony spinal column that rises as a series of bumps down the center of your back. In the space between each bone in your backbone there is a sack of pain neurons. You have one sack of pain neurons on each side of the spinal column at each segment in your articulated spine.

In four-legged creatures, there is ample room for a small sack of pain neurons stashed between each vertebral joint, but as humans rose up on two hind legs our vertebrae became stacked vertically, compressing the elastic disc of padding between each bone in our backbone. Our spine also became distorted away from the strong arched backbone of other four-legged animals—a horse for example—into a wispy bent S shape, creating kinks at our lower back and neck. These are the vulnerable points for neck and back pain that we humans endure in exchange for trading hooves for hands. Many of us suffer neck and back pain as a result of a herniated or compressed disk squeezing this sack of pain neurons between backbones. This pain is both agonizing and damaging.

The smashed nerve cells become inflamed and fire a barrage of nerve impulses in protest. Some nerve cells lose their pinched axons and die.

The nerve cells inside these sacks are unusual. Their cell bodies are round crystalline globes like gel-filled balloons. Pain neurons have no dendrites, but instead a slender string of an axon extends from each balloon and passes, bundled together with strings from other sense neurons, through our nerves to reach the skin or muscle somewhere on our bodies. Here the tiny nerve endings fray apart and become specialized into microscopic sensory organs that can sense touch, pressure, heat, cold, irritating chemicals, and substances released by skin cells damaged by sunburn, abrasion, or cuts.

So much of the world enters our mind through these tiny endings: from the silky textures of velvet, satin, glass, a lover's shoulder, or a puppy's fur, to the roughness of grit and sand, bark and stone. The sudden chill of an evening breeze, cool marble, and ice, to heat glowing on outstretched palms, warming rays of the sun bursting through a cloud, and scorching asphalt against bare feet in summer. The brittle dryness of an autumn leaf to the cloying wetness of a cold stream. The universe of knowing through fingertips touching another's skin, from the rugged firmness of a handshake to the thrilling intimacy of holding hands, to the soft, moist warmth of a gentle kiss. These are the neurons that give us sensation, both pleasurable and painful.

These nerve sensors also monitor the inside of our bodies. Some of them send their axons into muscle, where their tips spiral around individual muscle fibers, wrapping them like clinging vines. The tiny tendrils feel the stretch and strain of fibers in our muscles. They report back vital information to our unconscious brain on the tension and position of every muscle in our body. Without this delicate and intricate unconscious sensation, we could not move or even stand up balanced on two legs.

Every neuron from a particular sensory nerve sack has its own sharply defined domain on your body. The neurons sprouting out of each sack between every vertebra survey a narrow band of your skin and muscle, striping your body from head to toe, as though you were banded like a king snake. These stripes of sensation are revealed as a band of insensibility when you injure one of your nerve sacks, or in the case of chronic pain, as a band of intense hypersensitivity and pain. (A compressed nerve sack between vertebrae in my neck once painted an invisible stripe of numbing and tingling pain inside a band slashing across my pectoral muscle, down the triceps of my right arm, to encase the ring and pinky finger of my hand in an invisible but sharply defined glove.)

Each globular nerve cell also sends another axon into the spinal cord. Thus, sensory neurons have two axons, like two arms, one extending to the periphery where its fingers react to stimulation, and the other penetrating the spinal cord, where it signals what it has detected. The axons penetrate the top (or dorsal surface) of the spinal cord, giving these sensory neurons the name dorsal root ganglion (DRG) neurons. The left and right halves of your body are mirrored, so you have a dorsal root ganglion nerve sack on the left and right sides of your spine between each bump in your backbone.

The axons entering the dorsal surface of your spine then communicate through synapses to neurons inside your spinal cord. (The spinal column is your backbone, which shields your spinal cord. The spinal cord is an extension of your brain tissue running like a cord down your back inside these back bones. As mentioned earlier, it is part of your central nervous system, or CNS.) These spinal neurons are linked by a chain of neurons to the opposite side of your spinal cord. Sensations from the right side of your body are channeled across to the left side of your spinal cord, just as commands from the right side of your brain control movement of limbs on the left side of your body. Spinal cord neurons on this side then send axons up to your brain. After reaching the thalamus, the major switch box for information flow into and out of your cerebral cortex, neurons carry the pain signals to your cerebral cortex and to emotional processing centers of your brain, where you perceive the sensation of pain and associate the appropriate emotional reactions with it.

Pain can be stopped by silencing the pain neurons in the skin, as a dentist does when he injects Novocain into your gums. Novocain blocks the ion channels that spark nerve impulses. If there are no nerve impulses, no signals of pain will reach the spinal cord, and the tooth can be pulled painlessly.

Another strategy for pain relief is to inject anesthetics inside the spine, as doctors often do when giving a woman an epidural injection ("spinal block") to assure a painless labor. The sensory neurons in the woman's pelvis still fire intensely, but the signals are cut off by the anesthetic flooding the axons entering the spinal cord. Stopping pain here also has the advantage over general anesthetics of leaving the woman fully conscious to experience and assist in the process of delivering her child. General anesthetics numb neural circuits in the cerebral cortex of the brain as well, rendering people unconscious.

Painless: Anesthesia

The discovery of general anesthesia is an American success story and an intriguing example of how science proceeds inevitably from the deep human instinct to experiment—in this case to experiment with recreational drugs.

Around 1800, English scientist Humphry Davy inhaled nitrous oxide and felt an exhilarating sensation. Not only had he discovered a pleasant recreational drug, he noticed that a wisdom tooth that had been causing him great pain ceased to hurt while he was enjoying his sip of nitrous. Strangely, he never took the next step to apply the gas as an anesthetic for medical purposes. Had he done so, the renowned scientist might have given the gas a royal medical term, but instead he called it "laughing gas," and its use as a recreational drug prevailed.

Parties called "frolics" became popular social occasions. At such gatherings guests inhaled laughing gas and made fools of themselves in a senseless state of uproarious intoxication. Partygoers sometimes fell and hurt themselves giggling in euphoria without feeling the slightest pain.

In America too, frolics rapidly grew in popularity, but here a more widely available substance was the drug of choice: ether. Ether had been around as a solvent for some time, and at one point or another someone caught a good whiff of its fumes and enjoyed the buzz. Much like twentieth-century glue sniffers, but without the present-day (and well deserved) social stigma, Americans partied on, huffing their ether until the cows came home.

It was not science but rather partying mixed with pragmatism that inspired Crawford Long, a twenty-seven-year-old physician in small town Jefferson, Georgia, to explore anesthesia. In 1842, Dr. Long, who was familiar with ether frolics from his wilder days as a medical student, exploited the substance to dull his patient's pain during minor surgery to remove a skin tumor from a man's neck. Dr. Long did not bother to report his party trick until six years later, but by then anesthesia was rapidly becoming recognized as a major achievement of historical proportions. Dr. Long failed to draw attention to his experiments with anesthesia, but the public demonstration of anesthesia in 1845 by a dentist, Horace Wells, became a sensation.

Wells, a twenty-nine-year-old Hartford, Connecticut, dentist, had an aching molar extracted painlessly by a fellow dentist while Wells huffed laughing gas. Wells was so delighted that in 1845 in Boston, he gave a celebrated public demonstration of the use of nitrous oxide in dentistry. This should have been the crowning point of his medical career and a scientific breakthrough, but instead, fate took an abrupt turn and Wells was crushed. The demonstration failed horribly, with the patient screaming out in pain, discrediting Wells publicly as a quack. Wells never recovered from the disgrace, ending his life by suicide in 1848.

A colleague of Wells, William Thomas Green Morton, is credited with giving the first successful public demonstration of anesthesia, in 1846, to remove a painful tooth from Mr. Eben Frost. Rather than administer nitrous oxide, Morton used the preferred party drug of American frolics—ether. A vivid account was published the next day in the Boston press, and ether anesthesia began to be practiced widely.[3]

While Americans simply swabbed a rag with ether and held it to a patient's nose, English scientists began systematic studies of ether's anesthetic effects. James Simpson developed an apparatus to deliver ether precisely and determined the dosage required for various anesthetic and analgesic effects. He carefully described the stages and signs of anesthesia and published his findings in the first scientific paper on the subject. Simpson, an obstetrician, was the first to use ether as an anesthetic to relieve the pain of childbirth.

Some say deep resentment for the scientific success of the upstart Americans fueled British interest in finding an anesthetic substance of their own. (In fact, they already had such a substance, nitrous oxide, but the failure of Horace Wells in his public debacle drained their enthusiasm for pursuing nitrous as an anesthetic.) The Brits found what they were looking for in chloroform, a solvent that had been synthesized more recently. Simpson introduced chloroform as an anesthetic in 1847, and in 1853, John Snow administered chloroform to Queen Victoria during the birth of Prince Leopold, stimulating wide acceptance of chloroform for obstetric anesthesia. Ether continued to dominate in the United States because it was simpler to administer safely.

Long before general anesthetics like nitrous oxide, ether, and chloroform were discovered, people had chewed on enough plants to discover that some, such as coca leaves and juice from poppy seedpods (sources of cocaine and opium respectively) and other plant substances greatly relieved the discomfort of pain. Ingesting or smoking these and other plant products also had other strange effects on the brain. Why should a substance in a plant have such pain-relieving effects on the human brain? The answer in these cases, and in many others, including nicotine and marijuana, is that chemical substances very similar to those in plants are made by our body naturally to regulate brain function. These plant analogs are similar enough to the natural compounds in our body that they can regulate the transmission of pain signals in the spinal cord and brain. Nicotine, marijuana, and opium activate the same receptors on human cells that are stimulated by the body's own natural narcotics.

Although this insight came as a revelation in the 1980s, morphine, heroin, opium, and cannabis were proof certain that the body had its own powerful medicines to block pain. Narcotics like morphine block pain exactly where the surgeon's knife cannot reach: where the pain signals are passed from the sensory neurons to the pain circuits inside the spinal cord. The sensory neurons and the spinal cord neurons have receptors for these narcotic drugs, which inhibit the activity of the neurons. The natural opiates made by the body are called endorphins. These powerful pain relievers regulate pain naturally, in effect closing the valve on the flow of painful signals traveling up the spinal cord to the brain.

Endorphins flood the bloodstream during periods of intense pain or stress. This fact accounts for how a soldier can suffer a bullet wound unaware during the stress of battle. During labor, a woman's body naturally synthesizes endorphins. This "endogenous" (meaning "made internally") morphine enables her to endure the pain of childbirth. Endorphins are released not only in pain, but also during very intense exercise and during sex. Deep pressure by a chiropractor, nerve stimulation in acupuncture, and the placebo effect also work by causing the body to release endorphins. Experiments using drugs that block endorphins have proven the role of these chemicals in pain-suppressing treatments.

It is popularly reported that the endorphin high experienced by marathon runners also accounts for the afterglow following sex, when endorphins are released into the brain and bloodstream. A recent study questions the relation between endorphins and orgasm, however. To obtain sperm for artificial insemination in livestock production, breeders train boars to copulate with an artificial sow. The study found that more endorphins were released when the boar's efforts with the artificial mate went unfulfilled than when he achieved porcine orgasm and ejaculated. This suggests that the endorphins released during sex may have more in common with the pain and stress of acupuncture than with romantic afterglow. Opinions, however, may differ as to whether the sexual behavior of men is adequately modeled in this manner.

Studies using the drug naloxone, which blocks opiate receptors, uncovered an interesting clue—people apparently have other natural mechanisms to quash pain. Even with the opiate receptors blocked by the drug, people under stress still showed diminished sensitivity to pain, suggesting that there must be other natural painkillers in the body in addition to endorphins. In another link between pain relief and illicit drugs, it was discovered that the endogenous pain-relieving substance stimulated the same receptors that are turned on by the active ingredient in marijuana.

One study showed that just like a pothead feeling no pain on a cannabis high, people under intense stress and pain experience significant relief from the natural cannabinoid compounds released in their bodies to cope with the pain. Recently, neuroscientists have been startled to learn that spinal cord glia also have receptors for endorphins and cannabinoids. This puzzling discovery was a strong clue that glia might have an unsuspected role in chronic pain.

WHEN PAIN BECOMES DISEASE:
GLIA IN CHRONIC PAIN

Pain caused by injury is protective, but there is another type of pain that develops mysteriously and intensifies even after the injury has healed. This "neuropathic" pain transforms protective pain into an agonizing disease. Chronic pain develops because of changes in pain circuits in the central nervous system after injury and healing. Neuroscientists know that pain neurons in people suffering chronic pain become hyperexcitable after healing from injury. The slightest touch or change in temperature sets these neurons firing barrages of nerve impulses that signal intense pain to the brain. The neurons also fire abnormally in bursts of high intensity all on their own. Like a broken record, cycles of nerve impulses scream pain over and over again without any stimulus to trigger them.

What causes these pain neurons to spin out of control? After an injury to the body or an infection, cells involved in the body's defense and healing release powerful chemicals called inflammatory cytokines and chemokines. Drugs like ibuprofen and aspirin bring relief by blocking the action of these cytokines. Inflammatory cytokines act in many ways, and one of the consequences for pain fibers is heightened sensitivity. This is nature's way of telling us to take it easy until we are healed. We are all familiar with this phenomenon when we feel the region surrounding the site of injury become tender and painful. Also, we have all experienced the painful reaction to light and sounds while suffering a severe headache. Our senses become heightened to the point of pain. If the inflammatory condition does not resolve, however, pain and hypersensitivity will persist even after the injury has healed.

People suffering chronic pain often describe their misery as raw nerves, but the latest research suggests that chronic pain goes beyond nerve cells. For many pain sufferers, the source of their chronic, unreachable pain is glia. Linda Watkins, a pain researcher from the University of Colorado at Boulder, began to suspect microglia as the source of chronic pain that develops after nerve injury. She recognized that glia have no involvement in transmitting normal pain, but chronic pain that develops after an injury heals might be another matter. Microglial cells are well known as the immune system cells of the brain and spinal cord. They are also known as a main source of cytokines and chemokines after nervous system injury or infection. Both of these substances can cause neurons to become hyperexcited. All of this evidence points to microglia as a prime suspect in causing chronic pain. Indeed several years earlier, in 1994, Stephen T. Meller and colleagues at the University of Iowa had shown the involvement of another type of glia in chronic pain by administering a poison to animals that selectively kills astrocytes. These animals suffered far less chronic pain after nerve injury.

To test this theory that microglia may cause chronic pain, all one must do is block the normal reaction of microglia to injury and then test the animals to see whether numbing the glial response to injury brings relief from chronic pain. Watkins and colleagues conducted tests on rats with chronic pain that had been caused by spinal cord injury. She then administered a drug, minocycline, that targets microglia, preventing their activation and cytokine release in response to injury. Watkins and several other researchers found that injecting this drug into the spinal fluid relieves rats of chronic pain caused by spinal cord injury almost immediately. The experiments proved that the rats receiving the drug experienced greatly reduced chronic pain as a result of blocking microglial response to injury.

Other work is extending the findings on microglia and chronic pain to astrocytes by using minocycline. After all, responding to injury and infection is one of the shared functions of astrocytes and microglia. After injury, astrocytes proliferate and begin to express the skeletal protein GFAP at high levels, a cellular remodeling that allows astrocytes to change their shape and motility. Like microglia, astrocytes have many receptors for injury-related signals released by neurons, and in response, astrocytes release growth factors, cytokines, and chemokines that can contribute to chronic pain in some circumstances.

Microglia and astrocytes perform many vital functions after injury, however. Eliminating glial reaction to injury entirely would likely have many undesirable consequences. If we knew more about how microglia sense neural injury and the detailed mechanisms that control their many responses during injury, healing, and development of chronic pain, potent new pain medications could be developed to bring relief to chronic pain sufferers without sacrificing the healing functions of glia.

One of the signals damaged nerve cells are known to broadcast when they are in distress is a chemical called fractalkine. This molecule is tethered on the surface of neurons and released like signal flares upon injury or distress. Microglia have special receptors to detect these fractalkine distress signals. When the sensory receptor proteins on microglia sense fractalkines, the glial cells quickly move toward the injury and flood the area with cytokines. This response normally resolves after a period of weeks as the injury heals, but in some unfortunate cases, the microglia do not stop saturating the tissue with cytokines. The injury may be healed, but the painful injury response continues at full rage.

After administering a drug that dampens the microglial sensors for fractalkine, Watkins and her colleagues tested how sensitive the rat was to pain. There are several well established tests for assessing pain accurately and humanely—for example, placing the rat's tail on a heating pad and turning up the temperature gradually until the rat flicks the tail away. In rats suffering chronic pain as a result of previous nerve injury, even the slightest change in temperature causes the animal to flick its tail away quickly. Other tests using fine bristles of precisely calibrated stiffness to touch the skin can accurately gauge sensitivity to pain. Normally these bristles are painless, but in rats (and people) suffering chronic pain the slightest touch becomes excruciatingly painful, like poking an open wound. After researchers treated the rats with the drug blocking the fractalkine receptors on their microglia, the tests showed that the rats were released from the grip of chronic pain. Since the microglia could not receive the fractalkine distress signal sent by injured neurons, they did not release the inflammatory and painful cytokines.

This is an astonishing transition in medical science, for here is a painkiller that does not act on pain neurons; it brings relief from chronic pain by acting on nonneuronal cells. It is as if a door has been cracked open into a room filled with an entirely new stock of drugs to cure chronic pain. Some of these glia-targeted drugs are currently undergoing clinical trials in people suffering constant, uncontrollable pain.

JUST SAY NO TO PAIN: MARIJUANA

Marijuana is reported to have pain-relieving effects, in some cases easing pain that cannot be touched by any other painkiller. This fact is the basis for legalization of marijuana for medical purposes in California and in certain countries. Smoking or eating the plant is a primitive and imprecise mode of administering the drug, not to mention the significant cognitive side effects. In the past, the restrictive laws on obtaining cannabis or the active ingredients in marijuana for scientific research made it very difficult for all but a few laboratories to investigate how marijuana acts in the brain and relieves pain. Fortunately this situation has improved.

Research by Massimiliano Beltramo and colleagues at the Schering-Plough Research Institute in Milan, Italy, found that when they gave rats drugs derived from cannabis that stimulate the receptors on neurons activated by marijuana, called cannabinoid receptors, chronic pain was indeed reduced dramatically. This may not be surprising, but there are two types of receptors for cannabinoids in the body, CB1 and CB2. The CB1 receptors are present on several types of neurons in the brain, which accounts for the mind-bending effects of marijuana. The CB2 receptors are not present on neurons. These cannabinoid receptors are mostly found on cells outside the brain that are involved in inflammation, but there is one exception: they are also present on microglia.

There is intense interest in developing pain-relieving drugs that act on the CB2 receptors of microglia and other immune system cells for pain relief, because they would not have the negative side effects on cognitive function of marijuana acting on CB1 receptors. Recently it has been found that the number of CB2 receptors in the dorsal root ganglia and spinal cord increase after spinal cord injury. This seems to be the body's way of increasing its sensitivity to endogenous cannabinoids in its attempt to relieve chronic pain. When these investigators tested drugs stimulating the CB2 receptor in mice that had been genetically modified to remove the CB1 receptor gene, the animals still benefited greatly from receiving the cannabinoids for pain relief. This proves that compounds in marijuana activating CB2 receptors on microglia could be used to relieve pain without any of the negative cognitive effects and makes the CB2 receptor a highly attractive target for relief of chronic pain. When scientists learn to design drugs to act on microglia and astrocytes, marijuana may show us how to "just say no" to chronic pain.

A Cornucopia of Glial Pain Relievers

Microglia have so many attractive targets for development of therapeutic drugs for pain relief because they have so many "ears" to listen for brain injury. These new glial-targeted drugs for treatment of chronic pain work in many different ways. The current list includes fluorocitrate, a metabolic inhibitor of astrocytes; minocycline, which inhibits microglial activation; ibudilast (AV411) and propentofylline, which inhibit reactive astrocytes; methionine, which inhibits glutamine synthesis; and JWH-015 and Sativex, which activate CB2 cannabinoid receptors. Some of these glial-targeted drugs are being tested experimentally only in animals, but others are currently undergoing clinical trials in people suffering constant uncontrollable pain. Sativex, a drug acting on cannabinoid receptors, has been used in Canada and other countries since 2005 for treatment of chronic pain in multiple sclerosis patients. The drug is currently in phase III clinical trials for use in the United States.

Another interesting membrane receptor on microglia involved in chronic pain is the P2X4 receptor. This protein receptor is activated by the adenosine triphosphate (ATP) released when nerve cells are damaged. When it detects ATP, the P2X4 receptor opens a pore and calcium floods into the microglial cell, triggering a host of responses to the injury. In chronic pain, these responses overreact, and microglia become the cause of spinal cord pain.

Kazuhide Inoue and colleagues at the National Institute of Health Sciences in Tokyo, Japan, have found that after nerve injury, microglia in the spinal cord increase the number of P2X4 receptors on their membranes, making them hyperalert. When these receptors are overactivated, spinal cord pain increases greatly, mimicking neuropathic pain. Although an entirely different mechanism is involved, the research on P2X4 receptors proves again that glia are critically involved in chronic pain.

But how do hyperactivated microglia cause pain neurons to become supersensitive? Research published in 2005 by Jeffrey Coull and colleagues at Laval University in Quebec, Canada, has identified one way that microglia cause hyperexcitability in spinal cord pain neurons after their P2X4 receptors are activated by ATP. When the researchers inserted microelectrodes into the spinal cord neurons so they could measure the electrical signals, and then applied ATP to stimulate the P2X4 receptors on microglia, the voltage in the neurons dropped suddenly. Somehow the microglia were reducing the voltage in neurons, thus triggering pain neurons to fire impulses excitedly, which the brain could only interpret as intense pain. A series of studies showed that ATP released by injured neurons stimulates receptors on microglia (called P2X4 receptors), which causes them to release a growth factor (called BDNF) to aid healing of neurons. However, this growth factor also acted on spinal cord neurons, triggering impulse firing that signaled intense pain. Many other molecules released by microglia and astrocytes causing hyperexcitation of axons are being identified.

So now we have two methods of attack against microglia-induced spinal cord pain: block the receptors on microglia and block the release of substances from microglia making neurons hyperactive. These new drug approaches are not based on the old narcotic path to pain relief and could help patients avoid the medical and social toll narcotic drugs can exact.

This work is finding quick application outside the field of spinal cord injury, because so many inflammatory reactions also accompany infection and disease. One of the most exciting areas is relief of chronic pain associated with cancer. As in brain injury, horrible chronic pain in cancer is caused in part by inflammatory responses. Dr. Rui-Xin Zhang and colleagues at the University of Maryland have found a connection between astrocytes and microglia in pain resulting from bone cancer, which can cause intensely painful aching. Zhang's team found that astrocytes and microglia became activated in the pain circuits of the spinal cord after the investigators caused bone cancer to develop in the tibia of experimental animals by injecting cancerous prostate cells into the bone. Little wonder that cancer patients experience neuropathic pain when the microglia in their spinal cord are overactivated and pumping out cytokines that hyperexcite spinal cord pain neurons.

Infection with the HIV virus causes pain, but the virus cannot have a direct effect on pain neurons, because neurons cannot be infected by the HIV virus. Recent studies show that when a protein from the HIV virus (named gp120) is injected into the spinal cords of rats, the animals develop hypersensitivity and chronic pain. Linda Watkins's experiments showed that the same HIV virus protein activates microglia and causes them to release inflammatory cytokines.

The last bombshell to hit the chronic pain field was the finding that microglia and astrocytes have receptors for morphine and other opiates. Without realizing it, people responding to the release of endorphins, patients receiving morphine to relieve their horrible pain, and drug addicts shooting heroin were also unknowingly activating opiate receptors on their glia. Glia are now firmly implicated by this new evidence in the pain-relieving effects of opiates, but their involvement has surprised everyone. This research points to glia as an ugly demon contributing to drug tolerance and addiction, a subject that will be explored in the next chapter.

CHAPTER 10

Glia and Addiction:
A Neuron-Glia Dependence

What do Keith Richards, Mick Jagger, Eric Clapton, Charlie Parker, Billie Holiday, Ray Charles, Janis Joplin, Jonathan Melvoin, Jimi Hendrix, Lou Reed, Sid Vicious, Hillel Slovak, and John Lennon have in common besides music? The answer is heroin addiction. Some of them—Hendrix, Joplin, Slovak (Red Hot Chili Peppers), and Melvoin (Smashing Pumpkins)—died from an overdose of the drug.

Heroin is an easy drug to overdose on, because one's tolerance increases over time, making ever stronger doses necessary to achieve the same effect. The effective dose for a hardened addict may be a fatal dose for someone trying heroin for the first time.

It may be incorrect to assume from this list that musicians have an unusual propensity toward heroin abuse. The surprising number of musician addicts may be accounted for by their wealth, which enables them to afford the costly drug habit, and the publicity that opens their private lives to public scrutiny. Congressman Patrick Kennedy (Ted Kennedy's son), radio host Rush Limbaugh, and the wife of senator and former presidential candidate John McCain are recent examples of people overtaken by addiction to powerful prescription narcotics (like OxyContin) after originally taking them to relieve pain. These painkillers are not injected like heroin, but they act in the body in exactly the same way.

After back surgery to treat a ruptured disc in 1989, Cindy McCain became addicted to the narcotic painkillers Vicodin and Percocet. She soon faced federal charges carrying a potential twenty-year prison sentence after a DEA investigation revealed that she had forged illegal prescriptions and diverted narcotics for her own use that were intended for a nonprofit organization of volunteer doctors and nurses providing free medical care in Third World countries. Dr. John Max Johnson, the doctor providing her narcotic prescriptions, lost his license to practice medicine. McCain, like many others, became trapped into chemical dependence by drug tolerance, which weakens the potency of narcotic painkillers over time and causes painful withdrawal symptoms if the medication is stopped.[1]

I believe I can understand something of heroin addiction from personal experience. I once had a nasty ski accident, which came with a helicopter ride and, upon landing at the hospital, a shot of morphine. Even though my left leg was pulled out of my hip socket and twisted halfway around so my foot was on backward and the ball joint stuck out of my hip like a doorknob, once I got that injection of morphine, I could not have been happier—literally. In retrospect, that event is deeply frightening. From this brief experience with morphine, I learned that addiction to opiates is not simply a matter of morals, it is physiology. The pursuit of happiness is life's ultimate goal, and no matter how gruesome reality may be at the time for an addict, the rainbow truly ends in a vial of heroin. But that drug-induced happiness is only an illusion that quickly evaporates into the imprisonment of addiction. John Lennon's song "Cold Turkey" vividly describes heroin addiction and the painful withdrawal from the drug, including the fever, excruciating pain, body aches, sleeplessness, and desperation, to the point he wished he were dead.

A heroin addict's agony is caused by changes in his brain and spinal cord brought about by the habitual use of heroin. Prolonged use of opiates leads to drug tolerance—that is, increasingly higher levels of the drug are required to achieve the same effect. This tolerance is the result of the brain fighting to adjust its circuits to restore the normal balance of activity against the pain-numbing effects of heroin (or any opiate) as it suppresses activity in brain circuits involved in pain. Thus, the neurons increase their excitability when exposed to opiates for prolonged periods, and therefore more morphine or heroin is necessary to suppress their activity again. Slowly this sequential balancing act ratchets up drug tolerance until the brain must have a constant supply of these drugs circulating in the blood.

Quitting opiates cold turkey throws the system suddenly out of balance. As pain circuits are abruptly released from the suppression of opiates they fire wildly and this causes pain and extreme hypersensitivity. Because of this hypersensitivity, normal sensations—light and sound, for example—become painful, unleashing the symptoms John Lennon describes in his song. Lennon "rolled in pain" for thirty-six hours, waiting for the pain circuits in his body to readjust so they could operate at normal levels in the absence of the suppressing effect of heroin.

Withdrawal is much like the suffering of chronic pain patients in the absence of any pain drugs. In chronic pain sufferers, normal sensations become painful because their pain neurons have grown extremely hyperactive as a result of a previous injury, causing these individuals to suffer intense and unrelenting physical pain.

Glia have opiate receptors too. What is the effect of opiates on glia? In 1988 Lars Rönnbäck and Elisabeth Hansson at the University of Gothenburg, Sweden, proposed that glia could have a role in morphine addiction. Their research had found that morphine activated opiate receptors on astrocytes and caused them to release a number of substances that could stimulate pain circuits. In theory, glia could be releasing the neural stimulants that counteract the nerve-damping effects of morphine, and thus contributing to morphine tolerance. By nullifying the soothing effects of morphine in their effort to reestablish the normal balance of neural sensitivity and activity, glia could be ratcheting up the addict's need for ever-increasing stronger doses of morphine, thereby building drug tolerance.

The substances released by astrocytes and microglia in response to morphine include many biological compounds that are known to increase activity of pain neurons, including glutamate, nitric oxide, prostaglandins, and inflammatory cytokines. Similarly, morphine also causes the release of substances from glia that inhibit the action of morphine receptors on pain neurons, including cholecystokinin and dynorphin. This second glial mechanism directly suppresses the pain-relieving effect of morphine. Moreover, after prolonged exposure to opiates, glia actually increase the number of opioid receptors they have on their cell membrane. This makes them even more sensitive to morphine, a possible direct mechanism for drug tolerance. Glia counteract the suppressant effect of opiates by making neurons more excitable, and as they become more sensitive to morphine, their power to undermine morphine's action increases.

In 2004, Linda Watkins and colleagues reported that release of inflammatory cytokines and fractalkine from glia is also involved in drug tolerance. Their studies found that prolonged treatment of rats with morphine increased the amount of inflammatory cytokines glia produce. This increased production could cause pain and increase the sensitivity of pain circuits in heroin addicts, in the same way that microglia cause chronic pain, as discussed in the previous chapter. In support of this finding, researchers discovered that when morphine was given together with drugs that blocked the inflammatory cytokine receptors, morphine suddenly became much more potent in relieving pain because glial countermeasures against the numbing actions of morphine were suppressed. In effect, with microglia's actions counteracted, drug tolerance was thwarted.

Research published in 2009 by Joyce DeLeo, a pain researcher at Dartmouth Medical School, reveals an interesting interaction between opiates and the P2X4 receptor on microglia implicated in chronic pain and morphine addiction. Recall that this receptor detects ATP released from injured and hyperactive neurons. This new research shows that disrupting P2X4 receptor function with specific drugs can weaken the ability of microglia to cause tolerance to morphine. This treatment prevents microglia from moving upstream toward the source of ATP, where the glial cells would release substances increasing the excitability of pain neurons in order to counteract the blunting action of morphine.

Throughout history, chronic pain and drug addiction have always been interrelated; now we know that glia are one of the main reasons for this. These new findings could lead to treatments that prevent morphine tolerance from developing and permit doctors to maintain their patients on lower doses of narcotic pain relievers to treat their chronic pain effectively. At the same time, blocking the development of morphine tolerance would prevent the downward spiral of patients into narcotic addiction caused by the need for ever-increasing dosages to achieve the same effect and also by the individual's inability to stop taking narcotics without suffering painful withdrawal. (The same medical benefits from microglial research would be available to help heroin addicts.)

Consider the implications for drug addiction and relief for those suffering chronic pain who now require extraordinary levels of opiate pain medications for relief in part because their glia are fighting the pain-soothing action of morphine. Richard Paey, the forty-seven-year-old father of three and otherwise law-abiding citizen who resorted to forging prescriptions to obtain an effective dosage of opiate pain medications to ease his intolerable chronic pain, would not have been condemned to a Florida jail for a quarter century. He would have been at home, working as a lawyer, able to carry out his responsibilities as a father. Neither would countless others addicted to heroin fill our prisons. Addiction to opiates has an enormous personal and societal cost. Tiny glial cells may soon lead us out of this vicious, unbreakable cycle of addiction to opiates, which has plagued man for generations. Sherlock Holmes, with his dependence on the "7-percent-solution" (a mixture of cocaine Sherlock injected to alleviate his mental anguish), would not have been more delighted by the sudden turn of events in this story, and especially by the implication of a most unlikely of all culprits—glia—in the "Mysterious Case of Chronic Pain."

The same treatment directed toward glia will reduce morphine tolerance in pain patients and also the excruciating pain caused by cold-turkey withdrawal from opiate addiction. When the cytokine receptors are blocked in animals addicted to morphine, and the opiate drug is withdrawn cold turkey, the animals suffer far less from the withdrawal. Animals with blocked cytokine receptors are less sensitive after drug withdrawal than untreated rats addicted to morphine, which had grown hypersensitive to the point that even normal sensations were painful after withdrawing from the drug.

Extrapolating these new experimental results in laboratory rats to humans, the painful withdrawal from opiate addiction in drug addicts could be relieved by blocking the receptors on neurons for the inflammatory cytokines released by astrocytes and microglia. Even though the morphine-addicted glia would still flood the neurons with these excitatory substances, their stimulating effect on neurons would be blocked.

John Lennon pleads and prays in his song describing heroin withdrawal, promising anything to anyone who can get him out of his hell. Now we know what he was waiting for—for glia to stop flooding his neurons with inflammatory cytokines and other neurostimulants.

ALCOHOLISM: GLIA AND ALCOHOL

A child with mental retardation fills the heart with sympathy and sorrow. If only the misfortune that affected the budding brain in the womb could have been prevented. The sad reality is that we know the villain inflicting mental retardation on most children. Moreover, we know how to stop its slaughter, yet it continues. The leading cause of mental retardation in children is fetal alcohol syndrome.[2]

Here is what we know. Alcohol poisons neurons. The effects on the brain are most devastating and permanent when alcohol enters the system while the brain is developing. Chronic alcohol intoxication is not needed to poison the development of a baby's brain. A single exposure at a critical time—one binge—can damage the fetal brain permanently.

Distinct brain regions and various types of cells in the brain develop at different times during gestation. In general, the worst time to expose the fetal brain to alcohol is during the third trimester, when the brain swells with an amazing growth spurt. Neurons and glia are dividing and proliferating rapidly during this phase to fuel the massive growth. At this time, neurons are also actively sprouting dendrites that branch profusely like new twigs on saplings in spring. Axons extend through the brain to form networks of connections, weaving an intricate complexity connecting countless new synapses sprouting from the branches like new buds. Alcohol toxicity during this period stunts the spurt of brain growth—irreversibly. Eighty-five percent of children with fetal alcohol syndrome have abnormally small brains—"microencephaly," to use the medical terminology. Prenatal exposure to alcohol quashes the brain's growth spurt in two ways: it inhibits the proliferation of neurons (slows cell division), and it kills them, particularly in the hippocampus and cerebellum (areas of the brain important for memory and physical coordination).

What about glia? In comparison to the volumes of data on the effects of alcohol on neurons, much less is known about glia, but considering that glial cells constitute roughly 85 percent of the total brain cells, it is reasonable to expect that microencephaly may result from the loss of glia as well as neurons. The evidence bears this hypothesis out.

In cell culture, alcohol treatment of rat or human astrocytes is a potent inhibitor of astrocyte cell division. The retarding effects of alcohol on astrocyte cell division are so powerful that alcohol overwhelms the stimulating action of most known growth factors produced by the body. None of the powerful human growth factors researchers tried adding to the cell cultures could overpower alcohol's depression of astrocyte cell division.

Alcohol also kills glia. Alcohol's effects have been studied in cultures of rat astrocytes from the cerebral cortex and in the brains of rats exposed to alcohol during gestation. In one study, 40 percent of the cellular corpses counted in the rat cortex following exposure to alcohol as a fetus were astrocytes.[3]

Beyond the simple consequence that a small brain with fewer neurons and glia will have diminished mental capacity, one need only consider the crucial role of glia in orchestrating nervous system development to foresee the multiple devastating consequences of alcohol attacking fetal glia. As we will discuss in the next chapter, glia are the infrastructure of the fetal brain, the scaffolding upon which the fetal nervous system is built. Glia provide trophic factors to promote transformation of immature cells into appropriate types of neurons and sustain those neurons after development. Glia lay down molecules in tracts to guide axons to form appropriate connections that will make functional circuits. Glia promote the formation of synapses. Glia take up neurotransmitters and maintain the vital salt and nutrient concentration surrounding neurons at the proper levels. Glia protect the brain from infection. Glia control migration of cells during development. The loss of glia poisoned by alcohol will have a multitude of negative consequences on the developing brain.

Glia and neurons must assemble in the proper places to build the fetal brain. Abnormal glial migration is observed in children with fetal alcohol syndrome, as it is in experimental animals. Even a moderate exposure of glial cells to alcohol can cause this damage in misrouting the intoxicated glia. The brain that results is malformed. The important role of glia in forming the corpus callosum, the bridge between cerebral hemispheres, will be discussed more extensively in the next chapter; without this glial bridge, axons from neurons cannot span to the opposite side of the brain. In children with fetal alcohol syndrome, this massive interhemispheric bridge remains underdeveloped.[4] This brain abnormality is one of the signatures of fetal alcohol syndrome. Fetal alcohol syndrome is a glial disease at least as much as it is a disease of neurons.

Alcohol is a versatile toxin, intoxicating and poisoning glia and neurons in many ways. Many of the intoxicating effects of alcohol are due to its breakdown product, acetaldehyde, formed from the action of an enzyme in the liver, alcohol dehydrogenase. This essential enzyme developed because our ancestors would have ingested some rotting fruit in their diet, and the process of digestion itself generates some alcohol in the gut. Purposely ingesting alcohol, however, overwhelms the body's resources to detoxify itself from alcohol poisoning.

Alcohol itself (not just its breakdown products) is also the direct agent of neuronal and glial death. Alcohol acts on receptors that normally dampen overexcitation in the brain (GABA receptors). These are the same inhibitory receptors activated by barbiturates and Valium, which accounts in part for alcohol's sedative effects. However, in response to habitual alcohol use, the brain decreases the number of its GABA receptors that drive inhibitory brain circuits and increases the activity of glutamate-driven excitatory circuits. This leads to alcohol dependence, because in the absence of alcohol, circuits controlling anxiety, mood, and pain become hyperactivated. Alcohol has similar effects on the neurotransmitters dopamine and serotonin, which are involved in regulating mood and the brain's "reward circuits," contributing to alcohol dependence. The same three neurotransmitters are known to regulate the development of neurons and glia from fetal and adult stem cells. These changes also impair memory circuits. By decreasing inhibition and increasing excitation, chronic alcohol use releases neurons from the important inhibitory influence of GABA in controlling the normal level of brain excitation. As the inhibitory brain circuits are weakened with alcohol, brain circuits are prone to hyperexcitability, which causes the death of neurons from overstimulation.

Alcohol also numbs NMDA neurotransmitter receptors, dulling the mind in yet another way, because these receptors are the first essential step in increasing synaptic strength in learning and memory. Chronic alcohol use, though, increases the number of these receptors; more receptors make the neurons and glia more prone to death by overexcitation. Fetal brain cells that will develop into oligodendrocytes that myelinate the newborn's brain after birth are also killed by alcohol. In fact, loss of white matter is one of the main hallmarks of fetal alcohol syndrome, and as we now appreciate, loss of myelin means loss of mental function.

In combination with these injuries, alcohol also contributes to dehydration, oxidative stress, vitamin deficiencies, and a host of metabolic and injury responses, which all work to undo the miracle of fetal brain development. Glia are among this toxin's chief victims.

CHAPTER 11

Mother and Child

BUILDING A BRAIN:
GLIA IN WIRING UP THE BRAIN

Nothing in biology is more miraculous than the transformation of a fertilized egg into a complete organism. The unfertilized egg, silver and glowing in a halo of brilliant refracted light seen through a microscope, lies dormant, like a cold moon; once fertilized, it suddenly casts a spherical shockwave that radiates outward like a supernova, sweeping the surrounding void spotlessly clear and sterile. The shockwave hardens into a protective shell sealing off the egg from intrusion and contamination, blocking thousands of sperm crashing and squirming against the impenetrable fertilization ring.

The fertilized egg scintillates and churns like the surface of the sun, streaked with sunspots and surface eruptions that betray a powerful energy surging at its core. There is a sudden upheaval and the entire mass of the sphere abruptly cleaves in two, pulling apart into two perfect spheres pressed together within the protective halo of the fertilization ring. If these twin orbs become separated at this point, either naturally or artificially, identical twins will develop from the single cloven egg. The divisions accelerate and repeat as the two spheres cleave into four, four cleave to eight, and so on until with exponentially increasing speed the sphere has transformed into countless tiny cells massed together like a ball of clay. Already the individual fates of many of these cells have been set. Scientists have revealed this by injecting tracer dyes into single cells at this early phase of development and watching to see the cells evolve into heart or bone or brain.

This mass of cells now begins the first movement of coordinated choreography: assembling together to form a hollow ball. From this ball the defining topology of the embryo is cast—inside and outside. Cells that will form skin and nerve shuffle to the outer layer and cells that will form bone and gut slither to the middle; cells that will form muscle and blood sandwich themselves between the inner and outer shells.

The first organ to form in the tiny embryo is not the beating heart. In fact, the nervous system is the first bodily organ to emerge. The fetal nervous system serves as the cornerstone upon which the entire embryonic body will be built. The embryonic nervous system begins as a streak or crease on the surface of the ball that then closes to form a tube. This will become the spinal cord. This primitive step in formation of the nervous system casts the geometry for development of the entire body to follow: head and tail and the bilaterally symmetrical left and right sides of our body. The slit of embryonic cells that will form the nervous system marshal themselves by multiplying and moving in formation to form an eyedropper-shaped assembly of cells called the neural tube.

The brain begins as a swelling of the neural tube at one end, so that the nervous system in its initial stage of formation looks like an eyedropper. The "rubber bulb" continues to swell into a brain and then as it balloons it divides into its three basic parts: forebrain, which will one day sustain the higher mental functions of our cerebral cortex; the midbrain, connected to the eyes and unconscious automated control systems of our body; and the hindbrain, which will develop into medulla and cerebellum at the base of our skull, where the spinal cord enters the brain.

Severe birth defects, some of them devastating, result from failure of the depression of embryonic cells to form into a tube. These neural tube defects can result in the birth of babies without heads, or with a single cyclopean eye, or spinal bifida, in which the tail of the spinal cord fails to close completely, resulting in paralysis and learning disabilities. To prevent these devastating heartbreaks, scientists must discover the deep secrets of these first stages in formation of our brain.

As the eyedropper grows and morphs into the mature brain, the hollow cavity inside the neural tube will become the ventricles of our brain and spinal cord. All vertebrate brains are hollow, filled with circulating cerebrospinal fluid (it is this fluid that is sampled by doctors in performing a spinal tap). During fetal development, the hollow cavity in the brain (ventricle) becomes squeezed and bulges like a clown's balloon into four chambers to accommodate the twists and expansions of the major lobes of our brain. If the passageway between these chambers becomes blocked, the cerebrospinal fluid that fills the brain cavity will build up pressure, literally inflating the brain into the pathological distended condition called hydrocephalus. Babies born with hydrocephalus can have enormously expanded heads as their skull bones separate and grow large to accommodate the fluid-inflated brain. Hydrocephalus damages the brain, and unless it is controlled, the child will be mentally retarded.

Current treatment is the most rudimentary approach conceivable: tap a drain line into the ventricle to bleed off the excess brain fluid. A more sophisticated treatment could be devised if we understood the fundamental mechanism of how cells in our developing brain organize themselves to form these brain cavities. Recent research from Ken McCarthy's lab at UNC Chapel Hill shows that if astrocytes are genetically modified in mice, hydrocephalus can result. This was a completely unexpected discovery, leaving scientists to wonder how glia could contribute to hydrocephalus in the fetal brain. With further research, glia may offer real hope for understanding and preventing some forms of congenital hydrocephalus.

Assembling and Wiring the Brain

The human brain, with its 100 billion neurons and countless synaptic connections between them, has been described as the most complicated structure in the universe. Designing and building an equivalently complex machine far exceeds the capabilities of the finest team of hardware architects and engineers; yet our brains begin with the simplest possible foundation: a line embossed into a ball of embryonic cells as if an impression of a thumbnail drawn across a ball of clay in the hands of an artist. How will all of the brain's 100 billion nerve cells find their proper place in the embryonic brain? With as many as 100,000 synaptic connections to be attached properly to a single neuron, how will every microscopic axon fiber find its one-and-only proper receptacle on the correct neuron? What directs the position that individual neurons must assume as the structure of the brain develops from a formless mass? What guides the unfathomable maze of connections between them? The answer is glia.

During fetal development a special class of master glia arise to orchestrate building the brain and then vanish when their work is done. Some will transform into astrocytes, and others will simply disappear. As a result of their direction, the entire human brain—the most complex structure in the universe—will emerge from a ball of cells in less than nine months.

Spanning the entire thickness of the brain from the hollow ventricle to the outer covering is a special type of glial cell called radial glia. They can be visualized in many ways. Their name refers to their geometric arrangement, like that of radial fibers providing the structural support in an automobile tire. The substance of the brain is reinforced by these massive radial glial cells spanning floor to ceiling throughout the walls of the embryonic brain. If our brain cortex were concrete walls, radial glia would be the embedded framework of rebar steel supporting it. The radial glia are firmly anchored with their cell body on the floor of the brain, and they extend branches to the roof like a trained vine, pruned flat to the appearance of fan coral. This structure provides scaffolding from foundation to roof, with each radial glial cell spaced out regularly like studs in the wall of a house.

Neurons are born from stem cells in the floor of the embryonic brain. They bud off like lumps of dough and begin to slither up the scaffolding provided by the radial glia. The surface of the radial glial cell is painted with protein marker molecules instructing the infant neurons where to go. Sticky macromolecules on their surface, called cell adhesion molecules, help the baby neurons grip, and thus they coax the cells up to their proper position in the brain. Specific types of neurons read molecular signposts on the radial glia telling them where to get off and begin assembling with others of their kind to form layers of brain tissue.

STEM CELLS:
GLIA GIVE BIRTH TO NEURONS

Research published only in the last few years has undone much of the previous scientific bias that led scientists to segregate neurons and glia. Using new techniques to track the origin of neurons born from cells on the floor of the fetal brain, researchers saw that master glial cells not only direct the path of neurons, but they also spontaneously generate them. In one experiment, a virus was constructed in the laboratory to act as a cellular Trojan horse, concealing inside itself the genetic instructions to make a fluorescent green protein. When these virus particles were used to infect radial glial cells in experimental animals, the infected glia glowed with green light. Stephen Noctor, Paolo Malatesta, Arturo Alvarez-Buylla, Magdalena Götz, and others watched through a microscope as a daughter cell budded off the base of the green monster glia. They watched this daughter as it began to move up the trunk of the radial glial cell and transform from glial cellular stepdaughter into the princess neuron. Still marked by the telltale green pigment, its glial heritage was exposed. One mother radial glial cell spawns and guides generations of neurons to their destiny in the fetal brain. In the brain of the human fetus at midgestation, as many as thirty generations of migrating neurons cling to the shaft of a single radial glial cell.

Soon after an infant is born, its radial glia begin to fade away. Some transform into astrocytes, which now grow in ranks by rapid cell division, yet a few radial glial cells live on in parts of the mature brain. Why?

Perhaps they serve as a reserve should the brain suffer disease or injury. In the adult forebrain of mice, remnant radial glia have been caught recently transforming into neurons that are then incorporated into olfactory and hippocampal regions of the brain (the hippocampus is vital for transforming our everyday experience into enduring memories). Thus these master glial cells orchestrate the formation of our brain in fetal life, then step back behind the scenes and watch as the brain performs its function, but they stand ready anytime during our lifetime to step back on stage and play the neuron's role should our brain experience problems or disease.

Cellular Siblings

Once it was presumed that glia and neurons must have derived from completely unrelated ancestors, but in fact, neurons and oligodendrocytes are children of the same mother cell. As the mother cell divides into two cells, one becomes a neuron and the other remains a stem cell or becomes an immature glial cell. The fetal neuron continues to divide, generating only neurons through subsequent generations. The immature glial cell can divide into cells that can become neurons or glia. Once a neuron matures it cannot divide to make new neurons. Only the immature neurons in the fetal brain can do this. (Recently it has been discovered that a few remaining "mother cells" persist in certain places in the adult brain, where they can produce new neurons under some circumstances. Notably, this is seen after injury, but also after physical exercise and learning.) But young oligodendrocytes are far more dynamic than neurons. A mature oligodendrocyte can revert to an earlier stage and generate more oligodendroglia (immature oligodendrocytes) to repair the brain after injury or disease. Curiously, some oligodendroglia persist in an immature stage even in the adult brain. This is odd, because these immature oligodendrocytes have a relatively simple cell structure, and they do not form myelin. What are these juvenile oligodendrocytes still doing in our head decades after leaving the womb?

A New Character: The NG2 Cell

In the mid 1980s, William Stallcup and Joel Levine, then at the Salk Institute in La Jolla, California, discovered a new type of brain cell. Few took notice of the remarkable finding, and you have almost certainly never heard of it. The information is not yet in textbooks and it will not be there until scientists get a firmer grasp on what these new brain cells are.

NG2 glia are causing as much excitement among cellular neuroscientists as the discovery of a new fossil missing link would generate among paleontologists. Unlike anything seen before, this NG2 cell is very odd. The research team discovered this new brain cell by making antibodies against various lines of brain tumor cells grown in cell culture. Then the team added the antibodies to thin slices of normal brain tissue on microscope slides to see what kinds of cells the antibody would stick to. The antibody would stick to cells in the normal brain that shared the same protein the antibody recognized on the surface of cancer cells.

"We didn't really have a hypothesis," Joel Levine confessed recently. "We were just looking to see what new things we could find. Back then you could still do stuff like that."

In reminiscing, Levine was contrasting the old days with the current situation where funding for scientific research is so tight that only studies that a committee of scientists can agree will be certain to provide a meaningful new finding are considered for funding. Needless to say, Darwin never would have competed successfully for government funding under conditions that required him to prove that his observations would produce meaningful findings. Grant Application by C. Darwin: "To sail around the globe on the HMS *Beagle* to study nature in all her wondrous variety." Neither would Ramón y Cajal or Golgi have qualified.

What Levine and Stallcup found was that one of their new antibodies (called NG2, after an arbitrary inventory number assigned to it in their lab notebooks) attached very selectively to a population of cells throughout the adult brain. These cells were peculiar. They lacked any of the cellular hallmarks of neurons. They had no dendrites or axons. They were not astrocytes or microglia, yet they constituted nearly 5 percent of the total number of cells in the adult brain. Oddly, these mystery cells were sprinkled uniformly like pepper throughout the entire brain tissue, with little regard for confining themselves to the usual compartments and anatomical boundaries in the brain.

"When we showed our colleagues the [microscope] slides, they would look through the microscope and say, 'That's weird!' and then go back to their work at the bench," said Levine.

Levine and Stallcup were left to wonder, What are these brain cells? Where do they come from? What do they do? Are they a fourth kind of central nervous system glia or simply a type of oligodendrocyte suffering arrested development and hanging out in the mature brain? Do they change with disease or injury? As a result of twenty years of research and two recently published papers from other labs, the answers are coming in rapidly now and they are changing the rules of neuroscience.

Sleeper Cells

NG2 cells were found to differ from neurons in one more very important way: unlike mature neurons, NG2 cells can divide and proliferate. In fact, Levine and others were startled to learn that the vast majority of cells in the mature brain that can divide are NG2 cells—as many as 90 percent of dividing cells in the normal adult brain are NG2 glia. This fact makes these cells of special interest. NG2 glia might be the seeds of brain cancer, because tumors are cells that divide uncontrollably. Neurons do not divide, all but eliminating neurons as the root of cancer. Remember, the NG2 antibody was isolated from cancer cells in the first place. Alternatively, the ability of NG2 cells to divide also makes them potentially important in the brain's valiant response to defeat disease and heal after injury.

Here the plot thickens, because not only can NG2 cells divide, but researchers soon found that NG2 cells have remarkable stem-cell-like properties. Under the right conditions, Akiko Nishiyama and others, along with Vittorio Gallo at the Children's National Medical Center in Washington, D.C., found that NG2 cells could transform into many different types of brain cells. This includes oligodendrocytes, astrocytes, and neurons. NG2 glia are the main type of cells seen dividing after stroke and other brain injury.

The long-awaited dream of transplanting cells into humans to treat spinal cord injuries became reality in 2009 when the U.S. Food and Drug Administration granted a biotech company, Geron in California, permission to begin cell transplantation experiments on patients with spinal cord injuries. The cells to be used are derived from human embryonic stem cells, which have the potential to develop into endless varieties of cells, but what is the ideal type of cell to transplant into a damaged spinal cord?

Looking at the experimental data on animals the company placed its bet on a cell with unique potential for responding to injury—the NG2 cell. After coaxing stem cells to differentiate into NG2 cells, the team plans to inject ten patients with the cells a week or two after they suffer a serious spinal cord injury. While NG2 cells can form astrocytes, neurons, or oligodendrocytes under proper conditions, the team expects the cells to transform primarily into myelinating glia. By forming new myelin on axons that lose their insulating sheath after spinal cord injury, the NG2 glia should restore impulse conduction through the axons, bringing increased sensation and motor control to these patients. Moreover, as is now well appreciated, neurons that lose the myelin sheath on their axons often die. If the NG2 cells can restore the myelin insulation on axons, this process should rescue these precious neurons that survive the initial injury.

The major concern with transplanting any cell with stem-cell-like properties into human patients is that they could continue to divide and develop wildly into tumors. Because NG2 cells were first identified by raising antibodies against brain tumor cells, this is a serious concern. Nevertheless, Geron CEO Thomas Okarma brushes off these legitimate concerns as a consequence of the general ignorance about these cells among most neuroscientists, who have neglected to study them adequately. "There is so little expertise in the academic world about cell therapy that these people [who worry about the cancer-causing potential of NG2 cells] are rightly nervous . . . We are so far ahead of them." The company is also pursuing the use of NG2 glia in treatment for Alzheimer's disease, stroke, and multiple sclerosis.[1]

NG2 cells are even more peculiar in ways that upend the basic tenets of cellular neuroscience. In 2000, Dwight Bergles, now at Johns Hopkins School of Medicine, and colleagues discovered synapses that formed on NG2 cells. This finding violates a fundamental distinction between neurons, which communicate through synapses, and glia, which do not. This discovery was particularly baffling because NG2 cells are a category of glia that most closely resemble immature oligodendrocytes. Why would a myelin-forming glial cell with stem cell properties be wired into the nervous system by synapses? The NG2 cell has no output connections to activate other cells as neurons do, so it cannot relay a message to another cell in response to synaptic inputs. What good are synapses on glia?

Perhaps being wired into neural circuits enables NG2 glia to tap into electrical activity in the brain and monitor transmission of information between neurons. If so, NG2 cell division or development into other cells might be regulated by impulse activity in the brain. Perhaps this impulse activity could instruct NG2 cells to make more myelin-forming oligodendrocytes in the brains of a multiple sclerosis patients, or possibly it could even transform NG2 cells into new neurons in the adult brain.

In 2008 David Attwell, Ragnhildur Káradóttir, and colleagues at University College London made a discovery that completely upends one of the most fundamental distinctions between neurons and glia: the ability to fire electrical impulses (action potentials). When they stuck a microelectrode into NG2 cells to monitor the electrical signals they were receiving through their synaptic inputs, the researchers discovered that half of the NG2 cells were able to fire electrical impulses using the same mechanism as neurons. Other scientists remain unconvinced that these cells firing impulses are glia but think instead that they are a type of neuron in transition from an NG2 progenitor.

Whether NG2 glia are a new, fourth type of central nervous system glial cell, developmentally arrested oligodendrocytes that persist in the adult brain, or new neuron/glial hybrids is uncertain at the moment, but this recently discovered brain cell is blurring the line between neurons and glia. It is opening neuroscientists' imaginations to new possibilities at the interface between the neuronal brain and the other brain. Now scientists have plenty of specific hypotheses to test, and the grant applications for NG2 glia research are flying in.

OF TWO MINDS: GLIA UNITE

Unlike most bodily organs our brain is a paired structure connected together as a sort of Siamese twin. Thus, inside our skull we each have, in fact, two brains, although only one of them can talk. This may account for the subconscious hunches or vibes that nudge our rational decision-making process in directions we cannot always explain. In general, the left cerebral hemisphere in most people is specialized for analytical functions and the right is more adept at the artistic. Speech resides on the left side. The two brains communicate with each other through a massive trunk of nerve fibers (the corpus callosum), which carries information between sister portions of the cerebral cortex on either side of the brain. Patients who have had these lines of communication severed to control seizures appear perfectly normal in every way, but clever tests reveal that the two brains occupying one head are completely unaware of each other after losing these connections. This can be shown by tests presenting information to only one side of a patient's brain, for example by blocking one eye when showing the patient a picture. As is widely known, information tends to cross from the left side of the body to the right side of the brain, and visa versa. It is the corpus callosum that puts this information back together again, uniting our two brains into a single conscious mind. Without this connection intact, the left side doesn't know what the right is doing.

Our left hand is controlled by our right cerebral hemisphere, because the fibers leaving the brain to control our hand muscles cross to the opposite side of the body in the spinal cord. Information in our visual world is also crossed. If you focus on a point in space, visual information from the right side of the focal point crosses to the left cerebral cortex. This crossing of information takes place at a point called the optic chiasm, where the optic nerve fibers from our two eyes are routed to connect to either left or right brain. Unless visual information reaches the left cerebral cortex, which can speak, a person will deny that an object in the right visual field has been shown to him.

Say, for example, a split brain patient (someone who has had the corpus callosum severed) looks a pirate squarely in the nose. If the pirate wore a single earring appearing in the right side of the patient's field of vision, the patient could easily describe the earring in detail. This is because the visual information from the right field of vision crosses to the left cerebral cortex, which also controls speech. However, if the pirate wore an earring in his opposite ear, the patient would deny that there was anything at all hanging from the pirate's ears. Information in that half of the visual field went to the right brain, which cannot talk. Nevertheless this half of the brain does know all about the earring. This subconscious knowledge can be revealed by asking the patient to draw the pirate's face with his left hand. When he does so, the earring will be drawn in its proper place. Asked why he or she had drawn the earring, the patient would be flummoxed and reply, "I don't know." Indeed the "talking brain" does not know about the earring if the corpus callosum is cut, because an earring in the left visual field only reaches the right cerebral cortex. Although this right brain cannot speak, it can reveal what it perceives by nonverbal communication, simply by drawing with the hand that is controlled by the same hemisphere.

Roger Sperry was awarded the Nobel Prize in 1981 for his research on split brain patients and other groundbreaking research into how the brain wires itself up during embryonic development. This interesting research on split-brain patients continues to be pursued by many neuroscientists, notably Sperry's former colleague Michael Gazzaniga.

How these critical long-distance trunks of communication cables are laid down as the fetal brain develops reveals a great deal about the mechanisms that guide connections between all neurons during development. Moreover, remodeling the adult brain after injury and during learning reactivates many of the same mechanisms that formed the brain in fetal life.

As mentioned earlier, this interhemispheric connection is called the corpus callosum, which is the massive cabling between the two hemispheres, the most prominent structure seen inside the brain when it is cleaved in two. Laying down of this massive interhemispheric communication cable depends on the temporary formation of a glial bridge to span between the two hemispheres of the fetal brain. If this astrocyte suspension bridge is cut in experimental animals, the axons cannot cross to the opposite brain. Growing out of neurons like vines, the axons reach the chasm separating the left and right cerebral hemispheres and pile up into frustrated coils on either side, uncertain where to go or how to traverse to the other side. An artificial bridge can be made of filter paper covered with embryonic astrocytes, or even coated with the cell membrane stripped from these glia. When this artificial bridge is inserted, the axons scurry across.

Similar astrocytic bridges and paths appear elsewhere in the fetal brain, guiding the connections between neurons that should be joined into a circuit. Astrocytes guide axons not only by clearing a physical pathway for nerve axons to follow, but also by releasing molecular signals to stimulate and direct the growth of embryonic axons. At critical decision points in the brain, astrocyte guide cells also broadcast chemical signals that turn back axons growing out from neurons that are on the wrong path.

Interestingly, some abnormalities in brain development can lead to extraordinary abilities. It has recently been found that some savants—people with astonishing abilities to remember factual information, such as the full text of books read years ago—lack a corpus callosum. These people tend to suffer other mental impairments, however. Often they are unable to abstract or fully appreciate the information they so easily memorize. It is as if the full power of the brain normally distributed to multiple intellectual functions gets concentrated into a small functional circuit as a result of this miswiring of cortical connections in development.[2]

The midline between the left and right brain is a critical watershed, strictly protected by embryonic glia stationed along the boundary line. You cannot have incoming or outgoing signals from the left and right sides of the body cross-wiring in the brain. Molecular signposts on these midline glia and chemical signals that they broadcast are interpreted differently by axons that should cross to the other side of the brain and axons that should stay on their own territory. By using genetic engineering to place the wrong molecules on these midline glia in the fruit fly, scientists have observed the larval nervous system miswire itself as axons from the left and right sides of the body cross back and forth in a haphazard, hopelessly tangled nonfunctional snarl.

Scientists have learned a great deal about the molecules these midline glia use as signals to guide axons growing out of neurons, and about the molecular machinery in the growing tip of the axon that receives and analyzes these signals. Mastering this information that glia provide in embryonic development will be vital if we are ever to exploit it to rewire a damaged brain or spinal cord. In principle, all that must be done is to insert the proper glial guideposts where they are needed to allow doctors to redirect the growth of axons recovering from injury.

NEURONS IN A GLIAL SPIDER WEB: THE SPACE BETWEEN

The space between cells in our brain is easily overlooked, yet all our thoughts and mental functions travel through this difficult-to-study realm. Neurotransmitters released from neurons and electric currents generated by them carry messages to other neurons through the space between brain cells; hormones, nutrients, and disease organisms all trek through extracellular space. But what does this cerebral intercellular space look like? How much of our brain tissue is empty space between cells? Does intercellular space change with neural activity, or in learning, disease, or aging? If it does change, what are the consequences? How is this space regulated? How quickly and how far can neurotransmitters travel through the brain in the spaces between cells? Do glia influence this intercellular realm of the brain?

One would have thought that pioneering scientists would long ago have explored this cerebral terrain when they scrutinized the intricate details of brain cells with their microscopes, but, unfortunately, they had inadvertently obliterated it. When body tissue is preserved in formaldehyde it shrinks like a hamburger on a grill, squeezing out all the space between cells. Biochemists also missed this part of the brain because they analyze molecules in cytoplasm that come neatly packaged inside cells, while molecules in the extracellular space—many of them vital for controlling chemical communication between nerve cells—slip away when biochemists isolate the brain cells for study in their test tubes.

Neurobiologists Charles Nicholson of New York University and Eva Sykova of the Institute of Experimental Medicine in Prague have developed ways to probe the unseen extracellular space in the brain. By injecting tracers and tracking their diffusion through living slices of rat brain they have discovered that about 20 percent of our brain is extracellular space. This is a much greater fraction of empty volume than the picture of brain structure one gleans from looking at chemically preserved tissue on a microscope slide.

Surprisingly, Nicholson found that fluorescent dyes released into the extracellular brain space of fresh brain tissue do not diffuse away as quickly as one would expect, compared say to a drop of cream dispersing in coffee. Something about this extracellular brain region slows diffusion. By coupling tracer dye measurements with mathematical modeling of diffusion, Nicholson determined that diffusion of molecules in extracellular brain space is slowed and restricted because this micro-universe in our head has such a highly complex physical structure. The many nooks and crannies between brain cells impede the flow of substances through extracellular space as the diffusing molecules enter microscopic blind alleys and become trapped. The beneficial result is that neurotransmitters released by a synapse build up to a higher concentration and do not disperse as rapidly or as far from the release point as would be expected. This improves communication between neurons by concentrating and sustaining the message.

Research by Eva Sykova and her colleague Lýdia Vargová revealed that the extracellular space in the brain is highly dynamic. It changes with disease and aging, and this affects learning. Conditions producing lack of oxygen, such as stroke, shrink the extracellular space. As the space constricts, the diffusion of substances between brain cells slows. Brain injury from a stab wound has a similar constrictive effect, but brain cancer opens the spaces between brain cells to nearly 43 percent of brain volume—nearly half the volume of cancerous brain tissue becomes fluid! This accelerates the spread of cancerous cells throughout the now mushy brain.

Vargová's research shows that as we age, we lose space between cells in our brain, and as a result, the diffusion of messenger molecules between neurons is impeded. Just like people, rats experience mental decline as they age, and the amount of mental decline they experience approaching their golden years varies individually. When Vargová compared aging rats that were the fastest learners in a maze test with their senior friends that were slower learners, she found that the quick learners had lost much less extracellular space than the slower learners.

What controls the amount of extracellular space in the brain? One of the main factors is glia. Shrinking and swelling of glia and the extension and retraction of glial cellular branches is the main factor regulating the volume of intercellular space and its structural complexity. In this way, glia (astrocytes), with their probing cellular fingers that separate neurons, control the traffic of neurotransmitters through extracellular space and thus also control the speed and distance of information carried by neurotransmitters between neurons.

Unlike the vacuum of interstellar space, the space between cells in the universe inside our skull is not empty. It is filled with stringy proteins studded with enzymes that regulate the synthesis and breakdown of neurotransmitters. In nearly every tissue of our body, this matrix of protein glues cells together, but in the brain this extracellular matrix also contains a complex assortment of different molecules and enzymes that do far more than stitch cells into tissue. These molecules regulate the cellular environment and control signals passed between cells. By connecting with the macromolecules that penetrate through the cell membranes of neurons and glia, the matrix of molecules outside the cell relays information about the outside environment to the inside of cells.

The extracellular matrix in the brain is synthesized during brain development not by neurons, but primarily by astrocytes. In the nerves in the trunk and limbs of our body, Schwann cells spin this matrix around nerve cells.

Like the scaffolding of a building, the extracellular matrix is crucial in supporting development of the fetal brain. The protein matrix contains molecules that support the outgrowth of axons and guide them to the proper points of connection with other neurons. Astrocytes insert different types of growth-guidance molecules into the matrix at appropriate times in fetal development. Much like a spider making its web with sticky and nonsticky fibers, some of the molecules astrocytes embed in the extracellular matrix are adhesive, and neurons stick readily to them and grow on them. At other places in the brain where axons should not enter (or at particular times when it would be inappropriate for them to enter), astrocytes add slippery molecules to the extracellular matrix. These greased barriers prevent axons and neurons from crossing. As discussed in chapter 5 on paralysis, glial barriers with slippery coatings also prevent regeneration of axons after spinal cord or brain injury. Vigorous research is under way worldwide to find ways to regulate the adhesive and antiadhesive properties of the extracellular matrix in the glial scar formed by astrocytes after spinal cord or traumatic brain injury. Understanding this scaffolding will be essential if we are ever to reach the important goal of curing paralysis.

The scaffolding astrocytes build during brain development continues to function in the adult brain, and as mentioned above, there is clear evidence for the involvement of extracellular matrix in learning and memory, as well as in disease. Astrocytes modify the adhesive molecules linking synapses during learning to permit the formation of new synapses, and they secrete enzymes (proteases) that tear down the matrix during wound healing. When astrocytes become cancerous, these proteases turn against the brain by breaking down the extracellular matrix and allowing the tumor cells to metastasize and invade other parts of the brain. Escaping control of the extracellular matrix, cancer cells spread their damage widely.

Understanding the role of glia in making and degrading the extra-cellular matrix will provide important insight into how the brain is built, how it changes with experience, and how it is weakened in injury and disease, such as the spread of cancer. But, the most critical of all spaces between two neurons is the tiny gulf between neurons at synapses. New research reveals that astrocytes are the architects in forming these vital electronic junctions.

GENESIS OF SYNAPSES: HEAVENLY STARS

Neuroscientists are now facing a startling realization that wiring the brain into neuronal circuits is not controlled solely by neurons, but also by glia. Glia can even control the budding of synapses on neurons.

In 1997 Ben Barres and members of his laboratory at Stanford University reported a curious observation. Neurons isolated from the rat retina and grown in cell culture did poorly if all astrocytes were eliminated. Similar findings had been observed in cultures of neurons made from other regions of the brain, including the hippocampus. When Barres's team stained neurons to reveal their synapses and used microelectrodes to record the electric signals sizzling across synapses in these cultures there was no doubt about it: neurons cultured in the absence of astrocytes had fewer functional synaptic connections.

Prior to that point it had always been assumed that the number of synapses a neuron formed was entirely under the control of the neuron itself. All theoretical and applied neuroscience operated on that assumption. Barres's findings undermined that assumption and suggested that astrocytes helped determined in some unknown way how many synapses a neuron would form. Imagine: formation of synapses—the fundamental unit of function in the nervous system—controlled by glia!

The researchers set out to determine how it was possible for astrocytes to control the genesis of synapses. Were astrocytes stimulating birth of synapses by touching neurons or by releasing some substance into the medium that caused neurons to sprout more synapses?

To find out, Barres's team collected culture medium from pure cultures of astrocytes and added it to pure cultures of neurons. The experiment revealed that the active ingredient contributing to synapse formation was something released by astrocytes because the culture medium collected from astrocyte cell cultures stimulated formation of synapses just as well as the astrocytes themselves when added to pure neuron cultures. (Many readers would no doubt like to know what this synapse sprouting substance might be, and where they can get it.) Through a series of biochemical isolations and tests researchers determined that astrocytes release at least two and probably many more chemical signals that stimulate the formation of synapses.

Frank Pfrieger of the Max Planck Institute in Germany isolated cholesterol as one of the active components. It is simply too early to fully understand how cholesterol derived from astrocytes is promoting synaptogenesis.

In 2004, Karen Christopherson, Erik Ullian, and colleagues identified a large protein named thrombospondin as another synapse-stimulating molecule released by astrocytes. This molecule was known to scientists working on blood cells, but it was not previously appreciated as having any involvement in synapse formation.

We now are faced with the realization that astrocytes not only control the strength of synapses by regulating neurotransmission across the synaptic cleft, they can also control the formation of synapses in the first place. The formation of new synapses in fetal development and in learning in children and adults is one of the most fundamental and intensely studied processes in neuroscience. The last five years have shaken the field with the discovery that astrocytes are at the controls.

This glial control of synapse formation operates in our peripheral nervous system as well as inside our brain and spinal cord, but in the peripheral nervous system one can watch glia remodeling synapses under a microscope. Wesley Thompson at the University of Texas and colleagues and Jeff Lichtman and colleagues at Harvard have found that at the neuromuscular junction, which is the synapse between motor neurons and muscle fibers, Schwann cells control the development, removal, and repair of the synapses controlling contraction of our muscles.

By placing a newborn mouse under a microscope and examining the same neuromuscular synapse just under the skin every day for several days, these researchers are able to watch the process of a synapse forming on muscles and the elimination of extra synaptic connections.

The surprise was that Schwann cells at the neuromuscular junction control exactly where synapses form on the muscle fiber. They also found that Schwann cells guide the removal of extra synapses from our muscle fibers.

Similar developmental mechanisms often reappear in mature animals to repair the nervous system after injury. When a motor nerve axon is cut, the regenerated axon will grow out and reform synapses with its proper muscle fiber to restore lost muscle function. Watching this process of healing under a microscope, Thompson and colleagues saw the terminal Schwann cell, which normally wraps the synaptic ending tightly, suddenly transform itself. After axon injury the terminal Schwann cell sprouted wildly and sent out long runners. Some of these runners eventually reached a synapse on a nearby muscle fiber. When they did, that undamaged nerve ending responded by suddenly sprouting a new axon offshoot that followed the Schwann cell bridge back to the muscle fiber that had lost its synaptic connection, thus restoring function to that muscle fiber. Schwann cells, closely monitoring the function of every neuromuscular synapse, heal us by repairing the connections to our muscles that are broken in injury or disease. One wonders how much synaptic remodeling inside our head is being controlled by glia monitoring the traffic of information flowing across each synapse.

WINDOWS OF OPPORTUNITY AND WRAPPING THINGS UP: GLIA AND THE CRITICAL PERIOD

Learn to speak a new language after puberty, and you will forever speak it with an accent. You just can't teach old dogs new tricks. We are all born with an ability to distinguish any sound in any language, but in streamlining the operation of the brain, we sharpen the ability to distinguish sounds that are critical for our native language and discard the neural circuitry that would be essential to distinguish other sounds that are unimportant for our language. This shows that brain wiring is directed by environmental experience, not simply defined by a genetic blueprint. Young Japanese children exposed to English will have no difficulty with the "L" and "R" distinction, but their parents will be at a loss to distinguish any difference in the sound of the words "breach" and "bleach." Like the hard of hearing in any language, adult Japanese who never heard English as infants must learn to use other cues to gain comprehension, such as the shape of the speaker's lips or the context of the sentence, to make the proper distinction between words differing in sounds their brain cannot tell apart. Your brain actually prunes away unused connections to increase its performance and efficiency according to where and how you use it.

The best-studied example of this is the process of vision. In examining development of three-dimensional vision in kittens opening their eyes to the world, the Nobel Prize winners David Hubel and Torsten Wiesel found that visual interaction with the environment was absolutely essential in wiring the visual cortex, the part of the cerebral cortex where vision is analyzed. If they placed a patch over one eye of a newborn kitten and a few weeks later removed it, the cat would be forever blind in that eye. There was nothing wrong with the optics of that eye, but the brain could make no use of the input. Without the functional visual experience necessary to provide the feedback to connect neural circuits, the eye was not wired properly into the cerebral cortex.

As in any computer, once this wiring is complete, it is nearly impossible to rewire the circuitry substantially. Hubel and Wiesel found that no amount of visual experience in the adult cat could compensate for the visual deprivation the kitten experienced during the critical time when its visual circuits were wiring up to the brain. They called this the "critical period," and it applies to people as well as to experimental animals. This is why doctors are eager to correct cataracts in an eye of a newborn as soon as possible and correct the alignment of the two eyes in children who are born cross-eyed. Similar critical periods are now known in the development of many different sensory systems.

In 1989, Cristian Müller and Johannes Best at the Max Planck Institute in Frankfurt, Germany, reported that if they transplanted immature astrocytes into the brain of an adult cat, the visual cortex could rewire itself—the critical period had been shattered. Somehow the immature astrocytes returned the visual cortex to the youthful stage when it was pliable and capable of rewiring.

For years, no one knew what to make of the experiment. Does this mean that glia control the critical period? If that were true, doesn't that mean that glia have a role—a commanding role—in wiring and unwiring synaptic connections in the brain? And if that were true, glia had to have some way to tap into neuronal information processing and transmission. This possible ability of glia made no sense to many neuroscientists, to whom the other brain was as dark and mysterious as the other side of the moon. The reason transplanted astrocytes can reopen the critical period is still not understood. It may have something to do with astrocytes secreting growth factors or exuding enzymes to break down the extracellular matrix and allow new synapses to form. No one knows for sure, but now neuroscientists are beginning to understand why glia help establish a critical period for learning. This subject will be reexamined in part 3 of this book, but now let's revisit clues to solve the puzzle presented by paralysis, which is providing new insight into the critical periods for learning. Myelin in the brain and spinal cord is studded with proteins that block the regrowth of axons, but axons sprout and grow freely on myelin made by Schwann cells in the peripheral nervous system. Why?

Brain Damage in Premature Babies: A Hidden Message from Myelinating Glia

Many babies suffer oxygen deprivation during birth, and this can result in serious brain damage. Neurons are certainly damaged by the lack of oxygen, but the cells that give rise to oligodendrocytes are even more vulnerable. The death of these glial cells at birth causes mental retardation. These children develop periventricular leukomalacia, or PVL, which is a loss of myelin caused by the death of the cells that mature into oligodendrocytes. Children with these myelin deficiencies develop mental retardation and attention deficit hyperactivity disorder because neural impulses are not able to pass properly through the uninsulated axons.

Curiously, many of the axons in the brains of these children, and also in experimental animals with PVL, sprout branches exuberantly, growing to places they do not belong. Normally, myelinating glia suppress axon sprouting during development, and this may be essential for properly wiring the brain.

Myelin has been compared to electrical insulating tape wrapped around wires, and insulating is indeed an important function of myelin, but myelin may also resemble electrical tape in another way. Some scientists have a strong hunch that the coating of myelin on axons is there not only to speed impulse conduction, but to wrap up the neural circuits once they are fully formed, putting a necessary finish on the process of brain wiring, much like an electrician taping together wires inside an electronic device for stability once it is all wired up properly.

Something must do this, otherwise axons might continue to sprout wildly in our brains. The theory is that sprouts that form in the adult brain get pruned back once they touch the myelin coating. The fact that they lack fully formed myelin is one reason young children and animals are better than adults at repairing damage to their nervous systems. Marsupials such as the opossum, whose babies develop in the mother's pouch like a kangaroo, are ideal for developmental studies, because late stages of fetal development take place outside the womb. The spinal cord will heal after it is cut in a young opossum, for example—provided the injury is produced at an age before the spinal cord myelinates.

After injury, myelin is one of the first things to break down, and axons can once again send out new sprouts unimpaired. Unfortunately, once they contact myelin that is suppressing sprouting and speeding conduction through other uninjured axons, the sprouts are stopped in their tracks. This property presents a biological dilemma: axon sprouting is both good and bad, depending on the circumstances.

Taking an evolutionary view, it is easy to understand why axons in the brain and spinal cord do not grow back when damaged: in nature, an animal with damage to its central nervous system gets eaten. In contrast, animals are likely throughout life to suffer minor damage to their limbs that could be repaired before it becomes fatal. For wild animals, it does not matter if myelin in the brain and spinal cord is booby-trapped with proteins that block axons from sprouting and growing back after injury. The inhibitory proteins in myelin are there for another purpose: to stop axon sprouting once the mature circuit has developed.

Martin Schwab, who discovered Nogo, the first myelin protein inhibiting axon sprouting, agrees. Asked why he thought Nature added Nogo and other proteins in myelin that inhibit axon sprouting, he responded, "I think that Nogo and other inhibitors [in myelin] have a role in stabilizing the extremely complex structure of the adult CNS of higher vertebrates." Schwab went on to observe, "Nogo-A was 'invented' by frogs; fish and salamanders do not have it, and the restriction of plasticity to small spatial domains [in advanced brains] once the CNS has matured is one of its functions."

While it makes good sense that something must terminate the period of axon growth and sprouting in early development, why, I asked, should axon sprouting be inhibited by myelin and oligodendrocytes instead of some other cell?

"Oligodendrocytes mature late; in fact in correlation with the maturation of the fiber pathways, axon growth is shut down at the time the pathway is mature . . . Oligodendrocytes are simply there at the right time in development, and as they myelinate axons, the critical period for that circuit draws to a close."

Professor Stephen Strittmatter, neurobiologist at Yale University, agrees that oligodendrocytes are the right cells to close the window on the critical period: "The previous work from Aguayo and Schwab made clear that something was inhibitory in CNS myelin . . . the timing of myelination fits best with the developmental stage when critical periods close. In broad developmental outline, neurons are born, they migrate to the right place, then grow axons and dendrites, then make neurotransmitters and synapses, and then fine-tune connections with experience. After all of this is accomplished, myelin forms and experience-dependent tuning ends. As compared to any other event, oligodendrocyte maturation is best matched with the close of experience-dependent tuning."

Strittmatter and his postdoctoral fellow Aaron McGee recently put this theory to the test. As explained earlier, Nogo works by binding to a receptor protein in the membrane of neurons. Strittmatter and colleagues deleted the gene for the Nogo receptor in mice. In these mutant mice, axons sprouted profusely after a brain stroke or spinal cord injury. What is more, the injured animals lacking the Nogo receptor recovered substantial ability to walk and grasp objects after nerve injury. Schwab's laboratory has shown much the same thing by blocking Nogo with an antibody. But, in addition to allowing neuron repair, disabling the Nogo receptor had another important effect on learning.

The researchers found that blocking the inhibitory Nogo protein in myelin restored the youthful flexibility lost in the adult brain. As Hubel and Wiesel showed in the 1960s, neurons in the brain that normally receive input from an eye that is closed experimentally for a few weeks in young kittens will lose connections from the useless eye and become rewired to the open eye. This neuron can now "see" through the good eye. But in older animals this flexibility is not possible. The age of the critical period for this rewiring coincides with the time myelination is finished in visual circuits. Strittmatter's experiments published in 2005 revealed that the critical period could be reopened when they removed the Nogo receptor from mouse neurons—old animals could rewire their visual neurons to accommodate the loss of vision in one eye.

These new discoveries concerning myelin's role during the wrapping up of the critical period answer the riddle of why myelin that insulates axons contains proteins insuring that spinal cord injury results in permanent paralysis. These insights offer exciting new potential for treatments for paralysis after spinal cord injury, but they also help explain why there are critical periods for learning, and why the ages for critical periods are not the same for all kinds of tasks. Different parts of the brain myelinate at different times. Pieces of the puzzle found in exploring the other brain are coming together to fill missing gaps in knowledge about the neuronal brain that have been long-standing mysteries. These new discoveries reveal that myelin-forming glial cells control learning in ways that once were assumed to depend entirely on neurons.

CHAPTER 12

Aging: Glia Rage against the Dying Light

Children dress up and shuffle with delight in their parents' shoes. Teenagers try to pass for twenty-one. Some women (and some men) resist their thirtieth birthday, sometimes for years. Those nearing "senior" status search out miracle tonics, vitamins, tai chi, cosmetic surgery, Botox, and health spas. Our age, no matter what it may be, challenges us, because it is a part of ourselves that we cannot control.

The search to recover lost youth is a universal, ancient quest. We pursue it vainly. As legend has it, rather than return to Spain, Ponce de León stared out at the shimmering Caribbean waters and watched as the ships carrying Christopher Columbus back home to Spain vanished on the vast horizon. Rejecting the world of old, he chose to stay behind and hunt for the elusive treasure of youth in a land where everything was new and fresh. He failed, of course, but modern science, as it pries open the secret clockwork inside our cells, may succeed in finding it, even though we know that if our species ever finds the fountain of youth in the laboratory, the consequences would be cataclysmic. There is a reason for the biological cycle of life, and disrupting that cycle risks destroying the entire machinery of Nature. Still, even though every living thing has a life cycle that meshes perfectly with its natural history and ecology, we're tempted to think that it couldn't hurt too much to cheat just a bit.

It is safe to say that the world of science knows vastly more about the aging of skin cells than it does about the aging of glial cells—an investment in the superficial, some might suggest. Some modern-day Ponce de Leóns are beginning to explore this uncharted area, and in this chapter we examine the second phase in the life cycle of glial cells as we glide from maturity to the grave.

YOUR AGING BRAIN

It is astonishing to think that the same nerve cell that began processing the barrage of bewildering information you experienced that day you were born will still be processing information in your brain eighty years later. Neurons do not divide like our other cells. By contrast, your red blood cells are replaced every three or four months, and skin cells are sloughed off and renewed constantly. Few things in life are as durable as a brain cell—automobiles, houses, electronic devices would do well to survive intact for eighty years or more—but neurons do die throughout our lifetime, and our brains slowly lose weight as we enter old age.

By the time you reach sixty-five years of age, your brain will weigh on average 7 to 8 percent less than it did in midlife. The frontal lobe of your cerebral cortex, the brain region vital for planning and executive decision making, will have shrunk in volume by 5 to 10 percent. Fortunately, a modest loss in these once supreme functions may not be all that detrimental, and some might argue could even facilitate quality of life in retirement.

Individual neurons begin to deteriorate and die as we age, but this die-off varies widely in different regions of the brain. Loss of neurons in the hippocampus and substantia nigra is typical for senior citizens. This loss parallels the weakening of your memory and a growing propensity toward Parkinson's disease with age.

Much like your increasingly blotchy skin, your aged neurons begin to develop dark deposits and inclusions formed from tangles of coagulated proteins. It is the abnormal accumulation of the filamentous tau protein that forms the neurofibrillary tangles inside the neurons of people with Alzheimer's disease. Pools of junk protein also accumulate outside the neurons of Alzheimer's patients, like refuse in an alley. These amyloid plaques make the brain tissue of an Alzheimer's patient as instantly recognizable to a neuropathologist as chicken pox is to a pediatrician.

Like increasing clutter in the attic of an old house, many other proteins accumulate inside neurons with age, and this protein litter is associated with neurodegenerative disorders such as amyotrophic lateral
sclerosis, Huntington's disease, and Parkinson's disease. Trash-dump
accumulations of the protein alpha-synuclein create clogs in neurons
called Lewy bodies, which can be seen easily inside aged neurons under
a microscope. Many people suffering from Parkinson's disease have
Lewy bodies stuffed inside their neurons, as do some people suffering
from other forms of dementia, such as Alzheimer's, and those diagnosed
with Lewy body disease. The latter condition erodes language, memory,
and reason and can cause body tremors and hallucinations.

Recently pathologists have experienced the revelation of the Sherlock Holmes deduction "There is nothing more deceptive than an obvious fact." The same dark spots of coagulated protein distending the
bellies of aged neurons and drawing all the attention of neuroscientists
studying neurodegenerative diseases have recently been found packed
inside astrocytes and oligodendrocytes. These proteins can be seen in
the same microscope slide inside glia sitting right next to the neurons of
these patients. Until recently, this fact was simply overlooked: "The eye
sees only what the mind is prepared to comprehend," as French philoso-
pher Henri Bergson so astutely commented. What role these inclusions
in glial cells may have in neurodegenerative diseases and what opportunities they may offer for new medical treatments are topics only just
beginning to be investigated.

Not all of the shrinkage in your brain by age sixty-five is caused by
loss of nerve cells. Some of the loss in brain volume results from a
decrease in the space separating the cells in your brain. This cellular
overcrowding slows down diffusion of biological chemicals, neurotransmitters, nutrients, and other materials passing through the constricted
alleys between cells. This sluggish flow can in turn affect synaptic transmission, upset the balance of ions around neurons, and impair removal
of toxic products and byproducts of metabolism from your neurons. The
tight space also restricts the channel of communication between glia.

Interestingly, white matter volume decreases even more than grey
matter in the aging brain. (Recall that white matter is the portion of our
brain where heavily myelinated axons are bundled together, while grey
matter describes the parts of our brain where nerve cells are packed and

form synaptic connections with one another.) The inescapable conclusion is that myelinating glia suffer the effects of aging as much as or more
than neurons.

It is not just myelin-forming glia that are affected by old age. Astrocytes become more numerous and gliotic, much as they do after brain
injury. Perhaps, like so-called liver spots on aged skin, gliotic astrocytes
(cells swollen in response to injury) are the result of a lifetime of accumulated stresses and minor injuries to your nervous system. Alternatively, the increase in gliotic astrocytes in the brains of seniors may be a
response to the neuronal degeneration going on around them; astrocytes are doing whatever they can to protect the declining neurons in the
aging brain.

All your life astrocytes in your brain respond to minor brain stresses,
infection, and injury by proliferating, engulfing damaged tissue, controlling inflammation, secreting neuroprotective molecules, and walling
off damaged regions, preventing destruction from spreading throughout
the brain. But in reacting to injury, astrocytes can also release substances
that cause dangerous levels of inflammation and other chemicals that
are neurotoxic. This response inadvertently inhibits the regrowth of
neurons or attacks oligodendrocytes to cause demyelination.

As we age more of the microglia in our brain become activated too,
transforming more of these guard cells into the alerted battle-ready state
for fighting infection. But with so many microglia deployed, there are
dwindling reserves to respond when and where extra microglia are
needed to fight injury and infection in the elderly brain.

Just like everything else about your aging body, your astrocytes are
older. The glia are getting tired and weak. This is one of the reasons
we respond to brain injury better when we are young. Older astrocytes
become less responsive to the signals neurons send out as alarms
when they are injured. In experiments performed in a laboratory dish,
researchers can see that older astrocytes still respond to distress signals
added to their cultures, but their response weakens and slows as the
astrocytes age. Researchers must use increasingly higher concentrations
of the alarm signals to rouse the aged astrocytes to the same state of
activity. This is why people older than eighty recover from brain stroke
much more slowly than middle-aged stroke patients. This connection

between sluggish astrocytes and poor response to injury is backed up by postmortem exams of those who die after a stroke at middle age rather than in old age. Examining microscope slides of brain tissue, neuropathologists see clearly that a more vigorous gliotic response is marshaled following stroke in the middle-aged brain than in the aged brain.

After we reach a certain stage of maturity, we would all like to slow the effects of aging. Sex hormones can do it, for a while, but later there is a cost, as many women learned after estrogen replacement therapy. Originally estrogen replacement was touted as a means to reduce heart disease, stroke, loss of bone density, and cognitive decline in postmenopausal women, but longer-term studies showed that these women developed more cancers without any benefit to cardiac health. Some studies showed increased risk of heart disease and Alzheimer's disease in combined estrogen and progesterone therapy for postmenopausal women. Alzheimer's disease was a particularly unexpected and intriguing complication.

It may surprise many to learn that astrocytes have the same protein receptors for sex hormones that are on cells in our sex organs. Like other growth hormones, sex hormones stimulate the growth of young glia. But the time for cellular growth is when we are young, and Nature seems to turn down sex and growth hormones and our responses to them as we age, just as a cook lowers a flame from boil to simmer when a dish is nearly cooked. Experiments show that estrogen promotes the growth of young astrocytes, but aged astrocytes actually reverse their reaction to it, and estrogen (a hormone of both males and females, though females have more) actually inhibits proliferation of aged astrocytes. One wonders if glia might not provide another reason to avoid estrogen replacement therapy. Supplemental estrogen after the time in life when it would normally be circulating in high levels might slow already sluggish astrocytes in responding to injury or combating neurodegenerative diseases.

There seem to be good reasons why Nature has throttled back on sex and growth hormones as we age. We no longer need to grow rapidly (an understatement for most middle-aged Americans). Also, the longer we survive, the greater risk we face from cancer, a disease that takes many of us in old age. Cancers are cells that have suffered damage to their controls on cell division, and as a consequence they divide and grow rapidly and uncontrollably. The last thing one would want to do is throw oil on the fire in the form of additional sex and growth hormones. Hormones may temporarily restore youthful appearance and vigor, but think about the consequences of a careless chef raising the flame too high near the end of cooking a dish.

To the harmful effects of boosting cellular growth and division with sex hormones in old age, we now add new evidence that sex hormones can actually act negatively on aged astrocytes and other cells, because the receptors for these hormones in old age respond differently to them from the way they did when we were in the spring of life.

SHADOW OF THE SNIPER: GLIA AND ALZHEIMER'S DISEASE

"I have lost myself."
"What is your name?"
"Auguste," she replied.
"Last name?"
"Auguste."
"What is your husband's name?"
"Auguste, I think."
"How long have you been here?"
She could not say.[1]

With a slow exhalation of cigar smoke, the vivid scene of five years earlier cleared from the doctor's mind as he contemplated the pale yellow and purple splotched slice of brain stuck like a postage stamp to a glass microscope slide. It was Augustine Deter's brain.

Looking at the slide he recalled the eyes of the bewildered fifty-one-year-old woman looking to him earnestly for help, but there was little he could do. As her doctor, Alois Alzheimer watched Augustine Deter swiftly slip from her connection to reality. Within a few short years she was gone. In the end she had become delusional, incoherent, bedridden, and incontinent. She would scream for hours on end, inconsolable. Unreachable.

He was powerless to help Deter, but together they would expose the monster that ate away her mind, robbing her of all phases of her life: past, present, and future. A disease that left her forever wondering, lost without the familiarity of friends or place, a stranger to her own identity. From his first look at the horribly damaged cells in her brain, Alzheimer fingered a prime suspect: glia.

Over the years that suspect would elude most neuroscientists who tracked the cause of Alzheimer's disease in the tangled debris clogging diseased neurons. Today the spotlight has again focused on glia as culprits and conspirators in the slow, agonizing death of one's mind caused by the incurable disease that now bears the name of this German psychiatrist.

Like an indiscriminate sniper, this disease creeps up on its unsuspecting victims. Are the lapses of recall and the momentary confusions that occasionally startle and frustrate us ordinary mental hiccups? Or the shadow of the killer? In the early stages, no one can say for certain who is being stalked by the disease. Only microscopic analysis of the brain substance itself will reveal the definitive marks of the disease: the plaques and fibrous tangles that Alzheimer saw in Augustine Deter's brain as he unhinged the pince nez spectacles from the bridge of his nose and squinted through the microscope at her damaged brain.

Even those blessed with gifted memories and positions of authority in life can be shot down cruelly by Alzheimer's disease. A B-movie actor whose mind excelled in memorizing lines rapidly and who became the fortieth president of the United States of America slowly succumbed to destruction that obliterated any memory he had of his esteemed and demanding position.

Did he see the shadow? On the November anniversary of his reelection in 1985 President Reagan proclaimed:

I, Ronald Reagan, President of the United States of America, do hereby proclaim the month of November 1985 as National Alzheimer's Disease Month, and I call upon the people of the United States to observe that month with appropriate observances and activities . . .

   For more than two million Americans with Alzheimer's disease, each day is fraught with fear and frustration. Fear of getting lost in one's own neighborhood; of not recognizing members

of one's immediate family; of not being able to perform simple,
familiar chores. For the victims of this disease, tying shoes or
setting a table can be overwhelming tasks.[2]

On another November, in 1993, the proclamation became premoni-
tion. Doctors at the Mayo Clinic diagnosed Reagan with Alzheimer's
disease. The following November, the former president made a public
pronouncement in an eloquent handwritten letter to the people of the
United States. It was his final curtain call.

I have recently been told that I am one of the millions of Americans
who will be afflicted with Alzheimer's disease.

Upon learning this news, Nancy and I had to decide whether
as private citizens we would keep this a private matter or whether
we would make this news known in a public way.

In the past, Nancy suffered from breast cancer and I had my
cancer surgeries. We found through our open disclosures we were
able to raise public awareness. We were happy that as a result
many more people underwent testing.

They were treated in early stages and we were able to return to
normal, healthy lives.

So now, we feel it is important to share it with you. In opening
our hearts, we hope this might promote greater awareness of this
condition. Perhaps it will encourage a clearer understanding of the
individuals and families who are affected by it.

At the moment I feel just fine. I intend to live the remainder of
the years God gives me on this earth doing the things I have always
done. I will continue to share life's journey with my beloved Nancy
and my family. I plan to enjoy the great outdoors and stay in touch
with my friends and supporters.

Unfortunately, as Alzheimer's disease progresses, the family
often bears a heavy burden. I only wish there was some way I could
spare Nancy from this painful experience. When the time comes, I
am confident that with your help she will face it with faith and
courage.

In closing let me thank you, the American people, for giving me
the great honor of allowing me to serve as your president. When

*the Lord calls me home, whenever that may be, I will leave with
the greatest love for this country of ours and eternal optimism for
its future.*

*I now begin the journey that will lead me into the sunset of my
life. I know that for America there will always be a bright dawn
ahead.*

Thank you, my friends. May God always bless you.

Sincerely,
Ronald Reagan[3]

Nancy Reagan did indeed face the ordeal with devotion, faith, and courage, sustaining her husband on his dark journey, but within a few short years, the former president would no longer know his beloved Nancy at all.

ALZHEIMER'S: A DISEASE OF GLIA

Alzheimer's disease destroys neurons and communication pathways in the brain. Certain parts of the brain are more vulnerable than others, notably those brain regions controlling thinking (cerebral cortex), memory (hippocampus), and fear, emotion, and aggression (amygdala). These are the main targets of Alzheimer's disease. The disease afflicts 10 percent of us who reach the age of sixty-five. As average life spans increase through improved medicine, even more people will suffer from it. If you should reach the age of eighty-five, you will then have a fifty-fifty chance of acquiring Alzheimer's disease. To put it another way, if both you and your spouse should survive to this ripe old age, one of you will probably suffer dementia or personality changes from Alzheimer's disease. Age is the single greatest risk factor for this disease, which provides some intriguing biological insight into the disease mechanism.

The hallmark of Alzheimer's disease is the senile plaques seen in brain tissue as first reported by Alzheimer in 1907 based on his examination of slices cut from Augustine Deter's brain. These plaques are aggregates of material, beta-amyloid, that surround damaged neurons. But surrounding the plaques are large aggregates of microglia. Alzheimer himself saw and described these cells at these sites of brain damage. The plaques are also spots of intensive chronic inflammation of brain tissue. For this reason alone, it is to be expected that we should find these glial cells on the scene in abundance, but are they a part of the disease or just a response to it? The plaques are not seen in the brains of elderly people who do not suffer dementia, so the plaques are the result of disease, not a natural process of the aging brain. The second diagnostic feature of Alzheimer's disease is that neurons are choked with tangled bundles of protein filaments inside their cell body known as neurofibrillary tangles.

We are now beginning to realize that Alzheimer's disease is a powerful example of what can happen to the brain and mind when the healthy interaction between neurons and glia goes awry. The neuronal damage in Alzheimer's disease results from an attack on neurons by microglia and astrocytes that is linked to chronic, local brain inflammation.

The excess beta-amyloid protein that forms the senile plaques in the brain of Alzheimer's patients can be caused either by overproduction of the material or impaired clearance of beta-amyloid. Dysfunctional glia can contribute to the plaques in both ways. Microglia actively gorge themselves on the excess beta-amyloid, clearing it from the spaces between the neurons. This is the main reason why microglia are found in such abundance surrounding the senile plaques. However, under conditions of chronic inflammation, the ability of microglia to clear beta-amyloid from the brain is greatly diminished, so the toxic peptide accumulates more rapidly. In addition, several neurotoxic factors are released from microglia when they are treated with beta-amyloid (the list includes reactive oxygen species, cytokines, chemokines, and others). Reactive astrocytes are also recruited to the amyloid deposits, and they probe their cellular fingers into the senile plaques to release enzymes that dissolve the beta-amyloid. Thus astrocytes aid microglia in clearing beta-amyloid from the brain.

But astrocytes also cause the death of neurons in Alzheimer's disease. Astrocytes can generate the beta-amyloid peptide, which forms the senile plaques, by pathological processing of a precursor molecule, a protein called amyloid precursor protein (APP). This protein is cleaved by an enzyme in astrocytes to generate the toxic beta-amyloid peptide. Thus astrocytes make APP, as do neurons, and they also make the enzyme that generates the same neurotoxic beta-amyloid that is made by diseased neurons.

The ability of astrocytes to protect neurons from damaging oxidation becomes impaired in astrocytes exposed to the beta-amyloid protein, which suppresses their ability to generate the powerful antioxidant glutathione discussed in chapter 5. At the same time, reactive astrocytes in Alzheimer's disease release nitric oxide, a toxic component of smog, which at high concentration kills neurons by oxidative stress. The uptake of the neurotransmitter glutamate by astrocytes is also impaired in patients with Alzheimer's disease. This not only degrades synaptic function, it allows glutamate to rise to toxic levels, killing neurons by overstimulation.

Another important function of astrocytes is to regulate blood flow to neurons according to their metabolic demands, but this function becomes impaired in Alzheimer's disease. University of Rochester neuroscientist Maiken Nedergaard and colleagues determined this by watching calcium oscillations in astrocytes with a microscope focused through a hole made in the skull of a mouse under anesthesia. The researchers found that the normal dilation of microvessels in the brain cortex that is induced by stimulating brain function is greatly impaired in mice with Alzheimer's disease. Touching the mouse's whiskers increases the blood flow in the cerebral cortex where the whisker movements are analyzed. As explained earlier, this vascular regulation happens in part because astrocytes surrounding the microvessels communicate by calcium signals. But the calcium responses in astrocytes of animals with Alzheimer's disease are impaired, undermining the ability of astrocytes surrounding microblood vessels to regulate blood flow, causing neuron death.

Even as they are participating unwittingly in the death of neurons in Alzheimer's disease, astrocytes attempt to protect neurons from death by releasing growth factors. Brain-derived neurotrophic factor (BDNF) is produced in increasing quantities by astrocytes treated with beta-amyloid, and this factor rescues neuronal cells from toxicity induced by beta-amyloid in cell culture. Unfortunately, astrocytes are also victims of this disease. Recent studies published in 2007 find that the beta-amyloid peptide is also toxic for astrocytes in cell culture.

Chronic local brain inflammation around the plaques sets up the unhealthy environment that leads to changes in microglial and astrocyte responses in Alzheimer's disease. Microglia together with astrocytes induce this inflammatory response by releasing inflammatory cytokines. Treating astrocytes and microglia with these inflammatory agents not only causes them to lose their ability to clear beta-amyloid, it forces them to begin releasing neurotoxic substances. Treatment with anti-inflammatory drugs (such as ibuprofen) has been reported in some studies to decrease the risk of developing Alzheimer's disease. The effects are only neuroprotective, however; the anti-inflammatory drugs are not therapeutic once a person has developed the disease and his or her brain has presumably suffered irreversible neuronal damage.

This association between chronic inflammation and Alzheimer's disease may explain why age is the prime risk factor. Minor insults to the brain such as minor brain trauma, intoxicating drugs, and ischemia (impaired blood flow) accumulate over a life span. These conditions promote glial activation, which in turn promotes the release of inflammatory cytokines. When this situation is coupled with genetic risk factors, these small brain lesions may sow the seeds for Alzheimer's disease. We must also recognize that as we age, our glial cells also age and become less robust. At the end of a life span, microglia and astrocytes may no longer be able to keep up their vigilant service to protect neurons. When these glial cells become weak, neurons die.

Catching Alzheimer's Disease

This new understanding of glial involvement in Alzheimer's disease explains the most recent peculiar evidence that the disease can be caused by infection. Viruses, such as HIV and herpes simplex, cause a pronounced inflammatory response in the brain that contributes to neurodegeneration. One of the most recent and interesting infectious agents implicated in causing Alzheimer's disease is the bacterium *Chlamydia*.

This is not the same *Chlamydia* bacterium that causes sexually transmitted disease (*C. trachomatis*). This is its cousin, *C. pneumoniae*, which may be responsible for more deaths in America than any other microbe. It causes pneumonia. We all encounter this bacteria in our lives through common respiratory infection, sinus infection, or bronchial infection. In the elderly, these infections can be life threatening. Infection with *Chlamydia pneumoniae* (recently renamed *Chlamydophila pneumoniae*) is most common among those who are between sixty and seventy-nine years of age.

H. C. Gérard of Wayne State University in Detroit reports that *C. pneumoniae* is commonly found in the brain tissues of patients with Alzheimer's disease, especially in the brain regions that show neuropathology. The bacterium is not found in the brains of elderly people who do not suffer from dementia. In an interesting example of the interaction between environment and genes, these studies found that patients with the gene *APOE epsilon4*, a genetic risk factor for Alzheimer's disease, had higher numbers of *Chlamydia* bacteria in their brains than other Alzheimer's patients.

The bacteria infects neurons and can kill them, but a 2007 study by E. Boelen and colleagues of the Maastricht University in the Netherlands reported that this bug also infects microglia and astrocytes. Boelen's studies show that the inflammatory response resulting from glial infection by *Chlamydia* contributes to neuronal death in Alzheimer's disease. The infected microglia release many cytokines. When culture medium taken from the infected astrocytes was added to neurons, it did them no harm, but culture medium taken from microglia infected with *Chlamydia* killed neurons. This neuron death was caused by the inflammatory cytokines released by the infected microglia.

Alzheimer's is a disease of disrupted neuron-glial interaction. Understanding what goes wrong with glia in Alzheimer's disease should lead to better treatments; indeed most treatments of neurodegenerative disease, including Alzheimer's disease, will involve microglia. Vaccines have been produced against the human beta-amyloid protein, and in 1999 it was reported that immunization with the vaccine reduced Alzheimer's pathology in mice. It is known from studies in cell culture that the antibodies produced by the vaccination stimulate microglia to ingest the harmful protein. However, under inflammatory conditions, the microglia fail to do so adequately. On the other hand, appreciating the importance of microglia in clearing beta-amyloid, K. Takata and colleagues at the Kyoto Pharmaceutical University in Japan injected microglia into a rat brain after injecting it with beta-amyloid protein. They found that increasing the number of microglia greatly increased the clearance of the toxic peptide from the brain. In their 2007 paper reporting these results, Takata suggests, that microglial transplantation may be a new and effective therapeutic approach to treating Alzheimer's disease.

One hundred years after slices of it defined this devastating, mind-rotting disease, Augustine Deter's brain is now beginning to make sense. The explanation for her rapid mental decline is found in the puzzling plaques and the glia that frame them.

Part III

Glia in Thought
and Memory

The Mind of the Other Brain: Glia Controlling the Conscious and Unconscious Mind

It was more than one hundred years ago when Ramón y Cajal teased out the intricate structure of microscopic nerve cells packed invisibly inside brain tissue and with brilliant insight conceived the neuron doctrine. That doctrine has provided a solid foundation supporting a century of researchers seeking to understand how the brain functions. But as he probed the mysterious structure of brain tissue, Ramón y Cajal also saw that it was crammed with many cells that were clearly nothing like neurons. In fact, there were far more of these cells than neurons in the brain. Ramón y Cajal could not understand these odd and varied cells simply by pondering their shape and organization in the brain as he peered through his microscope and sketched them out precisely with a sharp pencil.

At the turn of the twentieth century the cellular structure of the brain was analogous to a brick wall. The bricks (neurons) were held in place by the cementing mortar (neuroglia, or "neuro glue") that served the needs of neurons in health and disease. The peculiar thing about *this* brick wall was that the mortar accounted for 85 percent of the structure, and the bricks were embedded spottily, like decorations, throughout the remaining 15 percent. At least that is the way scientists and doctors chose to see it.

Near the end of the twentieth century, Marian Diamond, looking for clues to Einstein's genius in his stolen brain, found no clues in his nerve cells, but in tallying up her cellular census she was struck by the unusually large number of glia. Was this a meaningless coincidence, or could it be a clue that glia might do more than everyone had assumed? Could it be that Einstein's remarkable brain, even in death, could become a beacon lighting the way for science to an unimagined direction?

On a much smaller scale of importance, I encountered a perplexing problem in my own lab. I had seen Schwann cells, filled with a dye that glowed when calcium poured into their cytoplasm, shine brightly when we stimulated nerve axons to fire impulses. Why was electrical insulation on the brain's wiring responding to the electrical signals flashing through the neural wires?

Only in the last few years have neuroscientists begun to explore the other brain, and as they do, the implications of these pivotal discoveries are challenging our presumptions about how the brain works. Clearly, neurons are only half the story, but what is the other half? Glia could cooperate in regulating the function of a healthy brain, but do they? Do glia contribute to thought and memory? Does information pass from the "neuronal brain" into the "other brain," where it is processed inside glial circuits and then drives neuronal information processing in ways that chains of neurons linked in series using synaptic connections cannot?

New research revealing that the brains of people with certain psychological disorders—including schizophrenia, depression, and anxiety disorders—have cellular abnormalities in glial cells is circumstantial evidence that dysfunctional glia might in some cases sow seeds of mental illness. But by extension, doesn't this also suggest that glia might participate in thought and function in the healthy mind? The fact that glia can detect information flow in neuronal circuits and influence it, coupled with the knowledge that glia can communicate among themselves, lays bare a new world of possibilities for cellular information processing in our head.

How will exploring glia change our understanding of how the brain operates? Could exploration of the other brain illuminate secrets of the human mind? Could glia play a role not only in servicing neurons, but in moving our conscious or perhaps unconscious mind?

We still know little about glia. Even the basic facts, such as how many kinds of glia there are, where they originate during development, and what the various glial cells look like in detail, are unknown. The simple fact is that few students enter the field of neuroscience to study nonneuronal cells. Worse, research by glial biologists is easily dismissed as unimportant by the establishment view that neurons are the only cells important for information processing in the brain. As a result, research on the other brain is one hundred years behind research on the neuronal brain.

Falling Stars

Even the basic facts of what glia look like and how many different types of glia there are in the brain are unknown today. As a remarkable example, neuroscientists thought they knew what astrocytes looked like until 2001, when Mark Ellisman and his graduate students published a paper on astrocytes, exposing the true character of these brain cells.

Ellisman is a wiry-haired dynamo who in the 1990s built and still runs one of the country's most advanced centers for exploring the cellular and subcellular structure of brain cells using light and high-voltage electron microscopy amplified by harnessing the power of supercomputers. A technology junkie, Ellisman always packs the latest electronic gadgetry in a fanny pack strapped to his waist, stuffed with electronic marvels destined to become the next craze in the consumer electronics industry. The first iPod I ever saw emerged like a bouncing baby kangaroo from Mark's pouch one evening after dinner at my home.

Mark was my Ph.D. advisor at the University of California, San Diego, in the early 1980s, and I was happy that he agreed to attend a symposium I organized on neuron-glia interactions in 2001. I wasn't sure what he would speak about, but I knew his visually dazzling lectures would ignite the crowd and launch the symposium on a high trajectory. He did not disappoint, astounding the audience of scientists with three-dimensional images and video adventures inside the brain. He used a "powers of ten" journey that began by penetrating into the interior of a neuron to reach inside the molecules that make it work. Then with expanding scales of magnification we raced from the synapse as a cloud of neurotransmitter molecules shot past in a billowing plume. Traversing tangles of synapses and dendrites, we accelerated out of the

tangle and ascended scales of anatomical structure past glial cells to escape through a pore in the skin where a bead of sweat emerged from the
root of a human hair and onto the scalp. If you have seen the movie *Fight
Club,* you have seen this footage: it is based on a collection of real scientific data, processed by supercomputer and edited together into a con-
tinuous video sequence carrying the audience on a journey that defies
imagination.[1]

At the end of the movie, Ellisman showed a collection of 3-D slides
revealing the finest delicate anatomy of astrocytes. I looked out at the
audience transfixed in awe and couldn't resist chuckling at the absurdity
of this scene of 150 of the world's top neuroscientists all sporting 3-D
glasses like kids rapt with suspense in a theater screening a 3-D horror
movie. Then, in 3-D, he showed the audience something no one had
seen before: what an astrocyte in the hippocampus, the part of the brain
crucial for human memory, really looks like.

Up until then, anatomists had used various stains to reveal astrocytes in brain tissue. Because the fibrous protein in these cells took up
stains particularly well, astrocytes were immediately recognizable by
their star-like structure, which had inspired their name. Rather than
using a stain, Ellisman and his colleagues pierced an astrocyte in the
hippocampus of a rat with a fine glass capillary and filled the cell with a
fluorescent dye. They saw that all the images of astrocytes heretofore
had been mere ghosts, skeletons actually, for the stains anatomists had
relied upon to identify astrocytes exposed only the fibrous skeleton
within these cells. Astrocytes were not star-like at all; they were as bushy
as the hair on Ellisman's head, and they were roughly two times bigger
than they appeared with stains that revealed only their internal skeleton. Once again, Nature had fooled scientists into giving a class of brain
cells a name that described not reality, but rather a relic of incomplete
staining.

More startling, these images showed that these enormous bushy
cells populated the hippocampus in a tile-like manner. Rather than
being hopelessly entangled as they appeared to be in their star-like form,
every astrocyte maintained its own unique territory in the brain. Following every fine branch in detail, Ellisman saw that no astrocyte ever
penetrated a twig into the bush of an adjacent astrocyte. This part of the
brain (later extended to most other parts of the brain) was divided into

sectors, each one of which was the sole domain of an individual astrocyte. When asked at the end of his presentation what was the functional
significance of this anatomical arrangement, Ellisman replied with a
wry smile that he did not know, but in an obvious bit of understatement
said that he thought the audience might find it interesting.

Why indeed is the hippocampus carved up into domains ruled by
astrocytes? Could these glial cells segregate and control neural circuits
in this part of the brain that is critical for memory? There is new research
supporting this hypothesis, and most of it comes from recent discoveries. Before exploring the possibility of glia participating in our conscious
mental functions and memory, we will consider a more mysterious
question that emerges from a property of the other brain that is unlike
the neuronal brain. The other brain does not communicate with electricity, as I have explained. This fact, completely outside all prior understanding of brain function, leaves neuroscientists adrift without a
compass. One consequence of the nonelectric information processing in
glial circuits must be that the other brain cannot respond with lightning
speed. Glia operate slowly and influence large territories of the brain.
This discovery is forcing neuroscientists to broaden their view of brain
function beyond what happens at the tiny synapse and to expand their
time frame for nervous system function beyond the millisecond-to-
millisecond signaling controlling reflexes and rapid responses in the
brain. Now they must also consider slowly evolving changes in information processing in our nervous system. Perhaps glia may participate in
quite different aspects of the mind where speed is not the priority. Could
glia operate in our unconscious brain?

GLIA MOVE OUR UNCONSCIOUS URGES

With so much emphasis on our intellectual mental functions from the
time we are schoolchildren through our adult life, it is easy to overlook
the astonishing range of vital computation and regulation our brain carries out unconsciously. (The silly adage that we only use 10 percent of
our brain shows how easily we dismiss the essential complex unconscious labor carried out by the bulk of our brain.) The most sophisticated robot moves with a halty mechanical clumsiness that contrasts

comically with the fluid graceful movements of any healthy animal. Consider the thousands of microscopic muscle fibers and the touch and body-position sensors that must be analyzed and controlled in a precisely coordinated manner to execute such fluid movements, all of which we take for granted. This exquisite control requires extremely complex computation with exacting precision. All of this is carried out automatically and unconsciously in our brain. You use this unconscious motor control every time you turn a page of this book with your fingers.

Control of body movement is only the most visible example, but our unconscious brain is a marvel of effortless multitasking and precision control as it maintains our body within the narrow limits that sustain life. Even a single-degree deviation in body temperature is cause for concern. Five degrees' variation can bring coma and death. Like a massive automated computer system controlling a power plant, our brain constantly monitors and regulates body temperature, fluid levels, hunger, respiration, digestion, circulation, muscle tone, balance, reproduction, growth, cycles of alertness and sleep, and our gradual progression through childhood, puberty, and beyond. Every minute of our life depends on maintaining vital systems within extremely narrow limits, while at the same time modulating and coordinating them precisely according to the body's constantly changing needs.

The hypothalamus is the master control center in our brain regulating many automatic functions vital to life. The hypothalamus controls these systems on a minute-to-minute basis, as well as the daily, monthly, and lifelong cycles. In a new mother the hypothalamus also controls the birth of her baby and the mysterious synthesis of milk arriving on schedule to feed her new infant. All this is carried out by a part of our brain beyond our conscious awareness.

The anatomical structure of such an intricate central processor is remarkably banal, as reflected in its name, which means simply "below the thalamus." The thalamus is a massive switch box for sensory information streaming into the brain en route to the cerebral cortex. Looking like nothing more than a few small clusters of cells near the base of the brain—tiny air bubble imperfections in a slice of bread—the hypothalamus contains neurons that send their axons into the pituitary gland.

The role of the pituitary in secreting hormones into the bloodstream is well known. People with certain pituitary problems never grow beyond dwarf stature or they keep growing until they are transformed into giants, simply because this part of the brain fails to mete out human growth hormone into the bloodstream at the proper pace. Fewer people know that it is the hypothalamus that controls the pituitary, suspended just below it like a single cherry on a slender stem. This link between our neural and endocrine (hormone) system also ties emotions and sexual activity to our higher cognitive systems.

Glia Quenching Thirst

Despite extreme variation in the amount and frequency of fluid we drink, our brain controls the water content of our body within precise limits. More critical than food, water must be maintained constantly at proper levels in the body of any living thing. Lack of water will impair bodily and mental function within a matter of hours, and death from dehydration will follow within a few days—faster than from most diseases.

One of the ways our body copes with dehydration is to release into the bloodstream the antidiuretic hormone ADH. Secreted from hypothalamic neurons this polypeptide acts on our kidneys to reduce the output of urine and ration our vital dwindling stores of body water. Anatomists observed that glia at these synapses in the hypothalamus responded in a surprising way when animals became thirsty.

This recent exploration of glia at work in our unconscious brain highlights another revelation: glia can move. At this very moment, astrocytes are alive and probing their cellular fingers between the neurons in your brain. As their tentacles slither and shrink, ooze between neurons and retreat, they are changing the circuitry of your brain. Even before we realized that glia are doing this, something vital always seemed to be missing from our cellular concept of the brain: it was too static. Much like microcircuits pinned down by countless solder junctions on a circuit board, neurons are tethered into a fabric of synaptic connections in the brain. This immobilized static state makes neurons seem artificial and unnatural. In contrast, the cellular tendrils of outcast glia, free to wander and probe at will through the tangled knotted network of nerve fibers in our brain, make brain tissue come alive with cellular motion. As they probe, they physically remodel your brain, changing the connections between neurons. The other brain, operating entirely out-side our conscious mind, shapes the circuitry of the neuronal brain.

This cellular remodeling of synapses in the hypothalamus allows astrocytes to change the properties of these synapses in response to thirst. In withdrawing their glial fingers from around synapses, they unveil more of the neurons. The withdrawing astrocytes are also less able to take up the neurotransmitter as quickly as when they closely sur-round the synaptic cleft. This delayed clearance of neurotransmitter leads to buildup of neurotransmitter in the extracellular space and changes synaptic transmission. Using microelectrodes to study synaptic function in this region of the brain, neuroscientists Stéphane Ouellet and colleagues in Bordeaux, France, found that the synaptic voltage decreases as astrocytes reconfigure themselves around these synapses when an animal is deprived of water. They suggest that similar control of synaptic voltage by glial fingertips might be happening elsewhere in the brain.

The probing glial fingers moving into and out of synapses can regu-late synaptic transmission in another way: by releasing substances that act on these synapses. In the hypothalamus, astrocytes release several types of neuroactive substances, now called gliotransmitters, that stimu-late neurotransmitter receptors on neuronal synapses. In this way glia directly regulate synaptic transmission, simply by releasing a large vari-ety of neurotransmitters, including the amino acid taurine, ATP, and D-serine—the same substances that neurons use for synaptic transmis-sion. Each of these substances released by astrocytes has a different effect on synaptic transmission.

The next time you are thirsty consider that your survival depends on the delicately probing fingers of astrocytes at the synaptic controls in this spot of your unconscious brain.

Changing Your Mother's Mind

Experiments on cells in dishes and slices of rat brain are one thing, but implicating glia in human behavior is quite another challenge. Neuro-scientists have discovered that the most basic, innate behavior of hu-mans, the defining feature of all mammals, is controlled by glia. Tracing the pathway of brain circuits controlling breast-feeding from molecules to cells to behavior, scientists have discovered that when a woman becomes pregnant, glia surrounding synapses in the brain controlling lactation change the physical structure of this brain region, rewiring these synaptic circuits. From this clue, researchers are beginning to discover many other mental processes where glia are in control, including circuits regulating physical coordination or "muscle memory" for physical skills in a part of the brain called the cerebellum. These recent discoveries have opened a new vista on how glia, by changing shape and moving, can alter brain structure and therefore its function in association with unconscious mental processes. From these examples it is easy to imagine glia operating in the conscious mind.

Birth, Motherhood, Love, and Glia

Melanie had been in labor for more than twenty-four hours. As her husband tried to comfort her, she was becoming exhausted from the lack of sleep and prolonged effort. Her doctor decided it was time for medicine to assist Nature. He hooked a bottle of saline solution onto the stainless steel support rod and adjusted the drip rate of the IV line draining into her forearm. Within minutes her contractions began to intensify and surge in regular waves. A short time later their daughter, Morgan, entered the world.

The substance in that bottle of saline that produced these rapid effects was oxytocin. This is not an artificial substance, but rather a natural hormone synthesized in the hypothalamus of a woman's brain. As just described, the hypothalamus in both men and women controls the body's vital systems automatically and unconsciously. No woman "knows" how to give birth to her child; she simply experiences this miracle as her unconscious brain takes command.

Oxytocin is made by specialized neurons in the hypothalamus that release it into the bloodstream. These giant neurons, called magnocellular neurons because of their large size, send their axons from the hypothalamus into part of the pituitary, where they release their contents into the space around capillaries. The capillaries then absorb the oxytocin into the bloodstream and distribute it throughout the body. Oxytocin is a short polypeptide made of nine amino acids, and in the female body it has two special functions: stimulating milk release from the mammary glands and stimulating contraction of the uterus at birth. Both of these functions are the result of smooth muscle contraction responding to oxytocin in the blood.

There is another intriguing and more subtle function of this hormone. It regulates maternal behavior and attraction. Although evidence is less clear, oxytocin may also have a related behavioral function in males. In the cerebrospinal fluid, oxytocin plays the role of Cupid, uniting mother and child in the powerful maternal behavior of bonding to her offspring immediately after birth. In the biological context, this strong emotional attachment is required to insure that she will nourish and protect her vulnerable offspring. In experiments on rats, injections that neutralize oxytocin in the brain prevent mother rats from accepting their pups. On the other hand, injecting virgin rats with oxytocin triggers them to exhibit maternal behavior toward any pups placed in their cage. Oxytocin can enter the body through aerosols, a property that has been exploited for profit. Colognes can be purchased containing oxytocin with the aim of making members of the opposite sex bond more easily.

In virgin rats the neurons containing oxytocin are tightly packed together in clusters, but they are separated from one another by thin sheets of astrocytes, like paper between stacked china. For years, electron microscopists noticed that this part of the brain physically changed in pregnant animals. This was astonishing because it was one of the first instances where the mysterious workings of the brain could be seen reflected in a structural change in its wiring. Through studies over several years by Dr. Glenn Hatton and colleagues at the University of California, Riverside, Dionysia Theodosis and Dominique Poulain of Bordeaux, France, and other groups in the United States and Europe, we now know that in an animal giving birth or breastfeeding, the astrocytes between these neurons actually move and change the structure of this part of the brain. In pregnancy, the thin veil of astrocytic membrane withdraws, exposing more of the oxytocin-producing neurons and their dendrites. This in turn increases the number of empty sites on each neuron available for new synapses to form. This action doubles the number of synapses that form on an oxytocin-containing neuron after the astrocytes withdraw. With more synapses stimulating the neuron, more oxytocin is released, preparing the new mother for birth.

This is not the only way astrocyte movement rewires this part of the brain. In addition to regulating synaptic input to neurons, astrocytes also control the delivery of oxytocin from the axon tips of these neurons, where the peptide is dumped into the bloodstream. During birth and lactation the astrocytes also withdraw from the nerve terminals, acting like floodgates opening to allow more oxytocin to inundate the capillaries and enter the bloodstream. The only thing separating the nerve terminals from the capillaries is these astrocytes.

The next time you see a child feeding in the arms of her mother, you will be seeing glia at work, controlling the synapses of neurons and the output of oxytocin into the bloodstream. The birth of our offspring and their nourishment depend on these glia.

GLIA AND SLEEP: THE OTHER LIFE OF THE OTHER BRAIN

Midway between our unconscious and conscious minds there is the altered mental state of sleep. If you should live to the age of seventy-five, you will have spent perhaps twenty-five of those years asleep. What goes on in your head during that block of your lifetime is largely beyond your knowledge or comprehension. It is a mysterious and still mystical chunk of ourselves. If sleep were simply a nightly hibernation, a shutting down of our system in the dark, it could be understood as a reasonable strategy to save power for the daytime when we can be physically active. Sleep might be much like a laptop computer going into temporary hibernation to save resources during long periods of inactivity. But hibernation is hardly what goes on in the human brain during sleep. (Or in the animal brain for that matter.) Sleep is a vigorous period of brain activity. It is an altered state, not an inert state. Sleep is an active mental process in which some circuits of the brain paralyze the body to allow our mind to cavort in a wild nighttime fantasy. This paralysis prevents us from leaping out of bed to run from the pursuer in our dreams or chase whatever fantasy we may be living in our dreaming minds.

There are cycles and patterns of activity during our nocturnal unconscious life shuttling enormous amounts of activity through different brain circuits. Events of the day—conscious and unconscious—are reexamined, sorted, associated, puzzled, filed, or discarded. Memories are moved from one place in the brain and filed in different places in our cerebral cortex according to such factors as the type of information they contain, their connections to other events, and the internal emotional states of mind stamping them with significance. This altered state of mind, perhaps one-third of your existence, is still puzzling to science and difficult to study. What happens to glia when we sleep? A more intriguing question is, could glia participate in controlling this state of mind we call sleep?

One insight has come from using gene chips (a new method that allows scientists to monitor the activity of thousands of genes at once) to detect changes in genes in brain tissue being switched on and off during different phases of sleep. This work shows that hundreds of genes in the brain are synthesized at distinct phases of REM or non-REM sleep. (REM sleep, for "rapid eye movement," is the dreaming phase of sleep.) A recent surprise is that many of them are genes found only in glia. In fact, some of the most highly regulated genes in the brain during REM sleep are genes in oligodendrocytes making myelin. No one knows why. But it is convincing evidence that glia do not simply go to sleep when we do. They are busy at something we do not comprehend.

Cycles of Brain Activity

The surges of mental activity swirling through the human cortex during sleep are readily apparent with EEG recordings. So large are these electrical fields inside our heads that they can be picked up and amplified by wires placed against our scalp.

Does the executive cerebral cortex drive and manage these cycles of activity in sleep, or is the cortex simply responding to cycles of activity that originate deep in the more "primitive" areas of the brain? The main portal of information into and out of our cerebral cortex is the thalamus. This nugget of neurons looks like the pit of a peach, with the flesh of the peach representing the cerebral cortex. Vincenzo Crunelli and colleagues at Cardiff University in the United Kingdom published results in 2002 examining this question. In their experiments they placed sets of electrodes simultaneously inside the thalamus and in the cortex of experimental animals to see which area leads and which follows. Their conclusion was that the thalamus, not the cortex, drives the cycles of activity during sleep as well as the various states of arousal while we are awake. Their data detected slow one cycle per second oscillations of neural impulse firing during sleep that originated first in the neurons of the thalamus. But what drives these oscillations in the first place?

To drive the cortex in cycles, large groups of thalamic neurons must work together cooperatively in a coordinated manner, just as fans at a baseball game must stand and raise their arms together in synchrony to generate a wave sweeping through the stands. The study found that groups of thalamic neurons are coupled together directly by protein junctions (gap junctions) that form tiny pores in the cell membrane connecting adjacent cells. This allows electricity to spread passively and rapidly between neurons without synaptic interconnections. This in turn yokes together groups of thalamic neurons, making them operate jointly in phases of activity. The electrical voltage of one thalamic neuron spreads instantaneously to the many other neurons that are physically coupled to it, and this causes large groups of neurons to work together and fire in unison and cycles, which in turn drives waves of activity in the cerebral cortex.

Something else operating outside the neuron doctrine of synaptic interactions also couples assemblages of these neurons together into ensembles—astrocytes. Crunelli and his colleagues took slices of the thalamus and soaked them in a calcium-sensitive fluorescent dye that is absorbed selectively by astrocytes. The researchers watched as, without any outside stimulus, waves of calcium swept through networks of astrocytes in cycles inside the thalamus. When they positioned electrodes in the neurons of the thalamus and recorded the voltage changes inside, they saw the voltage in the neuron change just as the wave of calcium swept through an adjacent astrocyte. Astrocytes were coordinating the cycles of neuronal activity generating brain waves during sleep.

This electrical response in the neurons was caused by the neurotransmitter glutamate being released from the astrocyte when the calcium wave passed through it. The glutamate activated glutamate receptors on the neurons, and this activation in turn triggered a voltage response that stimulated the neurons to fire impulses.

The startling conclusion: not only is the cerebral cortex not in control of these global cycles of brain activity during sleep—neurons are not exclusively in control either. The waves of activity flowing through astrocytes couple large groups of thalamic neurons together, coordinating neuronal activity like fans in a sports stadium. Just as we see global changes in brain waves during epileptic seizure and disease, waves of calcium activity in astrocytes surge in parallel with electrical activity in neurons. Astrocytes communicate among themselves without electricity, but instead signal one another by broadcasting chemical messages, controlling neuron firing by releasing the same neurotransmitter that neurons use to communicate with each other through synapses. The other brain is at work controlling sleep every night as we rest our head on our pillow.

Recent research on mutant mice that have disrupted sleep cycles has awakened many to the realization that the other brain is not simply shadowing the neuronal brain or slavishly tending to its needs; in fact, the other brain can control the neuronal brain. Much like a conductor coordinating ensembles of musicians in an orchestra to make music, astrocytes coordinate ensembles of neurons into coherent assemblages to make neurons fire together in rhythm.

The mutation in these mice that have disrupted sleep cycles is not to be found in their neurons; it is a genetic defect inserted by genetic engineering directly into astrocytes, preventing them from releasing the neurotransmitter adenosine. This mutation in astrocytes disrupts their ability to communicate with one another and with neurons. You may not be familiar with adenosine, a neurotransmitter that regulates excitability in neurons, unless you drink coffee. Caffeine blocks the neurotransmitter receptors for adenosine that normally calm activity in the brain and induce sleepiness. When caffeine blocks these calming adenosine receptors on neurons, neural circuits raise their level of excitability, arousing heightened electrical activity and alertness in the brain. Astrocytes have the same adenosine receptors, and they also release adenosine to stimulate the receptors on other astrocytes and on neurons. In this way, astrocytes can control the level of mental alertness (or sleepiness) in exactly the same way as a strong cup of java.

Neurobiologist Robert Jackson of Tufts University and his colleagues have extended this research to the lowly fruit fly, which is one of the favored organisms for genetic research. By comparing genes that are active in flies during their daily wakeful and sleep periods, researchers discovered several genes that had a clock-like cycle of activity on a twenty-four-hour rhythm. One of these clock-like genes, called *ebony,* became active during the day and inactive at night. When this gene was mutated, the flies suffered a seriously disrupted sleep/wake cycle, nodding off randomly any time of day or night. The protein made by this gene was found in the right place—inside the fly's brain—but in the "wrong" cells. That is, *ebony* was found exclusively in glia. The researchers concluded that glia in the fly's brain act as clocks, regulating nearby neurons that make the neurotransmitters dopamine and serotonin, which control the cycles of sleep and activity. These same neurotransmitters also regulate sleep in people; interestingly, they are also associated with mood, depression, and schizophrenia.

Coupling groups of neurons that are not wired together through synapses and driving global waves of neuronal activity in the sleep/wake cycles seem to be among the most important functions of glia in the brain. But why is the brain divided into three-dimensional sectors ruled by individual astrocytes? Many believe that this added dimension of regulation by astrocytes boosts the complexity of brain function beyond what could be achieved simply through joining individual neurons together in series through their synapses.

SEXUAL BEHAVIOR: GLIA AND GENDER

It seems doubtful that sleep is the only behavior that is regulated by glia, but it is difficult to conduct behavioral experiments on humans with respect to glia. For example, the cellular and molecular mechanisms of mate selection and sexual behavior are more readily investigated in experimental animals. Turning again to the fruit fly because its simple nervous system and easy genetic manipulation make it a useful experimental organism, Dr. David Featherstone and colleagues of the University of Chicago reported in 2008 a link between aberrant sexual behavior in flies and a gene mutation in glia. Male flies with a mutation in a gene called *genderblind* engage in normal heterosexual courtship and copulation with females, but these mutants were equally attracted to males, pursuing and attempting to copulate with them as frequently as with females.

The protein made by the *genderblind* gene is present only in glia that ensheathe synapses, where its function is to remove the neurotransmitter glutamate from the synaptic cleft. Courtship and mating in flies is

regulated by pheromones (as it is in part for humans). These are chemical signals exchanged between organisms that control specific behaviors.
The courtship process begins when a male fly taps a female with his
foreleg. If she is sexually receptive, this tender touching will trigger an
elaborate precopulatory song and dance in which the male extends and
beats one wing as he dances close behind the female. If she is sufficiently
impressed by her suitor, this fly foreplay progresses to the next stage,
where the male licks the female's genitalia to sample her pheromones,
after which she either strikes him violently with her wings and legs to
reject him or she allows him to mount her and copulate. A successful
consummation turns the cycle of life through another revolution with
a new generation of maggots.

Male flies will mate with flies that do not produce the male pheromone (normally these would be females), but males that cannot sense
the masculine pheromone will mate with males or females indiscriminately. In the *genderblind* mutants, the neurotransmitter glutamate accumulates to excessive levels in the synapse, disrupting the circuitry that
discriminates whether a pheromone is attractive or repulsive. By altering the activity of this glial gene, the researchers were able to change
the mate choice of male flies from females exclusively to males and females indiscriminately.

Surprisingly, the sexual preference associated with the male pheromone is not hardwired; it is learned in young flies. Inexperienced males
who initially attempt to mate with other males and suffer rejection eventually learn that this pheromone is a sexual turnoff. Interestingly, this
suggests to Featherstone and colleagues that glia are controlling learning
by regulating levels of the neurotransmitter glutamate at synapses in the
appropriate circuits.

BEYOND THE UNCONSCIOUS
TO THE CONSCIOUS BRAIN

Could the types of other brain functions at work in our unconscious
and semiconscious mind also participate in our conscious mind? Could
the other brain be involved in learning, thought, or memory? The other

brain coordinates groups of neurons, regulates the excitability of neural networks, enhances or inhibits synaptic strength by releasing or absorbing neurotransmitters, and probes its cellular fingers through brain tissue to strip away synapses or clear away space for new ones to form. These are all processes central to changing the function of neural networks, but previously it was assumed that the neuronal brain did it all. But consider that if learning involves rewiring brain circuits, how will neurons make and break their own connections? In contrast to neurons, which are tightly knitted into the fabric of neural circuits through countless synaptic connections, glia are free to move, as we have seen in the hypothalamus. In any significant remodeling of our brain during development or learning, glia now seem to be the cells most likely to act in cerebral remodeling. In this sense, neurons might be thought of as the architectural substrate of glia. A mechanism of synaptic remodeling by glia makes more sense to many brain scientists enlightened about the other brain than placing all the responsibility on neurons. The ability of glia to move and secrete substances that remodel tissue or stimulate neuron growth during development and repair of the brain makes them ideally suited to rewire the neural circuits during learning in the healthy brain.

Microglia are equipped with powerful protein-dissolving enzymes to melt extracellular matrix proteins that stitch neurons together, so that microglia can race between tightly packed brain cells to the site of infection and kill invading organisms. Recently microglia have been seen using these infection-fighting tools to strip synapses off neurons to rewire circuits after injury and during early life, when connections from our eyes, guided by visual experience, are wired to the proper places in the brain. These tiny glial cells were dismissed by most neuroscientists, except those interested in disease, because they were regarded as nothing more than immune cells in our brain there to track down and devour germs. Now we see that in addition to this extremely important function, microglia help us rewire our neural circuits to enable us to learn.

Is there any reason to presume that brain remodeling by glia, so well documented and tied to human behavior in the hypothalamus, is limited to that particular spot in the brain or to the particular behaviors we have discussed: birth, lactation, and water balance? A more reasonable conclusion is that glia remodel synapses everywhere in our brain, but the ways glia control diverse mental functions are more subtle than our crude methods of observation can capture at the present time.

Physical remodeling of synapses is beginning to be discovered in other regions of the brain involved in learning. The cerebellum, situated at the back of the brain, is vital for controlling physical movement of the body and in learning physical skills, such as perfecting a golf swing. Neurons in the cerebellum are tightly ensheathed by astrocytes called Bergmann glia, and the finger-like tendrils of these astrocytes also move. Scientists are beginning to discover the signals transmitted between neurons and glia that control these glial movements at synapses in the cerebellum.

When neuroscientists Masae Iino and colleagues at the Gunma Medical School in Japan inserted defective glutamate neurotransmitter receptors into the Bergmann glia, they saw the astrocytes withdraw from around neurons like a receding tide. This low tide exposed more neuronal shoreline, and new synapses in this part of the brain controlling movement were soon established on the newly exposed areas. Just as in the hypothalamus, the withdrawal of Bergmann glia not only increased the wiring of new synapses onto the neuron, it amplified the physiological strength of each synapse by allowing more glutamate to build up and spill out of the synaptic cleft as the synaptic glia withdrew. Logically, then, in astrocytes with normally functioning glutamate receptors, the glutamate released from synapses must beckon the astrocytes to fill in tightly around the synapse. In so doing, astrocytes regulate the number of synapses and their strength according to the level of synaptic activity and how much glutamate is released.

In 1961 anatomists John Green and David Maxwell of the University of California, Los Angeles, described similar physical changes in astrocytes in the hippocampus, a brain region known to be critical for memory. The possibility that glia might modulate synaptic strength and remodel connections in this part of the brain, controlling human memory, is inescapable. Many laboratories are beginning to test this hypothesis using advanced imaging in combination with electrophysiological tools.

It seems odd that the earliest and strongest evidence for glial involvement in information processing should have arisen first in association with so many unconscious brain functions—thirst, birth, breast-feeding, sleep, motor control, and sexual behavior. The unconscious processes in the brain are far more mysterious and inaccessible to research than conscious mental activity. Is this curiosity a coincidence or does it illuminate a more general property of the other brain? I believe it is the latter. Glia are not equipped with the rapid-fire means of communication that neurons use. Glia communicate through the slow spread of chemicals and calcium waves, not by jolts of electricity. Yet the unconscious and slowly evolving changes in our mind are an important and easily overlooked part of brain function. Perhaps our ignorance of the other brain is one reason the unconscious mind has remained so mysterious.

CHAPTER 14

Memory and Brain Power beyond Neurons

GLUING MEMORIES IN OUR MENTAL SCRAPBOOK

The most intriguing mental function for many neuroscientists is the mechanism of memory. A crucial part of the brain for memory is the hippocampus, and more is known about synaptic transmission and synaptic plasticity in the hippocampus than in any other part of the brain. For those interested in understanding the cellular circuits storing memories—and the possibility that glia have a role—the hippocampus was a natural place to look first.

The white lab rat is put to sleep with the same barbiturates a vet uses to painlessly end a pet's suffering. I draw my scalpel blade down the crest of its skull from front to back. Working quickly I snip through the skull bone, crunching from the occipital foramen, where the spinal column joins the skull to the forehead. Grasping another surgical instrument—stainless steel pliers with a sharpened bird beak designed to chip through bone—I snip bone and then crack the skull into two halves, revealing the brain exposed like a fleshy pink walnut on the half shell. With a Teflon-coated mini spatula I lift the brain out of its skull, severing the cranial nerves that exit through small holes in the skull to energize the muscles of the face. The brain drops with a plop into an ice-cold beaker of artificial spinal fluid chilled in a bucket of chipped ice. An hour before, I weighed out a long list of chemicals and dissolved them in ultra-pure water to formulate the artificial cerebrospinal fluid. The beaker hisses with a stream of fine bubbles from an aquarium store air stone infusing a mixture of 95 percent oxygen and 5 percent carbon dioxide into the fluid. It is important to work quickly to get the brain out and into a cold state of suspended animation before it dies. Often while I am doing this I think of Civil War surgeons, who excelled in proportion to their surgical speed.

Lifting the brain out of the now pink-colored ice-cold solution with a plastic spoon (metal would shock the brain like aluminum foil on a tooth filling) I blot it on a piece of filter paper and then slice the brain in half with a sharp scalpel into its two bilaterally symmetrical lobes. The dissection becomes more delicate as I probe through the folds of cortex to find and surgically remove the hippocampus, the tiny seed of brain tissue resembling a miniature banana. In a rat's brain it is only about one-quarter of an inch long. Using a razor blade I trim off two ends of the banana and pick up the central third on the tip of a mini spatula. Tipping the bit of tissue on end I can see through my dissecting microscope a swirl of white cream within the pink, looking much like a cinnamon roll.

It is this curl of tissue that gives this bit of brain its name. It resembles the coiled tail of a seahorse, prompting anatomists to bestow the Latin name for that fish—*Hippocampus*—on this now-famous bit of brain. No part of the brain has achieved greater fame since science learned in the 1950s that the hippocampus is the key to memory.

A patient known in the medical literature only by his initials HM had the hippocampus on both sides of his brain removed to control the severe epilepsy he developed after a fall from a bicycle as a nine-year-old boy. The treatment cured his seizures, but in the process surgeons inadvertently severed the mysterious connection between present and past. HM lost the ability to convert short-term memories into enduring memories. The surgery did not affect his intelligence or personality, revealing that the mysterious mechanism of memory is something apart from intellect. HM could learn new things perfectly well, but within a few minutes of entering his conscious mind, new experiences simply vanished. Memories already stored in his brain before the surgery were perfectly intact, but he could no longer make new memories.

In December 2008, the world finally learned HM's name when Henry Gustav Molaison died at the age of eighty-two in Windsor Locks, Connecticut. Imagine what it must be like to live without the possibility of making new memories. Henry Molaison could recall in vivid detail the streets of his home town and the layout of his boyhood home, but the floor plan of his current home remained a mystery to him. He could reread the same article in a newspaper repeatedly and experience each reading as fresh as the first. Although he met his doctors on a regular basis, he had to be introduced each time as if it were their first meeting. This banana-shaped bit of brain is the vital mental link between our present everyday experience and our personal past. HM, as a result of losing both hippocampi (from his left and right brain), was condemned to live forever in the present.

I squeeze a single drop of cyanoacrylate glue (a popular brand name is Krazy Glue) onto the bottom of a small black shallow pan, about the size of a woman's makeup compact. Then I carefully transfer the mushy bit of brain tissue to the drop of glue, where it instantly sticks. This is always the most critical step, for the blob of brain must be oriented precisely with the "cinnamon curl" facing up the instant it touches the glue. Should the tissue contact the glue drop at any other angle or fail to stick, I must begin again with another rat.

The black tray is actually the stage of a motorized precise slicing machine, which I will use to cut paper-thin slices off the cerebral cinnamon roll, each one less than 0.5 millimeters thick. The tray rests on a bed of ice, and the tissue is bathed in the ice-cold artificial spinal fluid. I insert a single fresh razor blade into the precision vibrating arm and advance it with the flick of a switch to slice through the tissue. The arm advances slowly, vibrating back and forth in a blur, buzzing loudly like a noisy hair clipper. It is possible to slice off about twelve slices, each of which is collected by sucking it up into a plastic eye dropper and squirting it out gently onto the warm moist stage of a recording chamber, where I arrange the delicate slices of brain using an artist's fine paintbrush.

The recording chamber sits prominently at the center of a stainless steel table floating on air pistons isolating it from minute vibrations that exist in all buildings. The card-table-sized air table sits in a room the size of a walk-in closet crammed to the ceiling with electronic instruments: amplifiers, computer screens, and oscilloscopes. It resembles the cockpit of a spacecraft, packed with a maze of buttons, blinking lights, and dials. The scene is accompanied by the sound of bubbling solutions and periodic beeps from various electronic monitoring devices. These amplifiers and electronic stimulators magnify the minute electric signals sent between neurons in the slice of brain.

Atop the air table sits a three-foot-by-three-foot cube constructed of shiny copper screen, which is open at the front. This is a Faraday cage, which is attached to a thick copper ground wire independent from all other electrical ground circuits in the building. Put your cell phone inside and it would fail to pick up any signal at all, for this cage screens out all electromagnetic radiation that fills our environment. Electromagnetic radiation is constantly emitted from power lines running through the walls of buildings and signals broadcast by radio and TV. Countless electrical devices in our modern world, from refrigerator motors and compressors to buzzing fluorescent lights, contribute to the electrical static. Connected through amplifiers to a speaker, this electrical noise creates the radio static you hear when you search between AM stations. These emissions are thousands of times stronger than the feeble bioelectric signals we wish to tap into to hear neurons communicating with one another in this part of the brain that is forging memory.

The room lights are dimmed to limit the glare on computer screens and eliminate the electrical buzz of the fluorescent tubes. The recording chamber sits prominently center stage on the air table, a crystalline cylinder of Plexiglas the size of a coffee can, shining bright white in the brilliant spotlight of cold fiberoptic beams. Inside the transparent chamber fine bubbles boil in solution warmed to body temperature to revive the neurons from the cold that protected them during the slicing. Solution drips from an IV line into the chamber and flows over the brain slices to keep them warm, moist, and oxygenated. Looking through a microscope suspended on a boom, I carefully position extremely sharp glass electrodes into the region of the brain slice where I know axons meet dendrites. I place another fine-tipped metal electrode in the proper position to deliver an electric shock to spark a volley of impulses shooting down the axons to release neurotransmitter from the synapse. All of this is done by twisting the dials of remote-controlled micromanipula-tors that move the electrodes precisely in three dimensions. The micro-scopic distances traversed are well beyond the limit of the unsteady human hand to negotiate with precision.

The strength of the synaptic signal is displayed continuously on the computer screen like a heart monitor. Every time I flip the switch to shock the axons a small blip ripples on the flat line monitoring the volt-age output from the synapse. The taller the ripple, the stronger the syn-aptic connection, for the display is simply a running graph of voltage generated by the synapse over time.

This bit of brain is so revered by electrophysiologists because using the instruments just described, they can see a memory form on the screen of their oscilloscope. In 1973 in Oslo, Norway, Tim Bliss and Terje Lømo first observed that if instead of giving a single shock they delivered a brief burst of shocks, the synaptic blip suddenly grew large—almost doubling in height. The synapse was now much stronger, pro-ducing nearly twice as much voltage as before. Afterward, it maintained its taller stature even in response to single test shocks applied hours later. Immediately they realized that with their burst of impulses they had strengthened the synaptic connection—the essence of learning.

This phenomenon, called long-term potentiation, or LTP, is now understood to be the cellular basis for learning and memory. Memories are connections between neurons, and these strengthened connections could now be seen by electrophysiologists in a slice of rat brain. Vigor-ous research over the last thirty years around the world has determined in fine detail the mechanisms for LTP down to the molecular level. If we can understand the molecules that store memories, we should be able to make drugs to improve memory or help those with weakened memory due to age, disease, or defects of birth.

The brief shocks mimic natural input to the hippocampus that the rat would experience automatically in processing sensory information as it explores its environment. Synapses are strengthened in LTP by cellular changes that increase the amount of neurotransmitter released from the synaptic terminal, or by increasing sensitivity of the receiving dendrite. Together these changes produce a greater voltage at the syn-apse and a strengthened connection between two neurons.

Neurophysiologists who worked on hippocampal LTP in the 1980s and 1990s were the elite among memory researchers, for they had reduced memory to its elemental form: the strengthening of a synaptic connection between neurons. With their fine glass electrodes they had pinned down the elusive mystery of memory in a dish, and they succeeded in isolating it in a single synaptic connection from the vast complexity of the brain. Memory was theirs to control and monitor at a cellular level with precision electronic devices. Previously the science of learning and memory was the domain of psychologists, who probed the phenomenon of human memory with cleverly designed memory tests and by running rats through mazes. Strategies involved testing animals under the influence of drugs designed to numb certain brain pathways or after experimental brain surgeries in order to find where the storehouse of memory was hidden in the brain and how it might work at a cellular level. Now the electrophysiologists had overtaken psychologists in memory research. Even minute advances in LTP research were likely to find their way into high-profile scientific journals. Rapidly, the mechanisms for this cellular learning began to be revealed in fine detail.

But neurophysiologists were reluctant to accept that anything beyond electrical events at the synaptic membrane could be relevant. They were wrong. Research soon showed that even the nucleus of the neuron is critically involved in the mechanism of memory. Genes coded in DNA must be read out to make new proteins to cement short-term memories into long-term memories, for example. Without this process, memories would quickly fade away like the forgotten name of a person you met moments ago. This molecular biology was completely outside the realm of electrodes to probe, but at least it was still a process residing inside a nerve cell. The idea that hippocampal LTP or memory could have anything to do with something other than neurons was inconceivable. This neurocentric view is collapsing now as glia are becoming implicated in information processing in this most hallowed of all brain regions for neurophysiologists.

In 1992 Stephen Smith and graduate students John Dani and Alex Chernjavsky reported experiments showing that astrocytes in slices of rat hippocampus light up with calcium when nerve axons fire in this part of the brain so essential for memory. This meant that astrocytes "know" about electrical signals transmitted through neural circuits coding memory. Ken McCarthy and J. T. Porter of UNC Chapel Hill revealed exactly how astrocytes were monitoring synaptic transmissions. As they suspected, astrocytes in the hippocampus were responding to the neurotransmitter glutamate released from the hippocampal synapses. The researchers proved this by adding drugs that block neurotransmitter receptors for glutamate. They found that this action blocked the calcium response in astrocytes. As of the mid-1990s there was no longer any doubt that glia were listening to synapses firing by picking up the neurotransmitter signals. But why?

Most researchers discounted the speculation that astrocytes might be participating with neurons in information processing, and especially in memory formation. The prevailing view was that the rise in calcium concentration in astrocytes simply reflected housekeeping functions of these glial cells in tending to the needs of neurons and their synapses. Even today this remains the view of many scientists in the field. But other researchers were not so willing to discard the possibility of glia participating in information processing and memory formation.

GLIA IN INFORMATION PROCESSING: NEURON-GLIA YIN YANG

By the late 1990s research on the other brain had advanced from the neuromuscular junction to the retina to the memory center of the brain. This research showed that glia were listening, but there was still little evidence that the calcium responses in glia had any consequence for neural function. As discussed in part 1 of this book, researchers studying synapses outside the brain had succeeded in showing that glia at the neuromuscular junction and in the retina were not only eavesdropping on synapses, they were regulating them. But no such evidence had yet been found for this effect anywhere inside the brain, and certainly not in the brain's memory center.

In 1998, Maiken Nedergaard at New York Medical College placed an electrode on an astrocyte in a slice of hippocampus from a rat brain and stimulated it with a small voltage. The instant she did, she and her colleagues saw the synapses in hippocampal neural circuits respond by shrinking in voltage. A new frontier had been crossed. The "support cells" had taken control of neural circuits in the heart of the brain where human memory was formed. Astrocytes had a firm grip on synapses holding memories.

CROSSING THE LINE

Other evidence suggesting that astrocytes participate in learning and memory came in 2002, when Nobufumi Kawai and colleagues in Japan reported that after brain stroke, hippocampal LTP was impaired in genetically modified mice that had the astrocyte gene coding for GFAP removed. This mutation only affected a fibrous protein in astrocytes; neurons were unaffected. Yet memory was impaired. That same year, Hiroshi Nishiyama and colleagues working in Japan found that LTP was enhanced in mice with another glial gene (called *S100*) knocked out. Moreover, these mice became smarter! Mice with the glial gene removed by genetic engineering learned to run mazes faster than normal mice. These findings were completely independent lines of evidence undermining the preconception that brain function and indeed memory were the exclusive domain of neurons.

Exactly how, at a molecular level, are astrocytes intervening in synaptic communication and helping record memories? In 1994 Philip Haydon and colleagues found that stimulating a calcium increase in an astrocyte in culture caused it to release the neurotransmitter glutamate. This glial-derived neurotransmitter in turn stimulated glutamate receptors on neurons and boosted synaptic transmission.

Research by Haydon, his colleague Vladimir Parpura, and others would eventually show that astrocytes, just like neurons, have synaptic vesicles releasing neurotransmitter. Crossing this line of segregation between neurons and glia was controversial, and it required years of experiments and debate to settle the issue. But because astrocytes communicate by broadcasting their signals widely, there was no need for Nature to focus all the synaptic vesicles into a point of contact the way neurons do in transmitting information along a chain of synapses. This is why anatomists missed synaptic vesicles in astrocytes. The vesicles were there to be seen inside astrocytes in their electron micrographs, but because they were scattered throughout the cell and did not look the way they do in neurons, they were overlooked. Now we can see that astrocytes have synaptic vesicles sprinkled throughout their cell body, and they can release neurotransmitters of various kinds from anywhere on their cell membrane. Neurons are far more restricted in their communication. If neurons are telephones connected through hardwired lines of communication, astrocytes are cell phones, broadcasting their signals widely. Moreover, astrocytes can release neurotransmitters that excite neurons and neurotransmitters that inhibit neurons. They use these neurotransmitters to communicate with one another and with neurons.

There are many types of neurotransmitters used by different neurons, but astrocytes exploit an even more universal molecule for communication: adenosine triphosphate, or ATP. This is the same molecule every athlete and biology student knows as the energy power pack for all cells, but outside cells ATP is rare. This scarcity makes ATP an excellent molecule for sending signals between cells—like a flashlight on a dark night—and the fact that no cell is without ATP makes it a universal messenger. Neurons and glia release ATP from synaptic vesicles, and this release is detected by membrane receptors on nearby cells. These ATP receptors then cause a flood of calcium inside the cell, and this in turn causes the astrocyte to release more synaptic vesicles, broadcasting more neurotransmitter and ATP into the environment, spreading the signal in chain-reaction fashion throughout the population of astrocytes. ATP and glutamate are the primary signaling molecules that cause waves of calcium to sweep through a dish of astrocytes or pass through astrocytes inside the brain. But as scientists study astrocyte communication, it is becoming clear that these are by no means the only substances used in calcium signaling between astrocytes.

We now understand that astrocytes do not fire electrical impulses because they have no need to transmit information rapidly over long distances the way neurons do. So astrocytes forgo this part of neuronal communication, but they fully exploit and participate in the second and more interesting part of neuronal communication, namely, communication by neurotransmitters.

Astrocytes cover a relatively large territory of brain, and their influence over synapses can be extensive. Yet by the beginning of the twenty-first century, researchers still hadn't tested the possibility that information could be picked up at one synapse by an astrocyte, flow through a glial circuit in the other brain, and then cause the release of neurotransmitter from an astrocyte to regulate neuronal communication at a distant synapse not directly connected in the neural circuit. In 2005, Philip Haydon's group proved this hypothesis. Haydon's studies confirmed that astrocytes responded to synaptic activity in the hippocampus with a rise in calcium, but this in turn regulated the strength not only of the neighboring synapse that had fired, but also the strength of other distant synapses on that neuron. Astrocytes spanning large distances through the brain and throttling synaptic transmission outside the neuronal circuit were the controlling mechanism. Information in the neuronal brain had been picked up by the other brain and used to control synaptic function at a different place in the neuronal brain. What might this meeting of our two internal minds accomplish that a neuronal brain on its own could not?

This phenomenon of regulating the strength of remote synapses—known as heterosynaptic depression—is much like what we have all experienced in carrying on a conversation in a noisy restaurant. We not only attend more closely to the conversation of our dinner guest, we simultaneously block out the kitchen noise and conversations going on between other diners nearby. This mental focusing is essential for sorting out the important signals in our environment from all the clutter. The same thing occurs in our hippocampus. Synapses from inputs carrying new information you want to learn are strengthened via long-term potentiation, but you also suppress inputs from other synapses carrying distracting information to that same neuron. You will not remember this suppressed background information. This sharpened focus of attention in learning in the hippocampus was known to depend on a neurotransmitter called adenosine, but it had always been assumed that the adenosine was released from synapses of other neurons in a noise-suppressing circuit. The assumption was not entirely correct.

Remember that from Mark Ellisman's studies we know that astrocytes spread well beyond their star-shaped cytoskeleton and they tile the hippocampus into sectors (see chapter 13). The domain of one astro-cyte can cover one hundred thousand synapses. Haydon's group found that the calcium rise in an astrocyte triggered by a synapse firing and a release of glutamate in turn caused the astrocyte to release adenosine triphosphate, or ATP. ATP molecules form by attaching three phosphate molecules to a core molecule of adenosine. After being released, ATP can quickly lose its three phosphates and leave the adenosine core, which is an inhibitory neurotransmitter (as discussed in chapter 13 in regulating sleep). As adenosine spreads from the astrocyte stimulated by the synapse, it decreases the strength of other remote synapses on the same neuron. The other half of the brain is in control of this mental fo-cusing in the hippocampus. Almost simultaneously, Richard Robitaille's group showed much the same thing by injecting a drug into hippocam-pal astrocytes to prevent the calcium rise and subsequent release of ATP. With the astrocyte communication blocked, the remote synapses were not depressed.

It comes as a shock to many to learn that astrocytes are the cells depressing these surrounding synapses and thus sharpening a particular input to our memory center. What if astrocytes should fail to perform their critical synaptic focusing? How might this affect learning, atten-tion, or even psychiatric conditions? Equally startling is the recognition that astrocytes regulate synapse strength by communicating through their own glial network. This network operates outside the neuronal network, unconstrained by the hardwired lines of connections strung between neurons. We know next to nothing about these cell-phone-like astrocytic networks. What are their boundaries? Are they modifiable—in other words, do astrocytes change from mental experience, do they learn? New research is beginning to provide evidence that astrocytes do indeed change the strength of their connections in learning.

Loose Ends

One thing is clear: glia are movers in the mind. But what about Schwann cells? The cells of Theodore Schwann in our nerves and the oligodendro-cytes in our brain, which Ramón y Cajal left to his student Río-Hortega, have nothing to do with synapses. They simply cling to nerve axons and coat them with myelin insulation. Yet in my lab we had discovered these cells tapping into information flowing through nerve fibers. Looking to astrocytes for understanding we find important clues, but no answers. Schwann cells are entirely different. We sense that we are missing something important about the inner workings of the brain; something that lies far beyond the synapse.

CHAPTER 15

Thinking beyond Synapses

SMART GLIA: MYELIN AND LEARNING

"Joey's IQ is perfectly normal—109." The anxious parents receiving this news are not sitting in a school office receiving a multiple-choice test score from Joey's counselor, they are in their pediatrician's office looking at a picture of Joey's brain scan. As their doctor points to the thick white cables connecting the left and right sides of their son's cerebral cortex, they can see for themselves the strength of their son's intelligence as easily as an orthopedist sees the strength of a leg bone on an X-ray.

This science fiction scenario became fact with the publication of a study by a team of researchers led by Dr. Vincent Schmithorst at the Cincinnati Children's Hospital in 2005. Even more amazing than the ability to capture intelligence in a picture is the fact that the parents are not looking at a snapshot of the area of brain containing neurons, the so called grey matter that your teachers always ribbed you about. This is a picture of the white matter tracts in the brain: the major trunks of axons bundled together like telephone cables buried deep below the cerebral cortex. There are no nerve cells in white matter tracts, no dendrites, no synapses—only glia nestled among axons.

For centuries man has sought ways to gauge intellect from the dimensions and shape of the human brain or from the bumps it embossed on the skull. None of this "phrenology" (from the Greek meaning "mind study") has any validity in revealing mental ability or personal character. But a picture of the brain in a region where glia reign exclusively can tell a doctor exactly how smart you are. How can that be?

The brain imaging technique revealing IQ is a special application of magnetic resonance imaging. Regular MRI scans are used by doctors to reveal the interior details of the brain in thin optical sections for diagnosis; the special type of brain scan, called diffusion tensor imaging (DTI), reveals the finest structure of axon cables in white matter, enabling scientists to untangle the bundles of axons in their jumbled pathways tunneling through living brain tissue. This X-ray vision is accomplished by computer-assisted imaging that is sensitive to how easily water molecules in the brain seep in all directions. Strong magnetic pulses from the machine bombarding the head cause water molecules in the brain to oscillate, and the radio signals generated by the vibrating water molecules are detected by the MRI device. The more thickly axons are covered with myelin, and the more tightly packed the axons are, the more easily water will seep along them rather than across them, like paint drawn up the bristles of a brush. DTI detects the amount of asymmetry in this diffusion of water between axons. The study found that the more easily water moves along nerve fibers rather than across them, the higher your IQ.

In this type of brain scan, the various areas of the brain are rendered in color by computer analysis to indicate how symmetrically water moves at each point. The white matter tracts of children with higher intelligence appear red on the pseudocolored brain scan, like high-speed highways on a road map. Cooler colors in this white matter imaging, like slower blue highways on a roadmap, reflect water motion that is less well directed along the axons, allowing water molecules to meander off the main highway. The more "blue highways" than "red highways" on the brain scan, the lower the child's IQ.

How can white matter structure have anything to do with intelligence? There are no neurons or synaptic connections there. Myelin is made by oligodendrocytes, and the "packing" material between the axons in white matter is astrocytes. Could these glial cells be contributing to intelligence?

A New Way of Learning

Long before now, someone should have asked this obvious question, but obvious clues were simply ignored. It has been known for decades that although most myelination takes place in the first five years of life, the process continues into early adulthood. Why? If myelin is simply electrical insulation, why isn't the job done before we are born?

There is a curious pattern in the way myelination proceeds in the human brain after birth: the last regions of the brain to become fully myelinated are those involved in higher-level cognitive function. In the human brain, myelination proceeds in a slow wave from the back of the cerebral cortex (shirt collar) to the front (forehead) as we reach adulthood. This wave of myelination may account, in part, for the teenager's notoriously impulsive behavior. By adolescence, myelination is not yet complete in the forebrain. This last region of the brain to myelinate is the area of our cerebral cortex critical for judgment and complex reasoning. This is the same area of the brain disconnected by surgeons in prefrontal lobotomy. Lobotomy renders patients unable to make complex decisions, to plan or show forethought. If the transmission lines to these regions of the brain are not fully formed, adolescents do not have the full circuitry that enables the adult brain to make rational decisions in complex situations.

Interestingly, the age at which most societies bestow full legal responsibility on an individual is not the age of puberty, but instead a later age that coincides closely with the completion of myelination in the forebrain (about twenty years of age). Myelinating glia provide a biological rationale for the age of legal responsibility. Recently these forebrain glia have found their way into courts as expert witnesses in arguments against subjecting juvenile criminals to adult penalties in the judicial system. Staci Gruber and Deborah Yurgelun-Todd argued in 2006 in the *Ohio State Journal of Criminal Law* that incomplete forebrain myelination is a compelling neurobiological basis accounting for the poor judgment and impulsivity of adolescents, and that this developmental immaturity makes it unreasonable to hold adolescents accountable as adults under the law. Still, why is myelin, the insulation on our axons, not complete in the human brain until early adulthood?

To answer this question, let's first take a step back. Many scientists, notably Marian Diamond at the University of California, Berkeley, and William Greenough at the University of Illinois at Urbana-Champaign, have devoted their scientific careers to exploring the process of how our brains become molded by our environment. (Diamond is the same neuroanatomist who examined Einstein's brain as described in chapter 1.) They reasoned that if they raised animals in environments that differed in the amount of cognitive stimulation they provided, the cellular structure of the animals' brains might differ accordingly. This was a radical departure from the standard approach to brain science. Traditionally scientists would alter the brains of experimental animals with drugs or surgery and then test the animals in mazes and other devices to see how this alteration affected learning and behavior. Diamond and Greenough each independently took the opposite approach. Reasoning from the fact that learning is the result of our brain interacting with the environment, these researchers altered the environment and then looked for cellular and biochemical changes in the brain.

"I saw my first human brain when I was about fifteen years of age and was blown over that those cells in the brain could 'think,'" Marian Diamond related to me recently in recalling the source of inspiration that changed her life. "I knew I had to study brains but could not have access to them until graduate school. There were no neuroscience courses back then. I was completely entranced and mesmerized and magnetized by brains ever since. Very few had the passion I did and especially very few women."

When both research groups compared the brains of animals, such as rats, raised in enriched environments to those raised in normal conditions, they found striking differences in the cellular brain structure. Enriched environments differ from the normal environment of a rat cage by providing the animals with playthings and opportunities for increased social interaction. As a result of the increased cognitive stimulation, the brains of animals raised in enriched environments are slightly larger. The implications for early childhood experience on human brain development are obvious.

Diamond's and Greenough's studies showed that this increased brain size was due to a significantly thicker cerebral cortex, the portion of the mammalian brain involved in higher-level cognition. This makes perfect sense, since the bulk of the brain covered by the cerebral cortex is devoted to controlling bodily functions and senses, not intellect, but a very surprising outcome of this work was their observation that this brain remodeling included not only neurons but also blood vessels and glia.

"At the time of this early work in the 1960s, no one else had published in this field of neurohistology responding to the environment, so I did not know what to expect. When J. Altmann also found similar results showing glial cells increasing in number with [environmental] experience, I was ecstatic," Diamond recalls.

The effects of environmental stimulation on glia, first reported by these pioneering researchers, have been confirmed repeatedly in several different laboratories in the decades since, but few have pursued this curious finding or followed the implications through to their logical conclusion. Environmentally induced changes in glia have been in the scientific literature for decades, but still they simply failed to gain traction with the majority of "neurocentric" neuroscientists. Vital clues were overlooked or dismissed because, as in every mystery story, they were hidden in the blind spot of preconceived ideas.

Even though it was recognized that astrocytes exist to shelter neurons and serve their every need, the idea that astrocytes might have a role in processing information or in learning was beyond consideration. It was felt that any changes in the number of astrocytes in these animals simply reflected the same thing that the increase in blood vasculature represented: a response of supporting cells to supply the increased demands of neurons aroused by the increased mental stimulation provided by their enriched environment.

The idea that myelinating glia in particular could have anything to do with intellect or learning was so remote that it was not given any serious consideration. Neuroscientists understood what myelin did: it insulated axons. Just as few electrical engineering students are attracted to the field of electronics to study plastic insulation on copper wires, few students of neurobiology are interested in myelin; their passion is to crack the secrets of cognition, learning, and memory. Researchers studying myelin are primarily medical scientists working on demyelinating disease or biochemists. Half of the human brain is white matter, so the bulk of what biochemists extracted in their test tubes from homogenized brain tissue is myelin. For doctors certainly, myelin is always at the center of their work, because myelin must always be repaired after injury and disease to restore electrical communication and function. Damage of myelin through disease, toxins, and infection causes the bulk of neurological disability, but myelin was thought to be irrelevant to the core issue of how the brain processes information and learns. This is still the prevailing view, but it is changing.

Let's pick up the trail of discarded clues. It has been known for forty years that the number of oligodendrocytes increases 27 to 33 percent in the visual cortex of young rats raised in an enriched environment. This odd finding does not fit; oligodendrocytes have nothing to do with information processing in neurons. All they do is wrap around axons to seal them so that electrical current will not leak out. They do not associate with synapses, dendrites, or the cell body of neurons.

Crazy as this clue may seem, it is not an isolated thread of evidence; there is corroboration. The oddity is not limited to glia in the visual cortex: rats raised in enriched environments have an increased number of myelinated axons in their corpus callosum, the major bundle of axons that connects the left and right sides of our brain, as discussed in chapter 11. This interhemispheric communication across the corpus callosum is vital for integrating our cerebral dual processors into a single cooperative system. But why should the insulation around these cables connecting our left and right brains increase, and the population of insulating oligodendrocyte cells expand by nearly one-third, in animals raised in enriched environments?

This observation is not an oddity limited to the lowly rat: rhesus monkeys reared in enriched environments also develop more myelin in their corpus callosum. Moreover, this difference correlates with improved cognitive performance on tests of learning and memory given to the monkeys.

The same clue to glial involvement in information processing turns up again and again, even in human studies. MRI scans show that the corpus callosum area is decreased 17 percent in children who suffered childhood neglect. Most surprising is the recent finding that brain scans of people with certain mental disorders, including schizophrenia and depression, also show reduced white matter development. White matter—not the grey matter that one might expect to atrophy in mentally ill people or children neglected and deprived of normal stimulation necessary to develop their minds.

A vexing fact that cannot be explained feels like a major hint: if myelination is simply part of the process of building our brain, why does the process continue for decades after we are born? An even greater question to ponder is, how do glia know their environment is enriched or impoverished? The obvious conclusion is that myelinating glia in our brains somehow sense impulses flowing through the axons they ensheathe. In fact this deduction was the inspiration for the Schwann cell experiments at the opening of this book. We return now to that scene, armed with a fortified understanding of glia's role interacting with neurons in health and disease.

GLIAL INTERCEPTION: MYELINATING GLIA WIRETAPPING AXONS?

How could I begin to find out if myelinating glia could detect nerve impulses? New evidence just published in the late 1990s was showing that astrocytes and Schwann cells around synapses could respond to neurotransmitter molecules leaking from a synapse, but how would myelinating glia sense neural activity speeding through the axon so far removed from synapses? We had to develop an experiment that would give us control of impulse activity in axons and the ability to monitor myelinating glia simultaneously.

I decided that only reconstructing the situation *in vitro* would allow adequate access to monitor and control the axons and myelinating glia and isolate the molecular mechanisms. We dissected sensory neurons from embryonic mice, the same neurons that convey the sense of touch, but we grew them in specially designed cell culture chambers equipped with platinum electrodes. Using electronic stimulators, we were then able to deliver brief shocks of electricity to the axons, causing them to fire impulses in any pattern we desired. Next we isolated myelinating glia from the nerves and brain of rats or mice before they started to form myelin and grew them in cell culture. Then we added these Schwann cells or oligodendrocytes to the neuron cultures under appropriate conditions to allow them to begin forming myelin around the axons. The entire process of forming myelin in cell culture requires growing the cells for two months in the laboratory incubator, which maintains the constant conditions of oxygen, warmth, and humidity of a mother's womb.

Before using calcium imaging we had discovered other strong clues that myelinating glia could detect and respond to electrical activity in axons. We had found that Schwann cells slowed down their rate of cell division when we grew them on axons stimulated to fire impulses. Somehow, the Schwann cells were picking up signals from the axons and changing their behavior. Philip Lee, Kouichi Itoh, and others in my lab then analyzed specific genes in Schwann cells and found that some of them could be turned on or off by firing impulses in the axons. Amazingly, impulse activity in an axon was able to control genes in the Schwann cell! If electrical impulse firing in axons could turn on and off specific genes in Schwann cells, the inescapable conclusion was that these glial cells could be controlled by electrical activity flashing through the animal's nervous system. Immediately we saw that some of these genes controlled by impulse firing could regulate Schwann cell development and myelination.

Still, we had no direct evidence that the Schwann cells or oligodendrocytes were sensing electrical activity in the axons. Another possibility could be that causing axons to fire impulses simply changed the properties of the axons themselves, which in turn affected how the Schwann cells responded to them. In fact, we found that this did occur. Stimulating axons to fire at different frequencies turned on and off certain genes in the neurons. Some of these neuronal genes made various proteins coating the surface of the axons. Intriguingly, some of these proteins, called cell adhesion molecules, were known to affect the attachment of Schwann cells to axons. They were also known to affect Schwann cell development and myelination. A most interesting aspect of these studies was finding that the pattern and frequency of impulses in the axons was important. By dialing up different frequencies on our electronic equipment stimulating the axons, we could selectively tune in to different genes and turn them on or off. Like a message transmitted in Morse code, the frequency code of neural impulses in some way reached the nucleus of neurons to control individual genes. Considering that every feature of our environment and every thought in our head is coded in nothing more than the pattern of impulses in our nerve axons, this finding explains how the environment we experience can alter the structure of neurons by controlling a neuron's genes. But now we saw that the same coded signals could reach glial cells, which responded to the changing types of molecules coating axons that were altered in response to particular patterns of neuronal firing.

This fascinating discovery did not eliminate the other possibility that Schwann cells and oligodendrocytes might themselves perceive electrical impulses in axons in some direct way. Since so many other types of stimulation cause calcium responses inside cells, I decided to use fluorescent calcium imaging to see if impulse activity in axons signaled to Schwann cells. This was the same strategy that had been used by others to detect astrocytes responding to neurotransmitters at synapses. As described in the opening chapters, I hoped these experiments might reveal communication between axons and Schwann cells by increasing calcium in the Schwann cells when we used a brief electric shock to fire nerve impulses in axons.

When after stimulating the axons we saw the Schwann cells light up with fluorescence, there was no doubt that these myelinating glia were able to detect impulse activity in the axons. Next I tried the same experiment using oligodendrocytes, and the result was the same. Myelinating glia from both the central nervous system and peripheral nervous system could sense electrical activity in axons. But how?

Unlike astrocytes surrounding synapses, myelinating glia have no access to neurotransmitter molecules seeping from the synapses. Myelinating glia wrapping axons can be more than a foot away from synapses in some cases. Contrary to established models of information flow in the nervous system, we were seeing communication taking place between brain cells without synapses. We guessed that when the axon fired bursts of electrical activity, it might release some signaling molecule that Schwann cells and oligodendrocytes could detect. Our approach to finding this mystery molecule was not very sophisticated: we just started guessing.

We tested many types of chemicals that might be released from axons firing impulses or that might activate receptors on Schwann cells to evoke calcium responses. We applied these chemicals to Schwann cells and oligodendrocytes in cell culture and watched to see if any of them caused a calcium response in the cells like those we had seen in the glia on firing axons. Many likely suspects, including the neurotransmitter glutamate, generated no calcium response in Schwann cells at all, but we found many other substances that could stimulate a rise in calcium in Schwann cells. The problem was, this increase in calcium could be an artificial response to a substance applied in the laboratory but might never occur in the cell in nature. How could we know which of these possible signaling molecules was actually used by myelinating glia to sense impulses in axons?

Our work began to narrow in on adenosine triphosphate (ATP) as a likely signaling molecule between axon and myelin-forming glial cell. As we saw earlier, ATP is the energy source for all cellular life, abundant inside all cells but rare outside them. This, in theory, makes ATP a good molecule to use for sending signals between cells. ATP was beginning to attract attention as a signaling molecule between neurons and astrocytes, because ATP is packaged inside synaptic vesicles together with neurotransmitter and released with it at synapses. In addition there are channels in cell membranes that can release ATP, which might allow axons to release it without synaptic vesicles. Indeed, applying ATP could produce waves of calcium through astrocytes in culture or in brain slices. Even though it was not known exactly how ATP might be released from the axon far from synaptic endings, we decided to try the experiment.

We found that applying ATP to Schwann cells or oligodendrocytes caused rapid and large increases in calcium inside the cells, much like those we saw when we fired impulses through axons as the glial cells clung to them in our cell cultures. To test whether ATP was indeed the natural signaling molecule responsible for the axon-glial communication we would need some way to disrupt ATP. There is an enzyme called apyrase that rapidly degrades ATP outside cells. We reasoned that if we added this enzyme to our cultures of neurons and Schwann cells, apyrase should disrupt the communication from axon to Schwann cell by intercepting and destroying the intercellular messenger, if indeed that messenger molecule was ATP. We added apyrase to the cultures and stimulated the axons to fire impulses. As we watched, the Schwann cells remained deep blue on our computer screen, indicating no rise in calcium at all. The axons themselves showed their usual large increase in calcium, indicating to us that they were firing impulses normally. Lucky guess perhaps, but the breakdown in axon–Schwann cell communication in the presence of apyrase was strong evidence that ATP was the messenger molecule communicating between axons and Schwann cells. The same molecule that provides the energy source for all cells in our body also appeared to be used outside cells as a form of communication between axons and myelinating glia.

This was strong evidence, but we needed definitive proof that ATP was indeed being released by axons when they fired impulses. The cold green light generated in the tail of a firefly provided the illumination that solved this puzzle. The green flash of the firefly is the result of a chemical reaction stimulated by an enzyme, luciferase, which breaks down a firefly protein, luciferin, to liberate a photon of green light. This reaction requires one more critical component: ATP.

I added luciferin and luciferase to the cultures of sensory neurons and used extremely sensitive digital cameras attached to a microscope to detect the green photons generated in the cell cultures. The dim photons were amplified by extremely powerful night-scope photomultipliers. Working in a completely dark room with the microscope itself encased inside a cabinet-sized light-tight box, I saw these few random photons, invisible as light to our own eye, splashing against the computer monitor like bugs hitting a windshield. I hoped that this night scope trained on the luciferin tracer added to the culture would permit us to see if any ATP was released when I stimulated the axons to fire electrical impulses. If ATP was released from the axon, it would fuel the reaction between the enzyme luciferin and the protein luciferase to generate a photon of green light. When I stimulated the axons the results came in a bright flash of photons speckling the computer screen.

When the axon fired, ATP had been released from the axons into the culture medium to drive the same light-producing reaction that scintillates summer evenings to the delight of children throughout the eastern United States.

There is much to be learned about this process. We still did not know how the ATP was released from axons firing electrical impulses in the absence of synapses. There is no doubt that this ATP release occurs, however, and myelinating glia sense the liberated ATP to detect impulse activity in axons. There remained, however, the burning question of what the eavesdropping glia were doing with the information they intercepted.

GLIA PLAYING THE PIANO

The old debate as to whether we are born a *tabula rasa* (blank slate) or predestined will probably resolve itself somewhere between these extremes. Human brains are extremely versatile, allowing the mind of each individual to develop the unique skills of a concert pianist, doctor, electronics repairman, or nurturing parent. This specialized programming of the brain of every individual has enabled the complex behavior, social structures, and interactions that allow humans to dominate life on this planet. The human brain is supremely creative, adaptable to changing situations, and capable of foresight, which is the ability to filter our past experience through the present moment and project the trajectory of events into the future: in effect, consciousness. A brain that can do this cannot be hardwired at birth; it must be designed uniquely in each individual to create an instrument suited to the demands of that particular individual's environment.

Evolution works slowly by eliminating over generations the individuals in populations that lose in the competition for food and mates. Human beings cheat the process of evolution by evolving our brains *after* we are born. In this way we each develop a brain that best suits the particular environment we find ourselves born into before it is time to reproduce. This remarkable capability of our brains to form through childhood and into early adulthood maximizes the probability of each individual's survival, success, and reproduction in its present environment, rather than the environment that was present in prehistoric times and recorded in our genes through heredity. In the Ice Age as in the Space Age, it is this ability of the human brain to mold itself uniquely to the environment early in life that separates man from animals whose brains are cast at birth. The plasticity of our brain prior to adulthood is the reason human evolution has exploded so far beyond that of any other organism. But how is myelination involved in the environmental molding of our brain?

Fredrik Ullén is a virtuoso classical pianist from Stockholm, Sweden. His performance of *Vertige* by György Ligeti is my personal favorite, which I recommend to anyone interested in music and piano virtuosity. Like many elite performers in any human endeavor—sports, music, chess—Ullén wondered what sets elite performers apart from the less highly skilled and what changes take place inside the brain as one masters a complex skill, such as playing the piano.

"Something which always fascinated me is the neural basis of optimal performance: what neural mechanisms enable the top-level performances of outstanding experts?" Ullén related to me recently. But in this case, the question was more than a musician's idle pondering, because Dr. Fredrik Ullén is also a neurologist at the Karolinska Institute in Stockholm.

"Musicians are of course a very useful model group to study different aspects of this problem, and that is what my research group focuses on now. The structural differences between musician and nonmusician brains are striking."

Using the same brain imaging technique that correlated IQ with white matter development, Dr. Fredrik Ullén and his colleagues at the Karolinska Institute published a paper in 2005 describing experiments comparing white matter tracts in brain scans of professional pianists with the brains of people of the same age who never learned to play the instrument. They succeeded where others had failed in visualizing learning simply by looking in the right but most unlikely place: the areas of the brain devoid of neurons. The brain images from professional pianists showed that they had more thickly myelinated axons in one white matter tract (the posterior limb of the internal capsule) on the right side of the brain. This fiber tract carries axons from areas of the cerebral cortex controlling finger movement, which is exactly the functional skill that one learns in mastering the piano.

This finding unhinges preconceptions most neurobiologists have about the brain. How could glial cells have anything to do with the difference between a concert pianist and a novice? Was the destiny of a concert pianist set at birth by the number of layers of insulation a glial cell wrapped around axons in this part of the brain? As unsavory as that cellular mode of predetermination may be, equally outrageous is the alternative that the myelinating glia wrapped more insulation around the axons as the pianist mastered the instrument. How could glia possibly know they were playing the piano? Even if they did know, how could electrical insulation have anything to do with learning?

The theory of predetermination crumbled when the scientists analyzed their data in another way. They compared the development of white matter tracts in the brain of pianists after sorting the musicians into three groups: those who began their piano lessons as children, as adolescents, and as adults. Then they sorted each group according to how many hours the person practiced the piano at each age. This revealed a striking finding. The organization of white matter tracts increased in proportion to the number of hours the person had practiced.

The researchers could see this trend regardless of whether people took up the piano as children or adults. However, in people taking up piano as adults, differences were seen only in the regions of cerebral cortex not yet fully myelinated. In the brains of children, which are still undergoing myelination more broadly, piano practice increased white matter structure in many more areas of the brain. The most important finding from these studies is that these concert pianists' brains were not hardwired at birth to excel at the piano; they had developed their brains through practice. This correlation between myelin and learning may relate to why it is necessary to begin learning complex skills early in life. A person can learn to play the piano at any age, but to reach the highest level of performance in music, sports, or most complex cognitive processes, it is necessary to begin at a young age when the cerebral cortex is still being myelinated.[1]

In a fundamental but unknown way, myelinating glia were contributing to learning. But how? Learning is based on synapses coupling circuits together more strongly with experience. All these glia do is make insulation, right?

GLIAL REPEATER AND RELAY: THE NODE OF RANVIER

We are all captives of our analogies. As a result, I could not at first understand why Nature had provided the cells insulating axons with an ability to sense impulses. What would they do with the information? All the information processing in electronic systems takes place before it is sent out through transmission lines. Why fool with the insulation on transmission lines?

People tend to think of nerve fibers as wires. Those who think this way regard myelin as nothing more than insulation. But if myelin were simply insulation, why did Nature do such a poor job of it, leaving bare spots exposing the axon every millimeter or so all along its length? From the first time anatomists looked through their microscope and saw that the axon was coated with a string of hundreds of "flattened pearls," they were perplexed by those bare spots between each glial cell. Workmanship of electricians varies widely, but Nature is never clumsy. Were these breaks in insulation caused by damage in the delicate fiber, or could they be natural?

The anatomists did their job, artifact or not, and gave the bare spots a name. In 1878, the exposed segment of axon between each glial sausage was called a "node" by the French neuroanatomist Louis-Antoine Ranvier. Ramón y Cajal had no idea what these nodes of Ranvier were, but he likened them to garter belts that he surmised held the fatty myelin sheath in place around the axon.

We now know that each node of Ranvier is not a careless break in insulation or a sort of garter belt attachment; it is a highly sophisticated electrical device, more like a communication relay station. In our quest to understand the node of Ranvier and myelin, we must first understand exactly what a nerve impulse is.

A single nerve impulse is a rapid voltage change that propagates down an axon like a ripple along a taut string. The impulse is generated by a carefully timed sequence of molecular events that open and close protein "valves" in the axon membrane to allow the passage of charged ions. As the charged ions surge into and out of the axon, the voltage at that spot on the axon changes briefly, reflecting the flow of these charges. Positively charged sodium ions rush into the axon through proteins in the axon membrane called sodium channels, raising the voltage as it accumulates more positive charges. Immediately thereafter, positively charged potassium ions rush out through potassium channels to reduce the excess positive charges and restore the axon to its original state so that it can fire again. The ripple in this voltage disturbance across the membrane moves down the axon.

This cycle is repeated every time an axon fires an impulse. It takes only about one-thousandth of a second for one cycle of sodium and potassium ions to generate the nerve impulse at one spot on the axon. However, these repeating delays of even one-thousandth of a second accumulate as the wave is generated incrementally in moving down the axon. This limits the speed of information transmission through the axon.

Myelin and the node of Ranvier fundamentally change the way impulses are transmitted through axons. Rather than carrying out this cycle every minute step of the way down the axon, as in invertebrate axons that lack myelin, the nodes of Ranvier in myelinated axons act as repeaters, greatly speeding the transmission of the signal over long distances. In myelinated axons, the nerve impulse is generated only at the bare nodes; it is not generated in the regions of the axon wrapped by myelin (the internode). Myelin seals off this part of the axon, preventing any voltage from leaking out, forcing the electricity to jump from node to node. Each node passes on the signal like an electronic relay station. Information can be transmitted over a series of nodal communication relays up to one hundred times faster than it would without these repeaters. Myelin is more than insulation, and the nodes are not mistakes; each one is a highly sophisticated electronic device speeding the transmission of information. Much like the difference between skipping sprightly across a stream on stepping stones and slithering across on a log, electrical impulses skip along a myelinated axon from node to node much faster than they can slither down an unmyelinated axon.

But another clue has been overlooked. Axons do not all transmit information at the fastest possible rate, as you might expect they would by analogy to a high-speed Internet connection. The speed of impulse transmission through our nerves and brain circuits varies tremendously among different axons. Some axons transmit impulses at slow rates of only 1 meter per second (the pace of a slow walk). The fastest axons transmit impulses at 100 meters per second. Why? Nature uses the fastest means of information transmission for processes that must be carried out rapidly, such as in sending impulses through motor axons to move our legs as we hurl our bodies through air, catching ourselves on one foot in midstride with every leap, which we call running. But why don't all axons conduct at the same rapid rate? Moreover, what determines the rate of impulse transmission in axons?

Glia control the speed of communication through myelinated axons. They do this not only by determining how many wraps of insulation they should apply to a given axon, but also by determining where the nodes of Ranvier will be placed on the axon and by concentrating the sodium and potassium channels into clusters along the axon to form the nodal and internodal regions. The more layers of myelin a glial cell wraps around axons, the better insulated the axons are, and with less voltage loss the signal is transmitted faster. It should also be clear that if the node of Ranvier is a repeater, there must be some optimal number of nodes and optimal spacing between them to achieve the most rapid relay of impulses down the axon. Glia control the distance between nodes of Ranvier and thereby control the speed of impulse transmission.

Myelinating glia are the foremen controlling construction of the axon in development and repairing it after it is damaged. They determine where the nodes will be formed in two ways: by releasing chemical substances that instruct the axon to form either node or internodal type of axon membrane and by physically sweeping the sodium channels in the axon membrane into clusters at the nodes as the glia wrap the bare axon and begin to myelinate it. It is this glial control of impulse conduction that may involve them in the process of learning.

Neurons That Fire Together Wire Together

It is a fundamental rule of learning originating from Pavlov's experiments with dogs trained to salivate at the sound of a dinner bell: "Neurons that fire together wire together." Pavlov was able to wire neurons responding to sound to neurons that stimulate saliva production by presenting food to the dog simultaneously with the dinner bell. After these neurons repeatedly firing together had wired together, Pavlov could make his dog salivate without any food, simply by ringing the bell. Neuroscientists interested in the cellular basis of learning have intensely studied the molecular mechanisms wiring neurons together more effectively at synapses, but they have completely overlooked a fundamental question: what determines whether or not two neurons fire together?

Timing is everything. Coordinating the timing of information flow is absolutely critical for the operation of any communication network. We have all experienced the breakdown in communication that occurs when timing is off, as when delays between sending and receiving information across long-distance telephone lines disrupt our conversation.

The brain is no different. Delay in information flow through the cerebral cortex is one of the underlying causes of dyslexia and possibly ADHD. (Interestingly, new research finds that children with autism have too much myelin in certain brain regions.) We do not understand how the speed of information flow through axons is regulated to provide the simultaneous arrival of needed information to a given neuron. What is clear, however, is that achieving this precise timing is absolutely essential for the brain to function. Logically then, our attention must turn to the cells controlling the speed of impulse conduction through axons: myelinating glia.

By controlling the speed of conduction through axons, myelinating glia can determine if impulses from two axons converge on a neuron at the same time. If impulses from the two incoming axons arrive simultaneously, the synaptic voltage they produce in the dendrite will add together, producing a bigger response. However, if impulses from the two incoming axons arrive at slightly different times, the synaptic voltages produced by each input will not add. This is much like two people working together to push a car out of a rut. The two must push at exactly the same time. The result of impulses not arriving at a dendrite simultaneously will be that two small voltage changes will be produced in sequence of only half the strength they would have produced if they had arrived simultaneously. These individual voltage pulses may be insufficient to trigger any response at all in the postsynaptic neuron.

Inputs in critical neural circuits will not arrive simultaneously if they are sent through axons of different lengths, as is usually the case. Thus, the speed of impulse transmission must be increased through the long axons and slowed through the short axons to assure simultaneous arrival as required by the rule "Neurons that fire together wire together."

The voltage change produced by a synapse is extremely brief: only a couple thousandths of a second. This speed requires very high precision of impulse arrival times. Is it possible that the optimal speed of impulse conduction through axons was established entirely by genetic instruction during brain development in every axon in our brain? Or is it possible that the conduction speed is regulated according to functional experience to optimize the performance of the circuit? Considering all the factors that affect the conduction delays between neurons separated by substantial distances—across the corpus callosum connecting our two cerebral hemispheres, for example—it seems implausible that genetics alone could account for all the variables. Factors affecting the arrival time of an impulse through an axon include the particular path the growth cone took in growing out during embryonic brain development, the caliber of the axon, the number of myelinating glia forming nodes of Ranvier all along its length, the thickness of the myelin sheath, the type and number of ion channels responsible for generating the nerve impulse, and many more. A more likely possibility for matching the speed of axon cabling to the requirements of every brain circuit is that conduction velocity is somehow regulated by functional experience.

Beyond simple reflexes, this process of regulating the speed and simultaneous arrival of impulse traffic through axons may be particularly important for learning complex skills, such as playing the piano, which requires integrating information between multiple regions of the cerebral cortex involved in any complex task. It appears that we are on the threshold of discovering a new type of learning beyond the simple reflexes that have occupied neuroscientists up to now, a form of learning that takes us not only beyond the synapse, but beyond neurons.

This reasoning and new information might explain why changes in white matter are seen as a person learns to play the piano or why white matter structure increases in proportion to a person's IQ. But, even if glia can sense electrical activity in axons, it remained to be determined if this sensitivity would indeed affect myelination to optimize performance and how ATP could be released from axons far from synapses.

Tasaki

Every day an elderly Japanese gentleman shuffles along the sidewalk purposefully stitching a pathway forward with deliberate jabs of his cane. Deep in thought, the old man seems oblivious. People scurrying past scarcely give him a second's thought, unless it is to admire the old man's stamina and focused determination, intent on reaching his destination. Few could imagine what that might be.

The grandfatherly man, two years from his one hundredth birthday, is on his way to work. He traverses the two-mile journey twice a day, every day. Before his wife of sixty years, Nobuko, passed away in 2003, he made this trip four times a day so that the two of them could lunch at home. Then they would return to work side by side. Early in their relationship, Nobuko realized that if they were ever to spend time together, she would have to become his assistant. In 2002 he vowed, "I will continue to work until my wife says she cannot work. If she can work until she is one hundred years old, then I will keep working too."[2] But after Nobuko was gone, he could not stop working.

Turning the brass key with shaky hands, the old man opens the door to his workplace, revealing a scientific laboratory cluttered with glassware and electronic instruments dating from the 1960s. As he steps into the room, his clothing and eyeglasses of a style once favored by engineers in the early days of the space program suddenly blend perfectly with the scene. On the street he is an anachronism, but as he flips the light switch he blends harmoniously with the scene from a bygone era bursting to life as if on a stage. Much of the dated electronic equipment and scientific apparatus surrounding him was built by hand. At the time of their construction his vision and equipment needs outstripped the available technology.

The name Ichiji Tasaki is not well known, but there is no one who does not know the fruits of this man's labor. Everyone knows that our nervous system works by sending electricity through our nerves to control our muscles, and impulses are sent from our sense organs to our brain. But how are impulses conducted through axons? Dr. Tasaki is the person who answered this question.

It is obvious that nerve axons are not copper wires. Even though these slender tubes transmit electrical impulses rapidly, nerve impulses are not sent through axons as electrons are sent through a wire. That intuitive thinking is clearly wrong, even though we persist in using the metaphor. The reality of how this vital communication process works is explained in every biology textbook, yet the name of the person who solved this mystery is rarely mentioned.

The discovery was made by Tasaki in Japan in the 1930s. After World War II, Tasaki continued his research in England, and in 1951 he came to the United States. In 1953 Tasaki joined the National Institutes of Health in Bethesda, Maryland, where he has remained ever since. The institute has provided him with laboratory space off a remote corridor in building 13, where he is an active member of a research group headed by biophysicist Peter Basser. Beyond the remarkable fitness of this man nearly one hundred years of age, Tasaki's mind is extraordinary—still as sharp as those years younger with whom he works. Tasaki is blessed with the brilliant mind of a biophysicist and mathematician.

"Intuitive logic is like . . . beautiful woman," he warned me only a few weeks ago with his characteristic easy laughter as I tried to press my arguments without the support of mathematical equations. Walking his fingers in the air he explained in halting English, "You can be led astray."

Today scientists have stunningly sophisticated instruments to look inside the human brain at work; to capture on a chip all the genes expressed in a nerve cell; to watch signals being sent through living neurons; and to observe glia by using computer-driven microscopes illuminated by lasers. How, in the 1930s when the most advanced electronic device was the radio, could anyone determine the sophisticated mechanism that transmits electrical signals through our bodies?

Tasaki did it by hand, using simple tools and hand-built instruments he made, and then he applied mathematical equations to unlock the meaning of his measurements. Tasaki found that myelin changed the way electricity is transmitted down an axon. Rather than shooting down the fiber in an electrical wave as everyone had assumed, Tasaki found that myelin forced the electrical impulse to leap from one node of Ranvier to the next like a ballet dancer leaping across the entire stage in a few bounds. This explained how myelinated axons transmit information one hundred times faster than unmyelinated axons. This fundamental process underlies the design and operation of every myelinated circuit in the brain and body, and it is the reason for paralysis afflicting people with multiple sclerosis and other diseases that attack the myelin insulation on axons.

Much of Tasaki's early work was ahead of its time, and so many of his unique observations linger as obscure curiosities. In 1958 he pierced glial cells of the cat brain with electrodes and was the first to find that although glia did not produce electrical impulses like neurons, they did have a steady electrical voltage with properties peculiar to themselves.

He also reported that he could make an astrocyte twitch microscopically by injecting a pulse of electric current into it. This he explained was simply the consequence of electricity forcing charged ions into and out of the cell, causing water molecules to follow.

In the 1960s while he was studying the electrical excitation of nerve fibers, his careful observations revealed that a nerve impulse caused subtle optical changes in an axon and minute temperature changes in accordance with the energy expended by moving ions across the membrane. More astonishing, he built delicate devices that detected microscopic swelling and shrinking of the axon caused by ions and water molecules flowing across the axon membrane during the flash of a nerve impulse.

Twitching axons are, however, an oddity. Biophysicists appreciate that the physical responses of axons inform us about the fundamental mechanisms of electrical excitation, but Tasaki's work was dismissed by some biologists as nothing that could have any biological purpose. Much like vibrations of an engine, the minute twitches an axon made when it fired an impulse were viewed as an inconsequential byproduct of an underlying mechanism. For several years Tasaki continued to publish studies showing that all kinds of axons twitched microscopically when they fired electrical impulses, but he was essentially alone in pursuing this phenomenon.

One day in my lab, as I was trying to discover how ATP is released from axons to signal to Schwann cells, I played back a time-lapse video of an experiment showing axons at high magnification through a microscope. The axon twitched. The movement was triggered when I stimulated the axon to fire electrical impulses, but the movement was so microscopic it was imperceptible without high magnification and video recording at just the right rate to capture the movement. I visited Dr. Tasaki in his lab to ask if I was seeing some type of artifact.

"No," he said, and he proceeded to explain how he had seen the same thing years ago. He sketched out a figure with arrows and equations explaining how ions and water molecules enter the axon when it fires an electrical impulse, which makes the cytoplasm near the cell membrane swell a small amount. Suddenly I realized that this might explain how ATP is released from axons firing electrical impulses.

I knew that all cells face the problem of regulating their volume precisely in a changing environment. As the amount of salt in bodily fluids decreases, cells swell as water and ions redistribute across the cell membrane to reach a balance inside and outside the cell. Cells do not burst when they begin to swell, because they have channels in their cell membrane that regulate the flow of water and small molecules into and out of the cell to restore the normal cell volume. If electrical impulses cause the axons to swell, these channels might open to release small molecules and water to shrink the axon back to its normal size. If ATP could exit through these channels, this could allow glia to sense neural impulse activity in axons even when they were far from synapses where neurotransmitter is released.

For nine years I performed various experiments to test the hypothesis. Finally I completed the research verifying the hypothesis and revealing a new way for information to be transmitted from axons to other brain cells without synapses. Writing up the results for publication I thanked Dr. Tasaki in the acknowledgements section of the paper. I raced off to give him a copy, with eager anticipation of his delight when I showed him that his odd phenomenon of twitching axons had a biological function after all.

On the way to Tasaki's lab I stopped by the scientific director's office to give him a copy of my new manuscript.

"This all stems from Dr. Tasaki's observations of decades ago," I began as a preface to explain my new paper to the director.

He interrupted me in midsentence, "You know Tasaki died."

"No!" It was such a loss. Only weeks ago he had joked with me about intuitive logic being like a beautiful woman. He had been so vigorous at the time.

"What an amazing person he was. What an inspiration," I said. I felt grateful to have known him. Scientists rarely use the word "genius," because in our line of work we are likely to meet one or two, and afterward the word is never misapplied.

I walked to building 13 and handed the manuscript to Tasaki's boss and longtime collaborator Peter Basser.

"I still walk into his lab and expect to see him sitting there," Peter said. "I told him he would probably outlive me. His mother lived to one hundred and eight . . . I can't believe he is gone."

Like many who drive to work at NIH, I now find myself searching an empty sidewalk.

IMPULSES STIMULATING MYELINATION

In the last ten years research in my laboratory has determined that impulse activity in axons affects the development of Schwann cells and oligodendrocytes, and thus myelination. Our research has determined three distinct molecular mechanisms that operate at different stages in the myelination process and work through different molecules to regulate myelination in response to impulse firing in axons. Our research on myelin formed in cell cultures equipped with electrodes to stimulate impulses in axons makes it clear that myelin is regulated by impulse activity.

Some of the effects of impulse activity inhibit myelination and some stimulate it. First, neuronal firing can change genes in neurons that make the proteins that coat axons. These proteins are critical for attachment of myelinating glia and wrapping the layers of myelin membrane around the neurons. Second, ATP can be released from axons, and both ATP and its breakdown product adenosine can be sensed by receptors on Schwann cells and oligodendrocytes that in turn regulate development of myelinating glia and myelination. Third, after oligodendrocytes mature, astrocytes nestled among the axons can sense the ATP released by axons firing electrical impulses, and when they do, they relay the signals to oligodendrocytes by releasing another signaling molecule, leukemia inhibitory factor (LIF), which stimulates mature oligodendrocytes to form more myelin. Although we have found three ways that impulse activity in axons is sensed by myelinating glia and how this sensing of impulse activity in turn affects myelination, we know that there are many more such methods of sensation yet to be discovered. This form of myelin plasticity in the brain may be as important as synaptic plasticity in improving the performance of the nervous system based on experience.

These new insights into how we learn spring from studying the interface between the other brain and the neuronal brain. The curious Schwann cells lighting up like Christmas trees launched us on a quest that drew us to visit other explorers of the nervous system who also encountered the other brain and were equally puzzled by it: Stephen Smith and the droplet of light radiating from astrocytes, and the brilliant observations and new ideas of Theodore Schwann, Santiago Ramón y Cajal, Camillo Golgi, Louis-Antoine Ranvier, and Fridtjof Nansen. Neuroscientists do not yet understand the other brain, but as we learn more about it we are glimpsing a far greater universe of brain function than we had ever imagined.

CHAPTER 16

Into the Future: The New Brain

The last chapter of the story of the other brain is unwritten. How will understanding glia change our understanding of the mind? Today we know that glia constitute another brain that was ignored for a century or more, a brain new to science. There, all along, the other brain was simply overlooked. Why?

To begin with, the wrong tools were used to explore it. The electrodes of neuroscientists are deaf to glial communication. Yet the glial brain was indeed communicating; it just works differently from the neuronal brain, communicating in different ways and on different time scales. But the lack of tools is not the complete answer to why neuroscientists missed half the brain until now.

Man is above all a master tool maker. Had scientists perceived a need for special tools to probe the glial brain, human ingenuity would have forged them. No matter how crude these tools may have been at the beginning, they would have been serviceable. After all, we are the creatures on this planet who learned to prey upon and dominate all other animals with nothing more than rocks sharpened with our unique ingenuity.

It was our thinking that failed us. We thought we knew how the brain worked. Dazzled by the electric neuron, neuroscientists tightened their focus intensely on this one cell type, virtually ignoring all others even though the other cells are superior in number and diversity to neurons. Our unconscious biases clouded our perception. The glial brain simply went unseen.

Science is a uniquely human activity. The most collaborative of all human activities, it spans time and space, politics and race, but it is subject to the same limitations and weaknesses as any human endeavor. The difference in science is that there is an elusive and ultimate truth to be found in this pursuit, whether our research uncovers it or not.

Understandably, research on the "unimportant" cells did not fare well in the fierce competition for precious funds doled out by government committees to support scientific research. Findings on the "unimportant" cells also lacked the "significance" required for publication in the mainstream scientific journals.

Suddenly this situation has changed. We are experiencing a scientific revolution sparked by a revelation: it's time to rethink our conclusions and reexamine our assumptions. We now know that the other brain works independently but cooperatively with the neuronal brain. Where will this new insight lead? Right now the answer is unknown, but neuroscientists in laboratories around the world are working vigorously to answer this question. It is astonishing how far we have come and how quickly. Until recently, few could even imagine how glial cells in Einstein's brain might relate to his genius. Now we can see multiple ways glia might have done so. One hundred years after Ramón y Cajal asked, "What is this other half of the brain?" scientists are answering.

Today German pathologist Rudolf Virchow is frequently mocked for, in 1856, naively naming these cells "neural glue." Here are cells that can build the brain of a fetus, direct the connection of its growing axons to wire up the nervous system, repair it after it is injured, sense impulses crackling through axons and hear synapses speaking, control the signals neurons use to communicate with one another at synapses, provide the energy source and substrates for neurotransmitters to neurons, couple large areas of synapses and neurons into functional groups, integrate and propagate the information they receive from neurons through their own private network, release neurotoxic or neuroprotective factors, plug and unplug synapses, move themselves in and out of the synaptic cleft, give birth to new neurons, communicate with the vascular and immune systems, insulate the neuronal lines of communication, and control the speed of impulse traffic through them. And some people ask, "Could these cells have anything to do with higher brain function?" How could they possibly not?

Anytime you take an unexpected turn, it is important to stop, pull out the map and compass, determine where you went wrong, where you are heading, and fix exactly where you are at this moment. Let's review the facts about glia. Information is indeed spreading through these cells, but it is transmitted by a different mechanism than electrical signals and over a different spatial and temporal scale. Glia communicate broadly, not linearly like neurons. Glia communicate slowly. These are important facts in rethinking how our brains may operate and in plotting a course of scientific investigation into the future.

The waves of calcium sweeping through glia move not on the thousandths of a second pace of neural impulses sparking along axons and across synapses, but at the glacial pace of seconds or even minutes. This fundamental difference tells us that the other brain is processing information differently from the neuronal brain and for different reasons. Glia cannot be directly involved in the quick reflexive functions where neurons excel. Glia do control the transmission of nerve impulses at the neuromuscular synapse, but they cannot be the cells saving us from falling on our face in the split second it takes to prevent a stumble from a stubbed toe. Glia do not communicate the pain of a hot stove with lightning speed to the brain to save your fingers from injury, but they do communicate chronic pain that can follow after the brain has recovered, remodeled, and recalibrated following an injury. Glia take us beyond reflexes.

Glia must be regulating other cognitive functions that are slow to develop. Astrocytes can increase or decrease neuronal excitability and synaptic strength. Their slow, steady influence suggests a critical role in balance, or "homeostasis," which keeps brain function within optimal limits rather than spinning wildly and destructively out of control or withering feebly. The importance of this control in cognition is evident from neurological disorders such as epilepsy or psychiatric conditions such as schizophrenia and autism, where neuronal function is out of balance. The rapid "within an eye blink" functions of our nervous system are actually a narrow slice of cognition. Many brain functions develop and operate slowly. Emotions and feelings, cycles of attention, cognitive changes with growth and aging, acquisition of complex skills like playing the guitar operate over time scales where glia excel and control neuronal function. These slowly changing aspects of brain function are relatively unexplored. Some would argue that these are the most interesting aspects of the mind.

Our artificial conceptual division separating the other brain from the neuronal brain is eroding, and as it dissolves, we are recognizing a new brain. The links from glia to disease are obvious: seizure, infection, stroke, neurodegenerative disease, cancer, demyelinating disease, and mental illness all involve many different types of glia, but glia regulate and remodel the brain in health as well as in sickness. Here the questions that are central to research on the neuronal brain are only now being asked of the other brain. How plastic are glia? Do they learn, sleep, age, differ in males and females, become impaired by disease? How many different kinds of glia are there?

It is this last question that is most bewildering at present. Glia are enormously diverse cells, and we lack the vocabulary and knowledge to speak about them in anything but general terms. Is an astrocyte at a synapse the same cell as the astrocyte controlling blood flow through the capillaries in the brain? Are the astrocytes surrounding dendrites in grey matter the same as astrocytes surrounding axons in white matter? Or are these cells as individually distinct as the cone and rod cells in our retinas? Separating glial apples from oranges is the first priority of glial biologists in understanding what these different components of the other brain are doing and how they interact with the neuronal brain.

Astrocytes also cover enormous territories of the brain. An oligo-dendrocyte ensheathes scores of axons. Microglia move at will through large regions of the brain. A single astrocyte can engulf one hundred thousand synapses. It seems unlikely that one astrocyte is monitoring and dictating the transmission of information individually across the thousands of synapses it surveys. A more likely possibility is that astro-cytes (and other glia) couple large groups of synapses or neurons into functional groups. This would vastly increase the power and flexibility of information processing in our brains beyond the simple changes in strength of individual synapses along a neural circuit. Glia give the brain a new dimension of information processing.

The physical dimensions and also the mechanism of communication suggest that glia cover an enormous domain of operation. The chemical means of cell-cell communication used by glia diffuses widely and across the hardwired lines of neuronal connections. These features equip glia to control information processing in the brain on a fundamentally different and more global scale than the point-to-point synaptic contacts of neurons. Such higher level oversight is likely to have significant implications for information processing and cognition.

Glia dazzle us with their multitude of communication channels. Astrocytes are equipped to intercept all types of neuronal activity, from fluxes of ions to every neurotransmitter neurons use, to the neuromodulators, peptides, and hormones that regulate nervous system function. Glia communicate among themselves with multiple lines of communication, not only with neurotransmitters, but also through gap junctions and glial transmitters, notably ATP. Neurons are fastidious; glia are promiscuous. Neurons form synapses only with the appropriate partner neurons, but glia communicate with one another and with neurons. Astrocytes sense neuronal activity and communicate with other astrocytes. But astrocytes communicate with oligodendrocytes and microglia, with vascular cells and immune cells. Glia are engaged in a global communication network that literally coordinates all types of information in the brain: glial, hormonal, immunological, vascular, and neuronal. Such global oversight and regulation must be critical to brain function, and neurons are incapable of doing it. Glia, we are beginning to see, knit *all* the cellular components of the nervous system into a functional network. In this sense, glia are indeed as Rudolf Virchow perceived—*neural glue.*

GLOSSARY

acetylcholine An excitatory neurotransmitter released from synaptic vesicles that is involved in muscle contraction and many other functions. Acetylcholine neurons are impaired in Alzheimer's disease, for example.

action potential The electrical impulse that is generated in nerve axons and carries electrical signals to the synapse. The electrical impulse is generated by the rapid exchange of charged ions (sodium and potassium) across the cell membrane of the nerve axon.

adenosine triphosphate See *ATP*.

Alzheimer's disease A neurodegenerative disorder causing senile dementia. The disease causes memory loss, difficulties with speech and understanding, and erratic emotions. Amyloid plaques and tangles of proteins inside neurons (neurofibrillary tangles) are characteristic features of brain tissue in patients with Alzheimer's disease.

amoeba A single-celled organism (protozoan) with an amorphous shape that moves by oozing and undulating motions of its cell body.

amyloid An insoluble mass of proteins that aggregate outside brain cells in several neurodegenerative diseases.

antidiuretic hormone (ADH) A hormone, also called vasopressin, that reduces the secretion of urine.

APP Amyloid-beta precursor protein, which when cleaved forms the beta-amyloid protein seen in Alzheimer's disease.

artifact An observation or measurement in an experiment produced by experimental error.

astrocyte One of four major types of glial cells, astrocytes are found only in the central nervous system, where they exhibit a wide variety of cellular properties and shapes. These cells are involved in a broad range of functions in disease and normal brain function. Astrocytes are not present in the peripheral nervous system.

311

astrogliosis An accumulation of reactive astrocytes at the site of injury or disease in the brain or spinal cord. Reactive astrocytes increase their content of several proteins involved in nervous system repair, notably GFAP and cytokines.

ATP (adenosine triphosphate) The molecule of cellular metabolism that stores energy for enzymatic reactions through the three phosphate molecules bound to an adenosine core. ATP (and adenosine) also act as signaling molecules between cells when released through channels or synaptic vesicles in cell membranes to activate membrane receptors on other cells.

axon The slender wire-like cellular extension from a neuron that sends electrical impulses to the next neuron in a circuit.

beta-amyloid A polypeptide contained in the amyloid plaques seen in brains of patients with Alzheimer's disease.

bipolar disorder Also known as manic-depressive illness, bipolar disorder is a psychiatric illness causing episodes of abnormally elevated mood (mania) cycling with severe depression. Genes and environment both have an influence. It can be associated with hormonal imbalances and imbalances in neurotransmitters, notably glutamate, dopamine, and serotonin.

blood-brain barrier The barrier of separation regulating the exchange of materials and cells between the bloodstream and the brain. The barrier is formed by specialized cells in the walls of the blood vessels in the brain that prevent passage of molecules and cells. Astrocytes closely associated with these cells regulate the barrier and also regulate the constriction and dilation of blood vessels to control blood flow locally to neurons according to changing demands. These perisynaptic astrocytes are specialized for the exchange of ions, water, and glucose between the blood and brain.

brain slice A portion of brain dissected out of an experimental animal and sliced thin to maintain it alive in the laboratory for neurophysiological studies.

brain stem The lower part of the brain connecting it with the spinal cord. This region of the brain provides essential activities necessary for maintaining attention, arousal, and motor function.

calcium wave The rise in calcium concentration in a cell that propagates in a wave-like manner to other cells.

CAT scan (computerized axial tomography) A three-dimensional X-ray of the body performed by computer reconstruction of serial X-ray sections taken at various angles.

cell body The part of a nerve cell containing the nucleus, and which in neurons gives rise to the slender axon and branched dendritic processes. In glial cells the cell body gives rise to cellular extensions from astrocytes, microglia, and oligodendrocytes. In oligodendrocytes extensions from the cell body wrap around axons to form insulating layers of myelin.

central nervous system (CNS) The portion of the nervous system inside the skull and spinal cord.

cerebellum One of the three major regions of the brain, situated at the base of the brain where it is involved in motor function and memory. The other major regions of the brain are the cerebrum and medulla.

cerebral cortex The superficial layers of the forebrain containing tightly packed neurons and dendrites. The cerebral cortex is involved in perceptual awareness, language, sensory and motor function, and consciousness.

cerebrospinal fluid The specialized fluid bathing the brain and spinal cord and filling the hollow cavities inside them.

chemokines A family of signaling molecules that regulate attraction of immune system cells and development, survival, and proliferation of neurons and glia.

CNS See *central nervous system.*

complement proteins Components of the immune system that kill foreign or abnormal cells by disrupting the cell membrane.

computerized axial tomography See *CAT scan.*

corpus callosum The major bundle of axons connecting the cerebral cortex of the brain's left and right hemispheres. The corpus callosum contains a quarter of a billion axons.

cytoarchitecture The cellular structure of the cerebral cortex, which appears as layers and variable packing densities of neurons with specific characteristic features in different regions of the brain.

cytokines Cell-to-cell signaling molecules of particular importance in immune responses and that regulate migration and other cellular responses, notably inflammation after injury. Recently cytokines have been implicated in chronic pain.

cytoplasm The interior fluids, soluble proteins, and organelles inside a cell, but not including the cell nucleus.

dementia A progressive decline in cognitive function including impairments in memory and learning.

dendrites The root-like extensions from neurons where synaptic inputs are received.

deoxyhemoglobin Hemoglobin that has lost oxygen, as when red blood cells deliver oxygen to tissues. A difference in the magnetism of hemoglobin and deoxyhemoglobin provides the signals for brain imaging by MRI.

differentiation The process of cellular development in which an immature or nonspecialized cell develops specialized adult form.

D-serine A signaling molecule that facilitates transmission at synapses using glutamate receptors, including those involved in long-term potentiation and memory.

ectoderm The outermost of three cellular layers of the early embryo, which will give rise to skin and cells of the nervous system. The other two layers are the endoderm and mesoderm.

EEG (electroencephalogram) A recording of electrical waves that originate from the combined action of electrical impulses in the cerebral cortex. The word is often used loosely to refer to brain waves in general.

electroencephalogram See *EEG*.

electron microscopy A technique for the analysis of cellular structure beyond the resolution of the light microscope. An electron microscope uses focused beams of electrons in place of light.

encephalitis Infection and inflammation of the brain.

endoplasmic reticulum (ER) A subcellular organelle that forms a network of sacs and tubes around the nucleus of a cell. The ER is involved in synthesis of lipids and proteins, but it is also a storehouse for calcium ions which are released from the cell after the proper stimulus.

epileptic seizure An abnormal, excessive, and synchronized neural activity in the cerebral cortex that causes involuntary changes in body movement, sensation, and consciousness.

estrogen A steroid sex hormone associated with female reproductive function, but also present in males and involved in regulating many cellular functions in both sexes.

excitatory neurotransmitter A neurotransmitter chemical released from synapses that excite neural activity.

frontal lobe The lobes of cerebral cortex at the front of the brain that are involved in higher-level cognitive function.

GABA (gamma-aminobutyric acid) A type of neurotransmitter chemical that is normally inhibitory, acting to decrease activity in the postsynaptic neuron.

gamma-aminobutyric acid See *GABA*.

ganglia A collection of neurons outside the central nervous system.

gap junction A protein connection between adjacent cells that allows the passage of small molecules between them. Astrocytes are extensively coupled through gap junctions to distribute ions, nutrients, and metabolic waste products between neurons and the bloodstream.

GFAP (glial fibrillary acidic protein) A filamentous protein found in astrocytes, which is greatly increased in response to neural injury or disease.

glia Short for "neuroglia," glia are nervous system cells that are not neurons and also not a part of other tissues in the brain such as vascular or connective tissue. Glia do not fire electrical impulses and they lack the identifying features of neurons, including axons, dendrites, and synapses.

glial fibrillary acidic protein See *GFAP*.

gliosis The response of an astrocyte to injury or disease, characterized by increased production of GFAP, increased cell volume, and the production of many specialized molecules involved in nervous system repair.

gliotransmitter A signaling molecule released from a glial cell that is detected by other cells, including glia, neurons, immune system cells, and vascular cells. Analogous to neurotransmitters, the signaling molecules released from neurons.

glucose A sugar used as a basic source for cellular energy.

glutamate The most abundant excitatory neurotransmitter in the brain.

Golgi stain A cellular stain developed by Camillo Golgi that involves a chemical reaction of silver salts, much like the chemistry of black and white photography, to stain neurons completely, but selectively, and in vivid detail.

grey matter One of two tissues in the brain, grey matter is made of densely packed neurons and synapses, primarily found in the outer cortex of the brain and at the core of the spinal cord. See also *white matter*.

growth factor A protein that stimulates growth or survival of neurons and other cells.

hagfish One of the most primitive living fishes, which is at the base of the evolutionary tree of vertebrates (animals with backbones).

hemoglobin The iron-containing molecule in red blood cells that carries oxygen to tissues.

hippocampus The region of the cerebral cortex resembling the curled tail of a seahorse that is involved in declarative and spatial memory.

homeostasis The process of maintaining physiological systems at an optimum level of performance, without exceeding or dropping below the normal range of function.

hypothalamus A portion of the brain that is involved in maintaining blood pressure, body temperature, hormonal cycles, and a wide range of automatic bodily functions.

hypoxia A condition in which body tissue suffers lack of adequate oxygen.

inhibitory neurotransmitter A chemical released from inhibitory synapses that decreases excitability of the postsynaptic neuron by hyperpolarizing it and thus reducing the recipient neuron's ability to fire an action potential.

interleukins A group of cytokines involved in immune system signaling.

invertebrates Animals on the evolutionary tree that are more ancient than those with backbones; for example snails, insects, and worms.

ischemia The loss of blood supply to a tissue. One of the consequences of brain stroke.

labile Capable of easily undergoing change.

lactation The production of milk for breast-feeding.

leukocyte A white blood cell.

lipids Oily or fatty organic molecules that form cell membranes.

long-term potentiation (LTP) The increase in electrical voltage (synaptic strength) produced by a synapse after stimulating it to fire repeatedly at high frequency.

LTP See previous entry.

magnetic resonance imaging See *MRI*.

medulla One of three brain regions, situated at the base of the brain and connected with the spinal cord. A part of the brain stem, the medulla is involved in many automatic functions such as breathing and regulating blood pressure, and it is a major relay station for the crossing of motor neuron tracts between the spinal cord and brain. The cerebellum and cerebrum are the other two major brain regions.

microglia One of the four major types of glia, these cells are the smallest of the glia present in the central nervous system. Their primary function is as immune system cells of the brain, protecting the central nervous system from infection and disease.

monocyte A type of white blood cell that ingests foreign organisms and materials. After leaving the bloodstream and entering tissue, monocytes develop into macrophages.

MRI (magnetic resonance imaging) A type of brain scan performed for medical diagnosis and research that provides a detailed three-dimensional image of the brain.

MS See *multiple sclerosis.*

multiple sclerosis (MS) A disease caused by the loss of electrical insulation (myelin) on axons in the central nervous system, resulting in loss of vision, mobility, or other brain functions.

myelin The electrical insulation on axons that is formed by oligodendrocytes in the central nervous system and Schwann cells in the peripheral nervous system. Myelin is a multilayered wrapping of cell membrane around an axon.

myelinated fiber An axon insulated with myelin.

nerve The cable-like tissue carrying electrical signals to the body from the brain and spinal cord and conveying signals from the sense organs of the body into the spinal cord and brain. Nerves are not to be confused with neurons, which are individual cells that are electrically excitable. The axons of neurons are bundled together inside nerve, but nerve also contains connective tissue and vascular tissue. Nerves do not exist in the central nervous system; instead, bundles of axon fibers in the brain and spinal cord are called "tracts."

neuromuscular junction (NMJ) The synapse formed by motor neurons on muscle fibers that stimulate muscular contraction.

neuron doctrine The theory formulated by Santiago Ramón y Cajal at the beginning of the twentieth century which states (1) that neurons are all separate cells communicating through specialized points of contact (synapses) but not by direct continuity of cellular protoplasm as had been assumed and (2) that neurons are polarized, such that incoming signals are received through the dendrites of a neuron and messages are sent out through its axon.

neurons The electrically excitable cells of the nervous system that transmit electrical impulses and communicate with other neurons through synapses. Neurons have three distinct cellular features: cell body, axon, and dendrite.

neurotoxin A toxin that kills neurons.

neurotransmitter A chemical substance released from synapses that allows communication between neurons.

nitric oxide A gas generated by cells that functions as a neurotransmitter.

NMDA (N-methyl-D-aspartate) A type of glutamate receptor that is essential for memory formation and long-term potentiation.

NMJ See *neuromuscular junction*.

NO See *nitric oxide*.

node of Ranvier A gap between regions of myelin electrical insulation on axons where the electrical impulse is generated.

nucleus The single organelle in all cells where the genes are located.

oligodendrocyte One of the four major types of glial cells, oligodendrocytes form myelin electrical insulation on axons in the central nervous system, but they are not present in the peripheral nervous system.

oxytocin A peptide synthesized by the hypothalamus that acts as a hormone to regulate many bodily functions, notably those involved in birth and maternal behavior.

peripheral nervous system (PNS) The portion of the nervous system outside the brain and spinal cord. It includes the nerves of the trunk and limbs.

perisynaptic Schwann cells Schwann cells that tightly surround synapses connecting to muscle fibers. Also called terminal Schwann cells.

pituitary A gland located at the base of the brain controlling hormonal functions of the body. It is controlled by a part of the hypothalamus that connects to it through nerve axons.

plaque An abnormal localized region of the brain containing accumulations of proteins, as in Alzheimer's disease, or cellular debris, as in multiple sclerosis.

PNS See *peripheral nervous system*.

postsynaptic terminal The receiving side of the synapse, where receptors on the dendrite detect neurotransmitter released from the partner axon's presynaptic terminal.

potassium channel A type of protein channel in cell membranes that allows potassium ions to pass into or out of cells. The release of positively charged potassium ions from axons through potassium channels is essential for generating the electrical impulse (action potential).

presynaptic terminal The side of the synapse where neurotransmitter is released from axon endings.

prodromal The period prior to the active phase of schizophrenia characterized by early, vague symptoms or psychosis.

progesterone A steroid sex hormone involved in the female menstrual cycle, but with regulatory effects on many other cells in the bodies of both men and women.

protoplasm See *cytoplasm*.

reactive astrocyte An astrocyte that undergoes changes in cellular structure and increases production of proteins involved in repairing the nervous system after injury, notably the cytoskeletal protein GFAP and cytokines.

schizophrenia A serious mental illness that undermines a person's perception of reality through hallucinations, delusions, and disordered thinking and behavior. Drugs affecting the balance of dopamine, serotonin, and glutamate neurotransmitters are used together with behavioral therapy to treat patients with this disorder.

Schwann cell One of the four major types of glial cells, Schwann cells are the glia found in nerves of the peripheral nervous system. These cells vary widely in shape and carry out many different functions. Schwann cells, in all their various forms, must perform all the functions of the three types of glia recognized in the central nervous system (microglia, astrocytes, and oligodendrocytes), as well as forming the myelin electrical insulation around axons in peripheral nerves.

sex hormones Steroids that regulate gender and reproductive functions of the body. Testosterone, estrogen, and progesterone are the most widely known sex hormones, but there are many others derived from them. All three of these major sex hormones are present in both sexes, but in variable amounts, with testosterone predominating in males and estrogen and progesterone predominating in females.

silver chromate stain A compound used in black and white photography applied by Camillo Golgi to stain neurons in fine detail. The method involves treating brain tissue with potassium dichromate and silver nitrate. A chemical reaction producing silver chromate fills the neuron with fine black pigment.

sleep apnea A sleep disorder characterized by prolonged pauses in breathing. Sudden infant death is sometimes caused by sleep apnea.

smooth muscle One of three types of muscle, smooth muscle is involved in contraction of involuntary organs, for example the uterus, gastrointestinal and respiratory tracts, and arteries. The other types of muscle include skeletal muscle, responsible for bodily movement, and cardiac muscle, which is specialized muscle tissue of the heart.

sodium channel A protein ion channel in the membrane of cells that allows sodium ions to pass into or out of cells. Sudden passage of sodium ions into an axon raises the voltage inside because the ions carry a positive charge; this process causes the electrical voltage spike called an action potential.

spinal cord An extension of the brain down the back through the hollow vertebrae of the spine (backbone). The spinal cord is where all sensory and motor nerves of the body enter the central nervous system, except for those from the face and head.

spreading depression A wave of electrical activity spreading through the cerebral cortex, which can be measured with an EEG. The depression results from the synchronized loss of voltage in large groups of cortical neurons that spreads through the cortex. It can be initiated by lack of blood flow and is associated with migraine headache and epilepsy.

synapse The point where communication takes place between neurons. The synapse is formed by nerve terminals that release neurotransmitter from synaptic vesicles; these neurotransmitters activate receptors on the dendrite of the next neuron in the circuit. Depending on the type of neurotransmitter released, the recipient neuron is excited or inhibited from firing an action potential. Thus synapses connect neurons into circuits and restrict the flow of information to one direction: from the axon terminal to the dendrite.

synaptic plasticity Changes in the strength of a synaptic connection. Strengthening or weakening synapses is the cellular basis for learning and memory. The formation and elimination of synapses are important in wiring brain circuits during development, in learning, and in recovery from injury.

synaptic vesicle An organelle shaped like a 25-nanometer-diameter sphere that contains neurotransmitter molecules for release from the presynaptic membrane of a synapse. Scores of synaptic vesicles are accumulated next to the presynaptic membrane. In response to the arrival of an electrical impulse, individual vesicles fuse with the cell membrane and release their contents onto the postsynaptic membrane of the synapse.

taurine A naturally occurring amino acid in the body that is not used in forming proteins but has various roles in chemical reactions and neurotransmission.

T cell A type of white blood cell identified by having CD4 receptors, T cells are involved in immune responses to infection. These cells are infected and killed in AIDS.

temporal lobe The region of the cerebral cortex near the temples that is a common site for epilepsy. Its primary role is in auditory and visual processing. The temporal lobe contains the hippocampus, which is essential for spatial and declarative memory.

terminal Schwann cell The specialized Schwann cell that tightly surrounds a synapse onto muscle.

testosterone One of the major steroid sex hormones, primarily associated with male reproductive function but also present in females and involved in regulating many cellular processes in both sexes.

thalamus Located at the center of the brain, the thalamus (a paired structure) is a tightly packed mass of neurons that relays information into the cerebral cortex from many different regions of the brain. Axons from every sensory system (except the olfactory) relay their signals through the thalamus to reach the cerebral cortex.

transporter A molecule in the cell membrane that transports other molecules into or out of the cell. Neurotransmitter transporters, which remove neurotransmitter from the synaptic cleft, have a critical function in maintaining synaptic transmission in the nervous system. Neurotransmitter transporters in astrocytes are the major mechanism regulating neurotransmitter uptake from synapses.

unmyelinated fiber An axon that is not insulated with myelin and thus lacks nodes of Ranvier. Unmyelinated axons are small in diameter and cannot conduct electrical impulses at high speed.

vertebrate Vertebrates are the animals with backbones, including fish, amphibians, reptiles, birds, and mammals.

white matter One of two types of brain tissue, white matter is the densely packed core of nerve axon fibers located beneath the cerebral cortex, connecting neurons in different regions of the brain into circuits. The white color is due to the electrical insulation on axons (myelin) that is formed by oligodendrocytes. Large bundles of white matter are also prominent in the spinal cord, where massive cables of axons carry signals to and from the brain. Unlike in the brain, white matter is concentrated on the surface of the spinal cord, with the neurons forming the core. See also *grey matter*.

NOTES

Chapter 1: Bubble Wrap or Brilliant Glue?

1. Paterniti, M. (2000) *Driving Mr. Albert: A Trip across America with Einstein's Brain*. Random House, New York.

Chapter 2: A Look Inside the Brain

1. Adapted from Fields, R. D. (2006) Beyond the neuron doctrine. *Scientific American Mind* June/July 17: 20–27.
2. Florkin, M. (1975) Theodore Ambrose Hubert Schwann, in *Dictionary of Scientific Biography*, Charles C. Gillispie, ed., vol. 12. Scribner, New York.
3. Ibid., p. 242.
4. Ibid., p. 244.
5. Multhauf, L. S. (1978) Fridtjof Nansen, in *Dictionary of Scientific Biography*, Charles C. Gillispie, ed., vol. 15, supplement. Scribner, New York, p. 430; see also Nansen, F. (1897) *Farthest North: Being the Record of a Voyage of Exploration of the Ship* Fram. *1893–1896, and of a Fifteen Months' Sleigh Journey by Dr. Nansen and Lt. Johansen,* 2 vols. Harper, New York.
6. Cited in Galambos, R. (1961) A glia-neural theory of brain function. *Proc. Natl. Acad. Sci. USA* 47: 131.

Chapter 5: Brain and Spinal Cord Injury

1. The Hearing Loss Web: http://www.hearinglossweb.com/Medical/Causes/nihl/mus/mus.htm#many.

Chapter 6: Infection

1. Gajdusek, C. (1963) Kuru. *Trans. Roy. Soc. Tropic. Med. Hyg.* 57: 151–69; see p. 152.

2. March 15, 1957, letter, D. Carleton Gajdusek to Dr. Joseph Smadel, NIH Medical Library. Note that the NIH Medical Library has a collection of letters and field notebooks provided by Carleton Gajdusek spanning the years 1957–93. The spiral-bound collection "Kuru Epidemiological Patrols from the New Guinea Highlands to Papua, August 21, 1957–November 10, 1957" and "Correspondence on the Discovery and Original Investigations on Kuru, Smadel-Gajdusek Correspondence 1955–1958" cover the period of events in this chapter. A selection of these documents has been reprinted in a book edited by Judith Farquhar and D. Carleton Gajdusek, published by Raven Press (see next note).

3. Julias, Charles (1957) Sorcery Among the South Fore, with Special Reference to Kuru, in Farquhar, J., and Gajdusek, D. C., eds. (1982) *Kuru, Early Letters and Field-Notes from the Collection of D. Carleton Gajdusek.* Raven Press, New York, p. 287.

4. March 15, 1957, letter, D. Carleton Gajdusek to Joseph Smadel, NIH Medical Library.

5. April 3, 1957, letter, D. Carleton Gajdusek to Joseph Smadel, NIH Medical Library.

6. Ibid.

7. April 9, 1957, letter, Sir Macfarlane Burnet to Carleton Gajdusek, NIH Medical Library.

8. May 28, 1957, letter, D. Carleton Gajdusek to Joseph Smadel, NIH Medical Library.

9. April 9, 1957, letter, J. T. Gunther to Sir Macfarlane Burnet, NIH Medical Library.

10. Sir Macfarlane Burnet, handwritten summary of telephone conversation with R. F. R. Scragg, March 29, 1957, NIH Medical Library.

11. March 13, 1957, letter, Carleton Gajdusek to Sir Macfarlane Burnet, NIH Medical Library.

12. Ibid.

13. March 30, 1957, Radiogram, R. F. R. Scragg to Carleton Gajdusek, NIH Medical Library.

14. March 30, 1957, D. Carleton Gajdusek to R. F. R. Scragg, handwritten note at bottom of radiogram in note 13.

15. May 28, 1957, letter, Carleton Gajdusek to Joseph Smadel, NIH Medical Library.

16. August 29, 1957, letter, Carleton Gajdusek to Vincent Zigas and Jack Baker, NIH Medical Library.

17. October 8, 1957, Carleton Gajdusek field notes, Camp 4, Iwane hamlet, Simbari Kukuku, p. 190, NIH Medical Library.

18. May 21, 1957, letter, Dr. Thomas Rivers to Carleton Gajdusek, NIH Medical Library.

19. May 27, 1957, letter, Sir Macfarlane Burnet to Thomas Rivers, NIH Medical Library.

20. Mid-April 1957, letter, Sir Macfarlane Burnet to Dr. John Gunther, NIH Medical Library.

21. Early April 1957, letter, Carleton Gajdusek to Sir Macfarlane Burnet, NIH Medical Library.
22. May 19, 1957, Carleton Gajdusek to Sir Macfarlane Burnet and Dr. Anderson, NIH Medical Library.
23. Early April 1957, letter, Carleton Gajdusek to Sir Macfarlane Burnet, NIH Medical Library.
24. Late May 1957, letter, Carleton Gajdusek to Joseph Smadel, NIH Medical Library.
25. June 4, 1957, letter, Carleton Gajdusek to Sir Macfarlane Burnet and S. Gray Anderson, NIH Medical Library.
26. Gajdusek, C. (1963) Kuru. *Trans. Roy. Soc. Tropic. Med. Hyg.* 57: 168.
27. Ibid.
28. Glasse, R. (1967) Cannibalism in the kuru region of New Guinea. *Trans. N.Y. Acad. Sci.* series 2, 29: 748–54; see pp. 751–52.
29. September 18, 1957, letter, Carleton Gajdusek to Joseph Smadel, NIH Medical Library.
30. Gajdusek, D. C. October 5, 1957, entry from Field Journals, Kuru Epidemiological Patrols, 26 September–9 November 1957, Auroga village, Camp No. 2, Kukukuku, NIH Medical Library.
31. Ibid.
32. Gajdusek, D. C., October 6, 1957, entry from Field Journals, Kuru Epidemiological Patrols, 26 September–9 November 1957, Tchaiorogoro hamlet, Auroga village, Kukukuku, NIH Medical Library.
33. Gajdusek, D. C., October 7, 1957, entry from Field Journals, Kuru Epidemiological Patrols 26 September–9 November 1957, Camp No. 3, Tchaiorogoro village, Auroga Kukukuku, NIH Medical Library.
34. Prusiner, S.B. (1995) The prion diseases. *Scientific American* 272: 48–51.
35. Prusiner, S. B. (1997) Prions, Nobel Foundation Lecture, December 8, 1997. http://nobelprize.org.
36. Weinberg, Rick (1991) Magic Johnson announces he's HIV-positive. ESPN.com, November 7, http://sports.espn.go.com/espn/espn25/story?page=moments/7.
37. UNAIDS, Joint United Nations Programme on HIV/AIDS (2006) 2006 Report on the global AIDS epidemic, http://data.unaids.org/pub/GlobalReport/2006/2006_GR-ExecutiveSummary_en.pdf. See also Centers for Disease Control and Prevention, HIV prevalence, unrecognized infection, and HIV testing among men who have sex with men—five U.S. cities, June 2004–2005, http://www.cdc.gov/mmwr/preview/mmwrhtml/mm5424a2.htm.

Chapter 7: Mental Health

1. DeToledo, J. C., and Lowe, M. R. (2003) Epilepsy, demonic possessions, and fasting: another look at translations of Mark 9:16. *Epilepsy and Behavior* 4: 338–39.
2. Masia, S. L., and Devinsky, O. (2000) Epilepsy and behavior: a brief history. *Epilepsy and Behavior* 1: 27–36.

3. Brown, B. (1866) On the Curability of Certain Forms of Insanity, Epilepsy, Catalepsy, and Hysteria in Females. Cox and Wyman, London, p. 13.

4. Ibid., pp. 79–83.

5. Masia, S. L., and Devinsky, O. (2000) Epilepsy and behavior: a brief history. *Epilepsy and Behavior* 1: 27–36.

6. Ibid., p. 29.

7. Xiao, L., et al. (2008) Quetiapine facilitates oligodendrocyte development and prevents mice from myelin breakdown and behavioral changes. *Mol. Psychiatry* 7: 697–708.

8. Saakov, B. A., Khoruzhaia, T. A., and Bardakhch'ian, E. A. (1977) Ultrastructural mechanisms of serotonin demyelination. *Bull. Exp. Biol. Med.* 83: 719–23.

9. Fink, M. (2005) ECT: Serendipity or logical outcome? *Psychiatric Times* 21(1), http://www.psychiatrictimes.com/display/article/10168/47531.

10. Ibid. See also Meduna, L. (1985) Autobiography. *Convulsive Ther.* 1: 43–57, 121–35; Fink, M. (1999) Images in psychiatry, Ladislas J. Meduna, M. D., 1869–1964 *Am. J. Psychiatry* 156: 1807; Meduna, L. (1937) *Die Konvulsionstherapie der Schizophrenie.* Carl Marhold, Halle, Germany.

11. Ibid. See also Fink, M. (2007) Electroshock works. Why? *Psychiatric Times* 24(7), http://www.psychiatrictimes.com/display/article/10168/54461?pageNumber=1; Fields, R. D. (2008) White matter in learning, cognition and psychiatric disorders. *Trends Neurosci.* 31: 361–70; Miller, G. (2005) The dark side of glia *Science* 308:788–81.

12. Fink, M. (2005) ECT: Serendipity or logical outcome? *Psychiatric Times* 21(1), http://www.psychiatrictimes.com/display/article/10168/47531.

13. Millett, D. (2001) Hans Berger from psychic energy to EEG. *Prosp. Biol. Med.* 44(4): 522–42.

14. Annas, G. J., and Grodin, M. A. (1992) *The Nazi Doctors and the Nuremberg Code.* Oxford University Press, New York, p. 24.

15. Ibid., p. 25.

16. Lifton, R. J. (1986) *The Nazi Doctors, Medical Killing and the Psychology of Genocide.* Basic Books, New York.

17. Gloor, P. (1969) *Hans Berger on the Electroencephalogram of Man.* Elsevier Publishing Company, New York, pp. 11–12.

18. Redies, C., Viebig, M., Zimmermann, S., and Frober, R. (2005) Origin of corpses received by the anatomical institute at the University of Jena during the Nazi regime. *Anatomical Record* 285: 6–10.

19. The Hebrew University of Jerusalem, Institute of Chemistry, http://chem.ch.huji.ac.il/~eugeniik/history/berger.html; German Epilepsy Museum, Kork, http://www.epilepsiemuseum.de/english/diagnostik/berger.html p. 1, top ten scientists who committed suicide listverse, http://listverse.com/people/top-10-scientists-who-committed-suicide/.

Chapter 8: Neurodegenerative Disorders

1. Hawking, S. (n.d.) My experience with ALS, http://www.hawking.org.uk/index .php/disability/disabilityadvice.
2. Gehrig, L. (1939) Farewell speech, http://www.lougehrig.com/about/speech.htm.
3. American Heart Association statistics, http://www.americanheart.org.
4. The Ohio State University Medical Center, Statistics of Stroke, http://medical center.osu.edu/patientcare/healthcare_services/stroke/statistics/Pages/index.aspx.

Chapter 9: Glia and Pain

1. Associated Press (2004) Rare disease makes girl unable to feel pain. November 1, see http://www.msnbc.msn.com/id/6379795/.
2. Schorn, D. (2006) Prisoner of Pain, CBS 60 minutes Jan 29, 2006, http://www .cbsnews.com/stories/2006/01/25/60minutes/main1238202.shtml.
3. For an introduction to the history and controversy surrounding the discovery of anesthesia see Alfred Randy (2008) Sept. 30, 1846: Either he was the first or he wasn't. *Wired* magazine. www.wired.com/science/discoveries/news/2008/09/dayintech_0930.

Chapter 10: Glia and Addiction

1. Kindy, K. (2008) A tangled story of addiction. *Washington Post,* September 12. www.washingtonpost.com/wp-dyn/content/article/2008/09/11/AR2008091103928.html.
2. Abel, E. L., and Sokol, F. J. (1986) Fetal alcohol syndrome is now leading cause of mental retardation. *Lancet* 2: 1222.
3. Climent, E., Pascual, M., Renau-Piqueras, J., and Guerri, C. (2002) Ethanol exposure enhances cell death in the developing cerebral cortex: role of brain-derived neurotrophic factor and its signaling pathways. *J. Neurosci. Res.* 68: 213–25; see also Costa, L. G., Yagle, K., Vitalone, A., and Guizzetti, M. Alcohol and glia in the developing brain, in *The Role of Glia in Neurotoxicity*, Michael Ascher and Lucio G. Costa, eds. CRC Press, Boca Raton, FL, 2005, pp. 343–54.
4. Guerri, C., Bazinet, A., and Riley, E. P. (2009) Foetal alcohol spectrum disorders and alterations in brain and behavior. *Alcohol and Alcoholism* 44: 108–14.

Chapter 11: Mother and Child

1. Couzin, J. (2009) Celebration and concern over U.S. trial of embryonic stem cells. *Science* 568: 323.
2. Treffert, D. A., and Christensen, D. D. (2005) Inside the mind of a savant. *Scientific American* December 294: 108–13.

Chapter 12: Aging

1. Shenk, D. (2006) The memory hole. *New York Times,* November 3, http://www
   .nytimes.com/2006/11/03/opinion/03shenk.html?pagewanted=print.
2. Reagan, R. (1985) Proclamation 5405 by the President of the United States—
   National Alzheimer's Disease Month, November 8, 1985, The American Presi-
   dency Project, University of California, Santa Barbara, http://www.presidency
   .ucsb.edu/ws/index.php?pid=38037
3. Reagan, R. (1994) Letter written by President Ronald Reagan announcing
   he has Alzheimer's disease, November 5, 1994, Ronald Reagan Presidential
   Library, University of Texas, http://www.reagan.utexas.edu/archives/reference/
   alzheimerletter.html.

Chapter 13: The Mind of the Other Brain

1. This video sequence can be seen at: http://latimesblogs.latimes.com/herocom
   plex/2009/06/edward-nortons-brain-up-close-and-personal.html.

Chapter 15: Thinking beyond Synapses

1. Coyle, D. (2009) *The Talent Code.* Bantam Books, New York; see also video clip
   by Bracken, K. (2007) The brains behind talent. *Play Magazine, New York
   Times,* March, http://video.nytimes.com/video/2007/03/02/sports/119481710
   8368/the-brains-behind-talent.html.
2. Marsiglia, S. (2002) NIH's senior scientist has no plans to retire. *The NIH
   Record,* June 11, 2002.

BIBLIOGRAPHY

PART I:
DISCOVERING THE OTHER BRAIN

Chapter 1: Bubble Wrap or Brilliant Glue?

Bentivoglio, M. (1998) Life and discoveries of Santiago Ramón y Cajal. The Nobel Foundation, April 20, 1998, http://nobelprize.org/nobel_prizes/medicine/articles/cajal/indes.html.

Diamond, M. C., Scheibel, A. B., Murphy, G. M. Jr., and Harvey, T. (1985) On the brain of a scientist: Albert Einstein. *Exp. Neurol.* 88: 198–204.

Fields, R. D. (2004) The other half of the brain. *Scientific American* April 290: 54–61.

———. (2006) Beyond the neuron doctrine. *Scientific American Mind* June/July 17: 20–27.

Golgi, C. (1906) The neuron doctrine—theory and facts. The Nobel Lectures, The Nobel Foundation, http://nobelprize.org/nobel_prizes/medicine/laureates/1906/golgi-lecture.html

Paterniti, M. (2000) *Driving Mr. Albert: A Trip Across America with Einstein's Brain.* Random House, Inc., New York.

Ramón y Cajal, Santiago (1928) *Degeneration and Regeneration of the Nervous System.* Oxford University Press, London.

———. (1990) *New Ideas on the Structure of the Nervous System in Man and Vertebrates.* MIT Press, Cambridge, MA.

———. (1995) *Histology of the Nervous System of Man and Vertebrates.* Oxford University Press, New York.

———. (1999) *Advice for a Young Investigator.* MIT Press, Cambridge, Mass.

Stevens, B., and Fields, R. D. (2000) Action potentials regulate Schwann cell proliferation and development. *Science* 287: 2267–71.

Witelson, S. F., Kigar, D. L., and Harvey, T. (1999) The exceptional brain of Albert Einstein. *Lancet* 353: 2149–53. See also *Lancet* 344: 1821–23 for reader comments and the author's reply.

Chapter 2: A Look Inside the Brain

Aldskogius, H., Liu, L., and Svensson, M. (1999) Glial responses to synaptic damage and plasticity. *J. Neurosci. Res.* 1: 33–41.

Alexander, W. S. (1949) Progressive fibrinoid degeneration of fibrillary astrocytes associated with mental retardation in a hydrocephalic infant. *Brain* 72: 373–81.

Aschner, M., Sonnewald, U., and Tan, K. H. (2002) Astrocyte modulation of neurotoxic injury. *Brain Pathol.* 12: 475–81.

Barger, S. W., and Basile, A. S. (2001) Activation of microglia by secreted amyloid precursor protein evokes release of glutamate by cystine exchange and attenuates synaptic function. *J. Neurochem.* 76: 846–54.

Barnett, J. A. (1998) A history of research on yeasts 1: work by chemists and biologists 1789–1850. *Yeast* 14: 1439–51.

Bergles, D. E., and Jahr, C. E. (1997) Glutamate transporter currents in hippocampal astrocytes. *Neuron* 19: 1297–1308.

Brenner, M., Goldman, J. E., Quinlan, R. A., and Messing, Albee (2008) Alexander disease: a genetic disorder of astrocytes, in *(Patho)Physiology of the Nervous System*, V. Parpura and P. Haydon, eds. Springer Science, New York.

Bruce-Keller, A. J. (1999) Microglial-neuronal interactions in synaptic damage and recovery. *J. Neurosci. Res.* 58: 191–201.

Bullock, T. H. (2004) The natural history of neuroglia: an agenda for comparative studies. *Neuron Glia Biology* 1: 97–100.

Fields, R. D. (2008) Schwann cell interactions with axons, in *The Encyclopedia of Neuroscience*, L. Squire, ed. Academic Press, Oxford.

Fields, R. D., and Stevens-Graham, B. (2002) New insights into neuron-glia communication. *Science* 298: 556–62.

Galambos, R. (1961) A glia-neural theory of brain function. *Proc. Natl. Acad. Sci. USA* 47: 129–36.

Harrison, R. G. (1904) An experimental study of the relation of the nervous system to the developing musculature in the embryo of the frog. *Am. J. Anat.* 3: 197–220.

Houades, V., et al. (2006) Shapes of astrocyte networks in the juvenile brain. *Neuron Glia Biology* 2: 3–14.

Jacobson, M. (1991) *Developmental Neurobiology*. Plenum Press, New York.

Johnson, A. B. (1996) Alexander disease, in *Handbook of Clinical Neurology*, H.W. Moser, ed. Elsevier Science, Amsterdam, pp. 701–10.

Kettenmann, H., and Ransom, B. R., eds. (2005) *Neuroglia*. Oxford University Press, New York.

Koelliker, R. A. (1852) *Handbuch der Gewebelehre des Menschen*. Engelmann, Leipzig.

Marin-Teva, J. L. (2004) Microglia promote the death of developing Purkinje cells. *Neuron* 41: 491–93.

Moriguchi, S. (2003) Potentiation of NMDA receptor-mediated synaptic responses by microglia. *Brain Res. Mol. Brain Res.* 119: 160–69.

Nansen, F. (1887) *The Structure and Combination of the Histological Elements in the Central Nervous System.* Bergen, Norway.

———. (1890) *The First Crossing of Greenland,* Hubert M. Gepp, trans. Longmans, Green, London.

———. (1891) *Eskimo Life,* William Archer, trans. Longmans, Green, London.

———. (1897) *Farthest North: Being the Record of a Voyage of Exploration of the Ship Fram. 1893–1896, and of a Fifteen Months' Sleigh Journey by Dr. Nansen and Lt. Johansen,* 2 vols. Harper, New York.

———. (1922) Autobiography. The Nobel Lectures, the Nobel Foundation, http://nobelfoundation.org/nobel_prizes/peace/laureats/1922/nansen-bio .html.

Oliet, S. H., Diet, R., and Poulain, D. A. (2001) Control of glutamate clearance and synergetic efficacy by glia coverage of neurons. *Science* 292: 923–26.

Ramón y Cajal, S. (1928) *Degeneration and Regeneration of the Nervous System,* Raoul M. May, ed. and trans. Oxford University Press, London.

———. (1989) *Recollections of My Life,* E. H. Craigie and J. Cano, trans. MIT Press, Cambridge, MA.

Ranvier, L. A. (1871) Contributions a l'histologie et à la physiologie des nerfs periphériques. *C. R. Hebd. Seances Acad. Sci.* 73: 1168–71.

———. (1872) Recherches sur l'histologie et la physiologie des nerfs. *Arch. Physiol. Norm. Path.* 4: 129–49.

Río-Hortega, P. del (1932) Microglia, in *Cytology and Cellular Pathology of the Nervous System,* vol. 2, W. Penfield, ed. Hoeber, New York, pp. 483–534.

Roumier, A., et al. (2004) Impaired synaptic function in the microglial KARAP/ DAP12-deficient mouse. *J. Neurosci.* 24: 11421–28.

Scemes, E., Suadicani, S. O., Dahl, G., and Spray, D. C. (2007) Conexin and pannexin mediated cell-cell communication. *Neuron Glia Biology* 3: 199–208.

Sherwood, C. C., et al. (2006) Evolution of increased glia-neuron ratios in the human frontal cortex. *Proc. Natl. Acad. Sci. USA* 103: 13606–11.

Somjen, G. G. (1988) Nervenkitt: notes on the history of the concept of neuroglia. *Glia* 1: 2–9.

Soury, J. (1899) *Le système nerveux central. Structure et fonctions. Historie critique des theories et des doctrines,* 2 vols. Carré et Naud, Paris.

Speidel, C. C. (1932) Studies of living nerves. I. The movements of individual sheath cells and nerve sprouts correlated with the process of myelin sheath formation in amphibian larvae. *J. Exp. Zool.* 61: 279–331.

Stagi, M. (2005) Breakdown of axonal synaptic vesicle precursor transport by microglial nitric oxide. *J. Neurosci.* 25: 352–62.

Vignal, W. (1889) *Développement des éléments du système nerveux cérébro-spinal.* G. Masson, Paris.

Virchow, R. (1856) *Gesammelte Abhandlungen zur Wissenschaftlichen Medizin.* Hamm, Frankfurt AM. (Note: The first use of the word "neuroglia" is on p. 890.)

———. (1858) *Cellularpathologie in ihre Begründung auf Physiologische und Pathologische Gewebelehre.* A. Hirschwald, Berlin.

Wegiel, J., et al. (2001) The role of microglial cells and astrocytes in fibrillar plaque evolution in transgenic APP(SW) mice. *Neurobiol. Aging* 22: 49–61.

Wekerle, H. (2004) Planting and pruning in the brain: MHC antigens involved in synaptic plasticity? *Proc. Natl. Acad. Sci. USA* 101: 17843–48.

Chapter 3: Transmissions from the Other Brain

Araque, A. (2008) Astrocytes process synaptic information. *Neuron Glia Biology* 4: 3–10.

Araque, A., Carmingnoto, G., and Haydon, P. G. (2001) Dynamic signaling between neurons and glia. *Annu. Rev. Physiol.* 63: 795–813.

Araque, A., Li, N., Doyle, R. T., and Haydon, P. G. (2000) SNARE protein-dependent glutamate release from astrocytes. *J. Neurosci.* 20: 666–73.

Araque, A., Parpura, V., Sanzgiri, R. P., and Haydon, P. G. (1999) Tripartite synapses: glia, the unacknowledged partner. *Trends Neurosci.* 22: 208–15.

Barres, B. A., Chun, L. Y., and Corey, D. P. (1988) Ion channel expression by white matter glia: 1. Type 2 astrocytes and oligodendrocytes. *Glia* 1: 10–30.

Beattie, E. C., et al. (2002) Control of synaptic strength by glial TNF alpha. *Science*, 395: 2282–85.

Coles, J. A., and Abbott, N. J. (1996) Signalling from neurons to glial cells in invertebrates. *Trends Neurosci.* 19: 358–62.

Cornell-Bell, A. H., Finkbeiner, S. M., Cooper, M. S., and Smith, S. J. (1990) Glutamate induces calcium waves in cultured astrocytes: long-range glial signaling. *Science* 247: 470–73.

Cotrina, M. L., et al. (2000) ATP-mediated glia signaling. *J. Neurosci.* 20: 2835–44.

Dani, J. W., Chernjavsky, A., and Smith, S. J. (1990) Neuronal-activity triggers calcium waves in hippocampal astrocyte networks. *Neuron* 8: 429–40.

Fields, R. D. (2004) The other half of the brain. *Scientific American* April 290: 54–61.

Fields, R. D., and Stevens, B. (2000) ATP: an extracellular signaling molecule between neurons and glia. *Trends Neurosci.* 23: 625–33.

Fields, R. D., and Stevens-Graham, B. (2002) New insights into neuron-glia communication. *Science* 298: 556–62.

Galambos, R. (2007) Reflections on the conceptual origins of neuron-glia interactions. *Neuron Glia Biology* 3: 89–91.

Gallo, V., and Ghiani, C. A. (2000) Glutamate receptors in glia: New cells, new inputs and new functions. *Trends Pharmacol. Sci.* 21: 252–58.

Georgiou, J., et al. (1994) Synaptic regulation of glial protein expression in vivo. *Neuron* 12: 443–55.

Guthrie, P. B., et al. (1999) ATP released from astrocytes mediates glial calicum waves. *J. Neurosci.* 19: 520–28.

Hassinger, T. D., Guthrie, P. B., Atkinson, P. B., Bennett, M. V., and Kater, S. B. (1996) An extracellular signaling component in propagation of astrocytic calcium waves. *Proc. Natl. Acad. Sci. USA* 93: 13268–73.

Haydon, P. G. (2001) Glia: listening and talking to the synapse. *Nat. Rev. Neurosci.* 2: 185–93.

Innocenti, B., Parpura, V., and Haydon, P. G. (2000) Imaging extracellular waves of glutamate during calcium signaling in cultured astrocytes. *J. Neurosci.* 20: 1800–88.

Jeftinija, S. D., Jeftinija, K. V., and Stefanovic, G. (1997) Cultured astrocytes express proteins involved in vesicular glutamate release. *Brain Res.* 750: 41–47.

Kang, J., Jiang, L., Glodman, S. A., and Nedergaard, M. (1998) Astrocyte-mediated potentiation of inhibitory synaptic transmission. *Nat. Neurosci.* 1: 683–92.

Kettenmann, H., and Ransom, B. R., eds. (1995) *Neuroglia.* Oxford University Press, New York.

Kimelberg, H. K. (2007) Supportive or information-processing functions of the mature protoplasmic astrocyte in the mammalian CNS? A critical appraisal. *Neuron Glia Biology* 3: 181–89.

Kirischuk, S., et al. (1995) ATP-induced cytoplasmic calcium mobilization in Bergmann glial cells. *J. Neurosci.* 15: 7861–71.

Lev-Ram, V., and Ellisman, M. H. (1995) Axonal activation-induced calcium transients in myelinating Schwann cells, sources, and mechanisms. *J. Neurosci.* 15: 2628–37.

Liu, Q. S., Xu, W., Kang, J., and Nedergaard, M. (2004) Astrocyte activation of presynaptic metabotropic glutamate receptors modulates hippocampal inhibitory synaptic transmission. *Neuron Glia Biology* 1: 307–16.

Nedergaard, M. (1994) Direct signaling from astrocytes to neurons in cultures of mammalian brain cells. *Science* 263: 1768–71.

Newman, E. A. (2004) A dialogue between glia and neurons in the retina: modulation of neuronal excitability. *Neuron Glia Biology* 1: 245–52.

Newman, E. A., and Zahs, K. R. (1997) Calcium waves in retinal glial cells. *Science* 275: 844–47.

Orkand, R. K., Nicholls, J. G., and Kuffler, S. W. (1966) Effect of nerve impulses on the membrane potential of glial cells in the central nervous system of amphibia. *J. Neurophysiol.* 29: 788–806.

Parpura, V., et al. (1994) Glutamate-mediated astrocyte-neuron signaling. *Nature* 369: 744–47.

Parpura, V., and Haydon, P. G. (2000) Physiological astrocytic calcium levels stimulate glutamate release to modulate adjacent neurons. *Proc. Natl. Acad. Sci. USA* 97: 8629–34.

Porter, J. T., and McCarthy, K. D. (1996) Hippocampal astrocytes in situ respond to glutamate released from synaptic terminals. *J. Neurosci.* 16: 5073–81.

———. (1997) Astrocytic neurotransmitter receptors in situ and in vivo. *Prog. Neurobiol.* 51: 439–55.

Ransom, B. R., and Orkand, R. K. (1996) Glial-neuronal interactions in nonsynaptic areas of the brain: studies in the optic nerve. *Trends Neurosci.* 19: 352–58.

Reist, N. E., and Smith, S. J. (1992) Neurally evoked calcium transients in terminal Schwann cells at the neuromuscular junction. *Proc. Natl. Acad. Sci. USA* 89: 7625–29.

Robitaille, R. (1998) Modulation of synaptic efficacy and synaptic depression by glial cells at the frog neuromuscular junction. *Neuron* 21: 847–55.

Sontheimer, H. (1994) Voltage-dependent ion channels in glial cells. *Glia* 11: 156–72.

Steinhäuser, C., Jabs, R., and Kettenmann, H. (1994) Properties of GABA and glutamate responses in identified glial cells of the mouse hippocampal slice. *Hippocampus* 4: 19–35.

Stevens, B., and Fields, R. D. (2000) Action potentials regulate Schwann cell proliferation and development. *Science* 287: 2267–71.

Stevens, B., Ishibashi, T., Chen, J.-F., and Fields, R. D. (2004) Adenosine: an activity-dependent axonal signal regulating MAP kinase and proliferation in developing Schwann cells. *Neuron Glia Biology* 1: 23–34.

Sul, J. Y., Orosz, G., Givens, R. S., and Haydon, P. G. (2004) Astrocytic connectivity in the hippocampus. *Neuron Glia Biology* 1: 3–11.

Verkhratsky, A., and Kettenmann, H. (1996) Calcium signalling in glial cells. *Trends Neurosci.* 19: 346–52.

Verkhratsky, A., and Steinhauser, C. (2000) Ion channels in glial cells. *Brain Res. Brain Res. Rev.* 32: 380–412.

Volterra, A., Magistretti, P. J., and Haydon, P. G., eds. (2002) *The Tripartite Synapse.* Oxford University Press, New York.

PART II:
GLIA IN HEALTH AND DISEASE
Chapter 4: Brain Cancer

Canoll, P., and Goldman, J. E. (2008) The interface between glial progenitors and gliomas. *Acta Neuropathol.* 116: 465–77.

Castellano, B., and Nieto-Sampedro, M. (2004) *Glial Cell Function.* Elsevier, New York.

Chekenya, M., and Pilkington, G. J. (2002) NG2 precursor cells in neoplasia: functional, histogenesis and therapeutic implications for malignant brain tumours. *J. Neurocytol.* 31: 507–21.

Claes, A., Idema, A. J., and Wesseling, P. (2007) Diffuse glioma growth: a guerilla war. *Acta Neuropathol.* 114: 442–58.

Collins, V. P. (1998) Gliomas. *Cancer Surv.* 32: 37–51.

Graeber, M. B., Scheithauer, B. W., and Kreutzberg, G. W. (2002) Microglia in brain tumors. *Glia* 40: 252–59.

Jackson, E. L., and Alvarez-Buylla, A. (2008) Characterization of adult neural stem cells and their relation to brain tumors. *Cells Tissues Organs* 188: 212–24.

Lee, P. R., Cohen, J. E., Tendi, E. A., DeVries, G. H., Becker, K. G., and Fields, R. D. (2005) Transcriptional profiling of an MPNST-derived cell line and normal human Schwann cells. *Neuron Glia Biology* 1: 135–47.

Martin-Villalba, A., Okuducu, A. F., and von Deimling, A. (2008) The evolution of our understanding on glioma. *Brain Pathol.* 18: 455–63.

McFerrin, M. B., and Sontheimer, H. (2006) A role for ion channels in glioma cell invasion. *Neuron Glia Biology* 2: 39–49.

Monje, M. L., and Palmer, T. (2003) Radiation injury and neurogenesis. *Curr. Opin. Neurol.* 16: 129–34.

Nanda, D., Driesse, M. J., and Sillevis Smitt, P. A. (2001) Clinical trials of adenoviral-mediated suicide gene therapy of malignant gliomas. *Prog. Brain Res.* 132: 699–710.

Perry, A. (2001) Oligodendroglial neoplasms: current concepts, misconceptions, and folklore. *Adv. Anat. Pathol.* 8: 183–99.

Pomeroy, S. L. (2004) Neural development and the ontogeny of central nervous system tumors. *Neuron Glia Biology* 1: 127–33.

Seaton, J. (2009) Kennedy taken from luncheon with Obama. Associated Press, January 20.

Sontheimer, H. (2008) An unexpected role for ion channels in brain tumor metastasis. *Exp. Biol. Med.* 233: 779–91.

Steindler, D. A., and Laywell, E. D. (2003) Astrocytes as stem cells: nomenclature, phenotype, and translation. *Glia* 43: 62–69.

Vescovi, A. L., Galli, R., and Reynolds, B. A. (2006) Brain tumour stem cells. *Nat. Rev. Cancer* 6: 425–36.

Watters, J. J., Schartner, J. M., and Badie, B. (2005) Microglia function in brain tumors. *J. Neurosci. Res.* 81: 477–55.

Chapter 5: Brain and Spinal Cord Injury

Bandtlow, C., Zachleder, T., and Schwab, M. E. (1990) Oligodendrocytes arrest neurite growth by contact inhibition. *J. Neurosci.* 10: 3837–48.

Benfey, M., and Aguayo, A. J. (1982) Extensive elongation of axons from rat brain into peripheral nerve grafts. *Nature* 296: 150–52.

Bradbury, E. J., et al. (2002) Chondroitinase ABC promotes functional recovery after spinal cord injury. *Nature* 416: 636–40.

Bregman, B. S., et al. (1995) Recovery from spinal cord injury mediated by antibodies to neurite growth inhibitors. *Nature* 378: 498–501.

Buchli, A. D., and Schwab, M. E. (2005) Inhibition of Nogo: a key strategy to increase regeneration, plasticity and functional recovery of the lesioned central nervous system. *Ann. Med.* 37: 556–67.

Cafferty, W. B., McGee, A. W., and Strittmatter, S. M. (2008) Axonal growth therapeutics: regeneration or sprouting or plasticity? *Trends Neurosci.* 31: 215–20.

Caroni, P., and Schwab, M. E. (1988) Antibody against myelin-associated inhibitor of neurite growth neutralizes nonpermissive substrate properties of CNS white matter. *Neuron* 1: 85–96.

Chen, M. S., et al. (2000) Nogo-A is a myelin-associated neurite outgrowth inhibitor and an antigen for monoclonal antibody IN-1. *Nature* 403: 434–39.

Chow, W. N., Simpson, D. G., Bigbee, J. W., and Colello, R. J. (2007) Evaluating neuronal and glial growth on electrospun polarized matrices: bridging the gap in percussive spinal cord injuries. *Neuron Glia Biology* 3: 119–26.

David, S., and Aguayo, A. J. (1981) Axonal elongation into peripheral nervous system "bridges" after central nervous system injury in adult rats. *Science* 198: 931–33.

Fischer, D., He, Z., and Benowitz, L. I. (2004) Counteracting the Nogo receptor enhances optic nerve regeneration if retinal ganglion cells are in an active growth state. *J. Neurosci.* 24: 1646–51.

Fouad, K., et al. (2005) Combining Schwann cell bridges and olfactory-ensheathing glia grafts with chondroitinase promotes locomotor recovery after complete transection of the spinal cord. *J. Neurosci.* 25: 1169–78.

Fournier, A. E., and Strittmatter, S. M. (2001) Repulsive factors and axon regeneration in the CNS. *Curr. Opin. Neurobiol.* 11: 89–94.

Friedman, B., and Aguayo, A. J. (1985) Injured neurons in the olfactory bulb of the adult rat grow axons along grafts of peripheral nerve. *J. Neurosci.* 5: 1616–25.

Friedman, H. C., Aguayo, A. J., and Bray, G. M. (1999) Trophic factors in neuron–Schwann cell interactions. *Ann. N.Y. Acad. Sci.* 14: 427–38.

GrandPré, T., and Strittmatter, S. M. (2001) Nogo: a molecular determinant of axonal growth and regeneration. *Neuroscientist* 7: 377–86.

Huber, A. B., and Schwab, M. E. (2000) Nogo-A, a potent inhibitor of neurite outgrowth and regeneration. *Biol. Chem.* 381: 407–19.

Kapfhammer, J. P., Schwab, M. E., and Schneider, G. E. (1992) Antibody neutralization of neurite growth inhibitors from oligodendrocytes results in expanded pattern of postnatally sprouting retinocollicular axons. *J. Neurosci.* 12: 2112–19.

Keirstead, S. A. (1985) Responses to light of retinal neurons regenerating axons into peripheral nerve grafts in the rat. *Brain Res.* 359: 402–6.

Keirstead, S. A., et al. (1989) Electrophysiologic responses in hamster superior colliculus evoked by regenerating retinal axons. *Science* 246: 255–57.

Kim, J. E., Li, S., GrandPré, T., Qiu, D., and Strittmatter, S. M. (2003) Axon regeneration in young adult mice lacking Nogo-A/B. *Neuron* 38: 187–99.

Kim, J. E., Liu, B. P., Park, J. H., and Strittmatter, S. M. (2004) Nogo-66 receptor prevents raphespinal and rubrospinal axon regeneration and limits functional recovery from spinal cord injury. *Neuron* 44: 439–51.

Kim, J. J., Gross, J., Potashner, S. J., and Morest, D. K. (2004) Fine structure of degeneration in the cochlear nucleus of the chinchilla after acoustic overstimulation. *J. Neurosci. Res.* 77: 798–816.

May, F., et al. (2004) Schwann cell seeded guidance tubes restore erectile function after ablation of cavernous nerves in rats. *J. Urol.* 172: 374–77.

McDonald, J. W. (1999) Repairing the damaged spinal cord. *Scientific American* 281: 64–73.

McGee, A. W., and Strittmatter, S. M. (2003) The Nogo-66 receptor: focusing myelin inhibition of axon regeneration. *Trends Neurosci.* 26: 193–98.

Moon, L. D., Asher, R. A., Rhodes, K. E., and Fawcett, J. W. (2001) Regeneration of CNS axons back to their target following treatment of adult rat brain with chondroitinase ABC. *Nat. Neurosci.* 4: 465–66.

Richardson, P. M., McGuinness, U. M., and Aguayo, A. J. (1980) Axons from CNS neurons regenerate into PNS grafts. *Nature* 284: 264–65.

Rossignol, S., Schwab, M., Schwartz, M., and Fehlings, M. G. (2007) Spinal cord injury: time to move? *J. Neurosci.* 27: 11782–92.

Schnell, L., and Schwab, M. E. (1990) Axonal regeneration in the rat spinal cord produced by an antibody against myelin-associated neurite growth inhibitors. *Nature* 343: 269–72.

Schwab, M. E. (2002) Repairing the injured spinal cord. *Science* 295: 1029–31.

———. (2004) Nogo and axon regeneration. *Curr. Opin. Neurobiol.* 14: 118–24.

Schwab, M. E., and Thoenen, H. (1985) Dissociated neurons regenerate into sciatic but not optic nerve explants in culture irrespective of neurotrophic factors. *J. Neurosci.* 5: 2415–23.

Strittmatter, S. M. (2002) Modulation of axonal regeneration in neurodegenerative disease: focus on Nogo. *J. Mol. Neurosci.* 19: 117–21.

Vukovic, J., Plant, G. W., Ruitenberg, M. J., and Harvery, A. R. (2007) Influence of adult Schwann cells and olfactory ensheathing glia on axon target cell interactions in the CNS: a comparative analysis using a retinotectal cograft model. *Neuron Glia Biology* 3: 105–17.

Chapter 6: Infection

Alper, T., Cramp, W. A., Haig, D. A., and Clarke, M. C. (1967) Does the agent of scrapie replicate without nucleic acid? *Nature* 214: 764–66.

Archer, F. (2004) Cultured peripheral neuroglial cells are highly permissive to sheep prion infection. *Virol.* 78: 482–90.

Bate, C., Boshuizen, R., and Williams, A. (2005) Microglial cells kill prion-damaged neurons in vitro by a CD14-dependent process. *J. Neuroimmunol.* 170: 62–70.

Bate, C., Kempster, S., Williams, A. (2006) Prostaglandin D2 mediates neuronal damage by amyloid-beta or prions which activates microglial cells. *Neuropharmacology* 50: 229–37.

Bate, C., and Williams, A. (2004) Detoxified lipopolysaccharide reduces microglial cell killing of prion-infected neurons. *Neuroreport* 15: 2765–68.

Beerhuis, R., Boshuizen, R. S., and Familian, A. (2005) Amyloid associated proteins in Alzheimer's and prion disease. *Curr. Drug Targets CNS Neurol. Disord.* 4: 235–48.

Bonthius, D. J., Mahoney, J., Buchmeier, M. J., Karacay, B., and Taggard, D. (2002) Critical role for glial cells in the propagation and spread of lymphocytic choriomeningitis virus in the developing rat brain. *J. Virol.* 76: 6618–35.

Brennemann, D. E., Hauser, J., Spong, C. Y., and Phillips, T. M. (2000) Chemokines released from astroglia by vasoactive intestinal peptide. Mechanisms of neuroprotection from HIV envelope protein toxicity. *Ann. N.Y. Acad. Sci.* 921: 109–14.

Burwinkel, M., et al. (2004) Role of cytokines and chemokines in prion infections of the central nervous system. *Int. J. Dev. Neurosci.* 22: 497–505.

Cheng-Mayer, C., et al. (1987) Human immunodeficiency virus can productively infect cultured human glial cells. *Proc. Natl. Acad. Sci. USA* 84: 3526–30.

Cronier, S., Laude, H., and Peyrin, J. M. (2004) Prions can infect primary cultured neurons and astrocytes and promote neuronal cell death. *Proc. Natl. Acad. Sci. USA* 101: 12271–86.

Ernst, T., Chang, L., and Arnold, S. (2003) Increased glial metabolites predict increased working memory network activation in HIV brain injury. *Neuroimage* 19: 1686–93.

Gajdusek, D. C. (1963) Kuru. *Trans. Roy. Soc. Tropic. Med. Hyg.* 57: 151–69.

———. (1976) Autobiography, The Nobel Foundation. nobelprize.org/nobel_prizes/medicine/laureates/1976/gajdusek-autobio.html.

———. (1976) Unconventional viruses and the origin and disappearance of kuru. Nobel lecture, December 13, 1976. The Nobel Foundation. nobelprize.org/ nobel_prizes/medicine/laureates/1976/gajdusek-lecture.html.

Garcao, P., Oliveira, C. R., and Agostinho, P. (2006) Comparative study of microglia activation induced by amyloid-beta and prion peptides: role in neurodegeneration. *J. Neurosci. Res.* 84: 182–93.

Gayathri, M. A., Taly, A. B., Santosh, V., Yasha, T. C., and Shankar, S. K. (2001) Vasculitic neuropathy in HIV infection: a clinicopathological study. *Neurol. India* 48: 277–83.

Gibbons, R. A., and Hunter, G. D. (1967) Nature of the scrapie agent. *Nature* 215: 1041–43.

Glasse, R. M. (1967) Cannibalism in the kuru region of New Guinea. *Trans. N.Y. Acad. Sci.*, series 2, 29: 748–54.

Gray, B. D., Skipp, P., O'Connor, V. M., and Perry, V. H. (2006) Increased expression of glial fibrillary acidic protein fragments and mu-calpain activation within the hippocampus of prion-infected mice. *Bochem. Soc. Trans.* 34: 51–54.

Griffith, J. S. (1967) Nature of the scrapie agent. *Nature* 215: 1043–44.

Gyorkey, F., Melnick, J. L., and Gyorkey, P. (1987) Human immunodeficiency virus in brain biopsies of patients with AIDS and progressive encephalopathy. *J. Infect. Dis.* 155: 870–76.

Hauwel, M., et al. (2005) Innate (inherent) control of brain infection, brain inflammation and brain repair: the role of microglia, astrocytes, "protective" glial stem cells and stromal ependymal cells. *Brain Res. Brain Res. Rev.* 48: 220–33.

Heppner, F. L., Prinz, M., and Aguizzi, A. (2001) Pathogenesis of prion diseases: possible implications of microglial cells. *Prog. Brain Res.* 132: 737–50.

Holley, J. (2008) Obituary D. Carleton Gajdusek: controversial scientist. *Washington Post,* December 16.

Hornef, M. W., et al. (1999) Brain biopsy in patients with acquired immunodeficiency syndrome. *Arch. Intern. Med.* 159: 2590–96.

Janssen, R. S., et al. (1992) Epidemiology of human immunodeficiency virus encephalopathy in the United States. *Neurology* 42: 1472–76.

Kato, T., Dembitzer, H. M., Hirano, A., and Llena, J. F. (1987) HTLV-III-like particles within a cell process surrounded by a myelin sheath in an AIDS brain. *Acta Neuropathol.* 73: 306–8.

Koenig, S., et al. (1986) Detection of AIDS virus in macrophages in brain tissue from AIDS patients with encephalopathy. *Science* 233: 1089–93.

Liberski, P. P., and Brown, P. (2004) Astrocytes in transmissible spongiform encephalopathies (prion disease). *Folia Neuropathol.* 42: suppl. B, 71–88.

Mahadevan, A., et al. (2002) Brain biopsy in Creutzfeld-Jakob disease: evolution of pathological changes by prion protein immunohistochemistry. *Neuropathol. Appl. Neurobiol.* 28: 314–24.

Marella, M., and Chabry, J. (2004) Neurons and astrocytes respond to prion infection by inducing microglia recruitment. *J. Neurosci.* 24: 620–27.

Mathews, J. D., Glasse, R., and Lindenbaum, S. (1968) Kuru and cannibalism. *Lancet* 2: 449–52.

Minagar, A., et al. (2002) The role of macrophage/microglia and astrocytes in the pathogenesis of three neurologic disorders: HIV-associated dementia, Alzheimer disease, and multiple sclerosis. *J. Neurol. Sci.* 202: 13–23.

Minagar, A., Shapshak, P., and Alexander, J. S. (2005) Pathogenesis of HIV-associated dementia and multiple sclerosis: role of microglia and astrocytes, in M. Aschner and L. G. Costa, eds. *The Role of Glia in Neurotoxicity,* second ed. CRC Press, New York, pp. 263–77.

Mirra, S. S., and del Rio, C. (1989) The fine structure of acquired immunodeficiency syndrome encephalopathy. *Arch. Pathol. Lab. Med.* 113: 858–65.

Prinz, M., et al. (2004) Intrinsic resistance of oligodendrocytes to prion infection. *J. Neurosci.* 24: 5974–81.

Prusiner, S. B. (1997) Prions. Nobel lecture, December 8, 1997. The Nobel Foundation. http://nobelprize.org/nobel_prizes/medicine/laureates/1997/prusiner -lecture.html.

Satheeshkumar, K. S., Murali, J., Jayakumar, R. (2004) Assemblages of prion fragments: novel model systems for understanding amyloid toxicity. *J. Struct. Biol.* 148: 176–93.

Schweighardt, B., and Atwood, W. J. (2001) Glial cells as targets of viral infection in the human central nervous system. *Prog. Brain Res.* 132: 721–35.

Shapshak, P., Fujimura, R. K., Srivastava, A., and Goodkin, K. (2000) Dementia and neurovirulence of HIV-1. *CNS Spectr.* 5: 31–42.

Shaw, G. M., et al. (1985) HTLV-III infection in brains of children and adults with AIDS encephalopathy. *Science* 227: 177–82.

Sponne, I., et al. (2004) Oligodendrocytes are susceptible to apoptotic cell death induced by prion protein-derived peptides. *Glia* 47: 1–8.

Whitfield, J. T., Pako, W. H., Collinge, J., and Alpers, M. P. (2008) Mortuary rites of the South Fore and kuru. *Philos. Trans. R. Soc. Lond. B, Biol. Sci.* 363: 3721–24.

Wojtera, M., Sikorska, B., Sobow, T., and Liberski, P. P. (2005) Microglial cells in neurodegenerative disorders. *Folia Neuropathol.* 34: 311–21.

Chapter 7: Mental Health

Angulo, M. C., Kozlov, A. S., Charpak, S., and Audinat, E. (2004) Glutamate release from glial cells synchronizes neuronal activity in the hippocampus. *J. Neurosci.* 24: 6920–27.

Batley, J., Wood, J., and Wellington, R. (1736) *Letters from a Moor at London to his Friend at Tunis Containing a Description of Bedlam, with Serious Reflections on Love, Madness, and Self-Murder.* Francis Noble, Thomas Davies, London.

Berton, R. (1974) Annals of medicine—as empty as eve. *New Yorker*, September 9, p. 84.

Binder, D. K., and Steinhauser, C. (2006) Functional changes in astroglial cells in epilepsy. *Glia* 54: 358–68.

Bowley, M. P., Drevets, W. C., Ongur, D., and Price, J. L. (2002) Low glial numbers in amygdala in major depressive disorder. *Biol. Psychiatry* 53: 404–12.

Brazier, M. A. B. (1961) *A History of the Electrical Activity of the Brain: The First Half-Century.* Macmillan Co., New York.

————. (1984) *A History of Neurophysiology in the 17th and 18th Centuries from Concept to Experiment*. Raven Press, New York.

Brown, B. (1866) *On the Curability of Certain Forms of Insanity, Epilepsy, Catalepsy, and Hysteria in Females*. Cox and Wyman, London.

Chambers, J. S., and Perrone-Bizzozero, N. I. (2004) Altered myelination of the hippocampal formation in subjects with schizophrenia and bipolar disorder. *Neurochem. Res.* 29: 2293–2302.

Coyle, J. T., and Manji, H. K. (2002) Getting balance: drugs for bipolar disorder share target. *Nat. Med.* 8: 557–58.

DeToledo, J. C., and Lowe, M. R. (2003) Epilepsy, demonic possessions, and fasting: another look at translations of Mark 9:16. *Epilepsy and Behavior* 4: 338–49.

Du, J., Quiroz, J., Yuan, P., Zarate, C., and Manji, H. K. (2004) Bipolar disorder: involvement of signaling cascades and AMPA receptor trafficking at synapses. *Neuron Glia Biology* 1: 231–34.

Fellin, T., et al. (2004) Neuronal synchrony mediated by astrocytic glutamate through activation of extrasynaptic NMDA receptors. *Neuron* 43: 729–43.

Fields, R. D. (2008) White matter in learning, cognition, and psychiatric disorders. *Trends Neurosci.* 31: 361–70.

Fink, M. (1999) Images in psychiatry. Ladislas J. Meduna, M.D., 1869–1964. *Am. J. Psychiatry* 156: 1807.

————. (2004) ECT: Serendipity or Logical Outcome? *Psychiatric Times,* January 21:1.

Friedberg, J. (1977) Shock treatment, brain damage, and memory loss: a neurological perspective. *Am. J. Psychiatry* 134: 1010–14.

George, J. A., and Michael, A. G. (1993) *The Nazi Doctors and the Nuremberg Code: Human Rights in Human Experimentation*. Oxford University Press, New York.

Gutnick, J. M., Connors, B. W., and Prince, D. A. (1982) Mechanisms of neocortical epileptogenesis in vitro. *J. Neurophysiol.* 48: 1321–35.

Hamidi, M., Drevets, W. C., and Price, J. L. (2004) Glial reduction in amygdala in major depressive disorder is due to oligodendrocytes. *Biol. Psy.* 55: 563–69.

Hsu, M. S., Lee, D. J., and Binder, D. K. (2007) Potential role of the glial water channel aquaporin-4 in epilepsy. *Neuron Glia Biology* 3: 2887–97.

Inazu, M., et al. (1999) Pharmacological characterization of dopamine transporter in cultured rat astrocytes. *Life Sci.* 54: 2239–45.

Inazu, M., et al. (2001) Pharmacological characterization and visualization of glial serotonin transporter. *Neurochem Int.* 39: 39–49.

Korn, M. L. (2001) *Historical Roots of Schizophrenia*. Medscape Portals, Inc., http://cme.medscape.com/viewarticle/418882_4.

Lee, P. R., and Fields, R. D. (2009) Regulation of myelin genes implicated in psychiatric disorders by functional activity in axons. *Front. Neuroanat.* 3(4): 1–8.

Li, T., et al. (2007) Adenosine dysfunction in astrogliosis: causes for seizure generation. *Neuron Glia Biology* 3: 353–66.

Lifton, R. J. (1986) *The Nazi Doctors: Medical Killing and the Psychology of Genocide*. Harper Collins, New York.

Manji, H. K., Drevets, W. C., and Charney, D. S. (2001) The cellular neurobiology of depression. *Nat. Med.* 7: 541–47.

Masia, S. L., and Devinsky, O. (2000) Epilepsy and behavior: A brief history. *Epilepsy and Behavior* 1: 27–36.

Meduna, L. (1935) Versuche über die biologische Beeinflussung des Ablaufes der Schizophrenie: Campher und Cardiozolkrympfe. *Zeitschrift für die gesamte Neurologie und Psychiatrie* 152: 235–62.

———. (1937) *Die Konvulsionstherapie der Schizophrenie.* Carl Marhold, Halle, Germany.

———. (1956) The convulsive treatment: a reappraisal, in *The Great Physiodynamic Therapies in Psychiatry,* Sackler, A.M., Sackler, M.D., Sackler, R.R., and Marti-Ibanez, F., eds. Hoeber-Harper, New York, pp. 76–90.

———. (1985) Autobiography. *Convulsive Ther.* 1: 43–57, 121–35.

Miguel-Hidalgo, J. J. (2006) Withdrawal from free-choice ethanol consumption results in increased packing density of glutamine synthetase-immunoreactive astrocytes in the prelimbic cortex of alcohol-preferring rats. *Alcohol and Alcoholism* 41: 379–85.

Miller, G. (2005) The dark side of glia. *Science* 308: 788–91.

Millet, D. (2001) Hans Berger from psychic energy to the EEG. *Perspect. Biol. Med.* 44: 522–42.

Mitterauer, B. (2004) Imbalance of glial-neuronal interaction in synapses: A possible mechanism of the pathophysiology of bipolar disorder. *Neuroscientist* 10: 199–206.

———. (2005) Nonfunctional glial proteins in tripartite synapses: a pathophysiological model of schizophrenia. *Neuroscientist* 1: 192–98.

Moises, H. W., Zoega, T., Gottesmann, I. I. (2002) The glial growth factors deficiency and synaptic destabilization hypothesis of schizophrenia. *BMC Psychiatry* July 3, 2:8.

Mustafa, A. K., Kim, P. M., and Snyder, S. H. (2004) D-serine as a putative glial neurotransmitter. *Neuron Glia Biology* 1: 275–81.

Ongur, D., Drevets, W. C., and Price, J. L. (1998) Glial reduction in the subgenual prefrontal cortex in mood disorders. *Proc. Natl. Acad. Sci. USA* 95: 13290–95.

Rajkowska, G., and Miguel-Hidalgo, J. J. (2007) Gliogenesis and glial pathology in depression. *CNS Neurol. Disord. Drug Targets* 6: 219–33.

Rajkowska, G., et al. (2002) Layer-specific reductions in GFAP-reactive astroglia in the dorsolateral prefrontal cortex in schizophrenia. *Schizophr. Res.* 57: 127–38.

Redies, C., Viebig, M., Zimmermann, S., and Frober, R. (2005) Origin of corpses received by the anatomical institute at the University of Jena during the Nazi regime. *Anatomical Record* 285B: 6–10.

Saakov, B. A., Khoruzhaya, T. A., and Bardakhch'yan, E. A. (1977) Ultrastructural mechanisms of serotonin demyelination. *Biulleten' Eksperimental'noi Biologii i Meditsiny,* 83: 606–10.

Schell, M. J., Molliver, M. E., and Snyder, S. H. (1995) D-serine, an endogenous synaptic modulator: localization to astrocytes and glutamate-stimulated release. *Proc. Natl. Acad. Sci. USA* 92: 3948–52.

Si, X., Miguel-Hidalgo, J. J., O'Dwyer, G., Stockmeier, C. A., and Rajkowska, G. (2004) Age-dependent reductions in the level of glial fibrillary acidic protein in the prefrontal cortex in major depression. *Neuropsychopharmacology* 29: 2088–96.

Stafstrom, C. E. (2006) Epilepsy: a review of selected clinical syndromes and advances in basic science. *J. Cereb. Blood Flow Metab.* 26: 9983–10004.

Stark, A. K., et al. (2004) Glial cell loss in the anterior cingulate cortex, a subregion of the prefrontal cortex, in subjects with schizophrenia. *Am. J. Psychiatry* 161: 882–88.

Tian, G. F., et al. (2005) An astrocytic basis of epilepsy. *Nat. Med.* 11: 973–81.

Tuke, D. H. (1882) *Chapters in the History of the Insane in the British Isles.* Kegan, Paul, Trench and Co., London.

van Calker, D., and Biber, K. (2005) The role of glial adenosine receptors in neural resilience and the neurobiology of mood disorders. *Neurochem. Res.* 30: 1205–17.

Wennstrom, M., Hellsten, J., Ekdahl, C. T., and Tingstrom, A. (2003) Electroconvulsive seizures induce proliferation of NG2-expressing glial cells in adult rat amygdala. *Biol. Psychiatry* 55: 464–71.

Xiao, L., et al. (2008) Quetiapine facilitates oligodendrocyte development and prevents mice from myelin breakdown and behavioral changes. *Mol. Psychiatry* 7: 697–708.

Chapter 8: Neurodegenerative Disorders

Akiyama, Y., et al. (2004) Remyelination of spinal cord axons by olfactory ensheathing cells and Schwann cells derived from a transgenic rat expressing alkaline phosphatase marker gene. *Neuron Glia Biology* 1: 47–55.

Allan, S. (2006) The neurovascular unit and the key role of astrocytes in the regulation of cerebral blood flow. *Cerebrovasc. Dis.* 21: 137–38.

Brown, D. R., Schmidt, B., and Kretzschmar, H. A. (1996) Role of microglia and host prion protein in neurotoxicity of a prion protein fragment. *Nature* 380: 345–47.

Bueler, H. R., et al. (1992) Normal development and behavior of mice lacking the neuronal cell-surface PrP protein. *Nature* 356: 577–82.

Burns, R. S., et al. (1983) A primate model of Parkinsonism: selective destruction of dopaminergic neurons in the pars compacta of the substantia nigra by N-methyl-4-phenyl-1,2,3,6-tetrahydropyridine. *Proc. Natl. Acad. Sci. USA* 80: 4546–50.

Collinge, J., et al. (2006) Kuru in the 21st century—an acquired human prion disease with a very long incubation time. *Lancet* 367: 2068–74.

D'Amato, R. J., Lipman, Z. P., and Snyder, S. H. (1986) Selectivity of the Parkinsonian neurotoxin PMTP: toxic metabolite MPP+ binds to neuromelanin. *Science* 231: 987–89.

Davis, G. C., et al. (1979) Chronic Parkinsonism secondary to intravenous injection of meperidine analogues. *Psychiatry Res.* 1, 239–54.

Depino, A. M., et al. (2003) Microglial activation with atypical proinflammatory cytokine expression in a rat model of Parkinson's disease. *Eur. J. Neurosci.* 18: 2731–42.

Dodick, D. W., and Gargus, J. J. (2008) Why migraines strike. *Scientific American* 299: 56–63.

Frank L., et al. (2001) Pathogenesis of prion disease: possible implications of microglial cells. *Prog. Brain Res.* 132: 737–50.

Furman, S., et al. (2004) Subcellular localization and secretion of activity-dependent neuroprotective protein in astrocytes. *Neuron Glia Biology* 1: 193–99.

Gordon, G. R., Choi, H. B., Rungta, R. L., Ellis-Davies, G. C., and MacVicar, B. A. (2008) Brain metabolism dictates the polarity of astrocyte control over arterioles. *Nature* 456: 745–49.

Gozes, I., et al. (2003) From vasoactive intestinal peptide (VIP) through activity-dependent neuroprotective protein (ADNP) to NAP: a view of neuroprotection and cell division. *J. Mol. Neurosci.* 20: 315–22.

Hamel, E. (2006) Perivascular nerves and the regulation of cerebrovascular tone. *J. Appl. Physiol.* 100: 1059–64.

Hawkins, B. T., and Davis, T. P. (2005) The blood-brain barrier/neurovascular unit in health and disease. *Parmacol. Rev.* 57: 173–85.

Heikkila, R. E., and Niclaas, W. J. (1986) Studies on the mechanism of the dopaminergic neurotoxicity of 1-methyl-4-phenyl-1,2,3,6-tetrahydropyridine (MPTP). *Curr. Sep.* 7: 52–54.

Koehler, R. C., Gebremedhin, D., and Harder, D. R. (2006) Role of astrocytes in cerebrovascular regulation. *J. Appl. Physiol.* 100: 307–17.

Lewin, R. (1984) Trail of ironies to Parkinson's disease. *Science* 224: 1083–85.

———. (1984) Brain enzyme is the target of drug toxin. *Science* 225: 1460–62.

———. (1985) Parkinson disease: an environmental cause? *Science* 229: 257–58.

Li, J., et al. (2005) Increased astrocyte proliferation in rats after running exercise. *Neurosci. Lett.* 386: 160–64.

Liberski, P. P., Yanagihara, R., Wells, G. A., Gibbs, C. J. Jr., and Gajdusek, D. C. (1992) Ultrastructural pathology of axons and myelin in experimental scrapie in hamsters and bovine spongiform encephalopathy in cattle and a comparison with the panencephalopathic type of Creutzfeldt-Jakob disease. *J. Comp. Pathol.* 106: 383–98.

Marchetti, B., et al. (2005) Hormones are key actors in gene x environment interactions programming the vulnerability to Parkinson's disease: glia as a common final pathway. *Ann. N.Y. Acad. Sci.* 1057: 296–318.

Mattson, M. P., and Chan, S. L. (2003) Neuronal and glial calcium signaling in Alzheimer's disease. *Cell Calcium* 34: 385–97.

Nouchine, H., et al. (2001) Mechanisms of migraine aura revealed by functional MRI in human visual cortex. *Proc. Natl. Acad. Sci. USA* 98: 4687–92.

Pardo, C. A., Vargas, D. L., and Zimmerman, A. W. (2005) Immunity, neuroglia and neuroinflammation in autism. *Int. Rev. Psychiatry* 17: 485–95.

Pearson, H. (2006) Experts comb tropics for clues to vCJD. *Nature* 44: 1033.

Peters, O., Shipke, C. G., Hashimoto, Y., and Kettenmann, H. (2003) Different mechanisms promote astrocyte Ca2+ waves and spreading depression in the mouse neocortex. *J. Neurosci.* 23: 9888–96.

Prusiner, S. B. (1982) Novel proteinaceous infectious particles cause scrapie. *Science* 216: 136–44.

Quaid, L. (2005) Possible case of mad cow investigated. *Washington Post,* July 27.

Raeber, A. J., et al. (1997) Astrocyte-specific expression of hamster prion protein (PrP) renders PrP knockout mice susceptible to hamster scrapie. *EMBO J.* 16: 6057–65.

Reichenbach, A., and Wolburg, H. (2008) Structural association of astrocytes with neurons and vasculature: defining territorial boundaries, in *Astrocytes in (Patho)Physiology of the Nervous System,* V. Parpura and P. G. Haydon, eds. Springer, New York.

Sailer, A., et al. (1994) No propagation of prions in mice devoid of PrP. *Cell* 77: 967–68.

Savidge, T. C. (2007) MIND the gap: an astroglial perspective on barrier regulation. *Neuron Glia Biology* 3: 191–97.

Schmidt, K., and Oertel, W. (2006) Fighting Parkinson's. *Scientific American Mind* February/March 17: 64–69.

Spencer, M. D., Kinght, R. S. G., and Will, R. G. (2002) First hundred cases of variant Creutzfeldt-Jakob disease: retrospective case note review of early psychiatric and neurological features. *BMJ* 324: 1479–82.

Timmons, H. (2003) Mad cow disease in the United States: a case history; Britain has learned many harsh lessons in a long effort to combat mad cow disease. *New York Times,* December 26. www.nytimes.com/2003/12/26/US/mad-cow -disease-united-states-case-history-britian-has-learned-many-harsh.html.

Vargas, D. L., Nascimbene, C., Krishnan, C., Zimmerman, A. W., and Pardo, C. A. (2005) Neuroglial activation and neuroinflammation in the brain of patients with autism. *Ann. Neurol.* 57: 67–81.

Vourc'h, P., et al. (2003) The oligodendrocyte-myelin glycoprotein gene is highly expressed during the late stages of myelination in the rat central nervous system. *Brain Res. Dev. Brain Res.* 144: 159–68.

Weingarten, H. L. (1988) 1-methyl-4-phenyl-1,2,3,6, tetrahydropyridine (MPTP): one designer drug and serendipity. *J. Forens. Sci.* 33: 588–95.

Chapter 9: Glia and Pain

Aldskogius, H. (2001) Regulation of microglia—potential new drug targets in the CNS. *Expert Opin. Ther. Targets* 5: 655–68.

Beltramo, M., et al. (2006) CB2 receptor-mediated antihyperalgesia: possible direct involvement of neural mechanisms. *Eur. J. Neurosci.* 23: 1530–38.

Cavaliere, C., et al. (2007) Gliosis alters expression and uptake of spinal glial amino acid transporters in a mouse neuropathic pain model. *Neuron Glia Biology* 3: 141–53.

Clark, A. K., Gentry, C., Bradbury, E. J., McMahon, S. B., and Malcangio, M. (2006) Role of spinal microglia in rat models of peripheral nerve injury and inflammation. *Eur. J. Pain* 11: 223–30.

Coull, J. A., et al. (2005) BDNF from microglia causes the shift in neuronal anion gradient underlying neuropathic pain. *Nature* 438: 1017–21.

Fields, R. D. (2009) New culprit in chronic pain. *Scientific American* 31, no. 5 (November).

Guo, L. H., Trautmann, K., and Schluesener, H. J. (2005) Expression of P2X4 receptor by lesional activated microglia during formalin-induced inflammatory pain. *J. Neuroimmunol.* 163: 120–27.

Hains, B. C., and Waxman, S. G. (2006) Activated microglia contribute to the maintenance of chronic pain after spinal cord injury. *J. Neurosci.* 26: 4308–17.

Hayward, P. (2006) Microglial role in neuropathic pain. *Lancet Neurol.* 5: 118–19.

Holguin, A., et al. (2004) HIV-1 gp120 stimulates proinflammatory cytokine-mediated pain facilitation via activation of nitric oxide synthase-I (nNOS). *Pain* 110: 517–30.

Inoue, K. (2006) The function of microglia through purinergic receptors: neuropathic pain and cytokine reasease. *Pharmacol. Ther.* 109: 210–26.

Inoue, K., Koizumi, S., Tsuda, M., and Shigemoto-Mogami, Y. (2003) Signaling of ATP receptors in glia-neuron interaction and pain. *Life Sci.* 74: 189–97.

Inoue, K., and Tsuda, M. (2006) The role of microglia and ATP receptors in a mechanism of neuropathic pain. *Nippon Yakurigaku Zasshi* 127: 14–17.

Inoue, K., Tsuda, M., and Koizumi, S. (2004) Chronic pain and microglia: the role of ATP. *Novartis Found. Symp.* 261: 55–64.

Inoue, K., Tsuda, M., and Koizumi, S. (2004) ATP-and adenosine-mediated signaling in the central nervous system: chronic pain and microglia: involvement of the ATP receptor P2X4. *J. Pharmacol. Sci.* 94: 112–14.

Ji, R. R., Kawasaki, Y., Zhuang, Z. Y., Wen, Y. R., and Decosterd, I. (2006) Possible role of spinal astrocytes in maintaining chronic pain sensitization: review of current evidence with focus on bFGF/JNK pathway. *Neuron Glia Biology* 2: 259–69.

Ji, R. R., and Strichartz, G. (2004) Cell signaling and the genesis of neuropathic pain. *Sci. STKE* 252:reE14.

Lacroix-Fralish, M. L. (2006) Differential regulation of neuregulin 1 expression by progesterone in astrocytes and neurons. *Neuron Glia Biology* 2: 227–34.

Ledeboer, A., et al. (2005) Minocycline attenuates mechanical allodynia and proinflammatory cytokine expression in rat models of pain facilitation. *Pain* 115: 71–83.

Ledeboer, A., et al. (2006) The glial modulatory drug AV411 attenuates mechanical allodynia in rat models of neuropathic pain. *Neuron Glia Biology* 2: 279–92.

Lindia, J. A., McGowan, E., Jochnowitz, N., and Abbadie, C. (2005) Induction of CX3CL1 expression in astrocytes and CX3CR1 in microglia in the spinal cord of a rat model of neuropathic pain. *J. Pain* 6: 434–38.

Milligan, E. D., and Watkins, L. R. (2009) Pathological and protective roles of glia in chronic pain. *Trends Neurosci.* 10: 23–36.

Milligan, E. D., et al. (2006) Intrathecal polymer-based interleukin-10 gene delivery for neuropathic pain. *Neuron Glia Biology* 2: 293–308.

Moalem, G., and Tracey, D. J. (2005) Immune and inflammatory mechanisms in neuropathic pain. *Brain Res. Brain Res. Rev.* 51: 240–64.

Narita, M., et al. (2006) Direct evidence for spinal cord microglia in the development of a neuropathic pain-like state in mice. *J. Neurochem.* 97: 1337–48.

Saab, C. Y., Wang, J., Gu, C., Garner, K. N., and Al-Chaer, E. D. (2006) Microglia: a newly discovered role in visceral hypersensitivity? *Neuron Glia Biology* 3: 271–77.

Sawynok, J., Liu, X. J. (2003) Adenosine in the spinal cord and periphery: release and regulation of pain. *Prog. Neurobiol.* 69: 313–40.

Suter, M. R., Wen, Y. R., Decosterd, I., and Ji, R. R. (2007) Do glial cells control pain? *Neuron Glia Biology* 3: 255–68.

Tanga, F. Y., Nutile-McMenemy, N., and DeLeo, J. A. (2005) The CNS role of Toll-like receptor 4 in innate neuroimmunity and painful neuropathy. *Proc. Natl. Acad. Sci. USA* 102: 5856–61.

Tsuda, M., et al. (2003) P2X4 receptors induced in spinal microglia gate tactile allodynia after nerve injury. *Nature* 424: 778–83.

Tsuda, M., Inoue, K., and Salter, M. W. (2005) Neuropathic pain and spinal microglia: a big problem from molecules in "small" glia. *Trends Neurosci.* 28: 101–7.

Verge, G. M., et al. (2004) Fractalkine (CX3CL1) and fractalkine receptor (CX3CR1) distribution in spinal cord and dorsal root ganglia under basal and neuropathic pain conditions. *Eur. J. Neurosci.* 20: 1150–60.

Vit, J. P., Jasmin, L., Bhargava, A., and Ohara, P. T. (2006) Satellite glial cells in the trigeminal ganglion as determinant of orofacial neuropathic pain. *Neuron Glia Biology* 2: 247–57.

Watkins, L. R., et al. (2001) Glial activation: a driving force for neuropathological pain. *Trends Neurosci.* 24: 450–55.

Watkins, L. R., Hutchinson, M. R., Johnston, I. N., and Maier, S. F. (2005) Glia: novel counter-regulators of opioid analgesia. *Trends Neurosci.* 28: 661–69.

Watkins, L. R., Milligan, E. D., Maier, S. F. (2003) Glial proinflammatory cytokines mediate exaggerated pain states: implications for clinical pain. *Adv. Exp. Med. Biol.* 521: 1–21.

Wieseler-Frank, J., Maier, S. F., and Watkins, L. R. (2004) Glial activation and pathological pain. *Neurochem. Int.* 45: 389–95.

———. (2005) Central proinflammatory cytokines and pain enhancement. *Neurosignals* 14: 166–74.

Wotherspoon, G., et al. (2005) Peripheral nerve injury induces cannabinoid receptor 2 protein expression in rat sensory neurons. *Neuroscience* 135: 235–45.

Wu, Y., Willcockson, H. H., Maixner, W., and Light, A. R. (2004) Suramin inhibits spinal cord microglia activation and long-term hyperalgesia induced by formalin injection. *J. Pain* 5: 48–55.

Zhang, J., Hoffert, C., Vu, H. K., Groblewski, T., Ahmad, S., and O'Donnell, D. (2003) Induction of CB2 receptor expression in the rat spinal cord of neuropathic but not inflammatory chronic pain models. *Eur. J. Neurosci.* 17: 2750–54.

Zhang, R. X., et al. (2005) Spinal glial activation in a new rat model of bone cancer pain produced by prostate cancer cell inoculation of the tibia. *Pain* 118: 125–36.

Chapter 10: Glia and Addiction

Aberg, E., Hofstetter, C. P., Olson, L., and Brene, S. (2005) Moderate ethanol consumption increases hippocampal cell proliferation and neurogenesis in the adult mouse. *Int. J. Neuropsychopharmacol.* 8: 557–67.

Albertson, D. N., et al. (2004) Gene expression profile of the nucleus accumbens of human cocaine abusers: evidence for dysregulation of myelin. *J. Neurochem.* 88: 1211–19.

Collins, M. A., Zou, J.-Y., and Neafsey, E. J. (1998) Brain damage due to episodic alcohol exposure in vivo and in vitro: furosemide neuroprotection implicates edema-based mechanism. *FASEB J.* 12: 221–30.

Costa, L. G., Yagle, K., Vitalone, A., and Guizzetti, M. (2005) Alcohol and glia in the developing brain, in *The Role of Glia in Neurotoxicity*, Michael Ascher and Lucio G. Costa, eds., CRC press, Boca Raton, FL. pp. 343–54.

Cullen, K. M., and Halliday, G. M. (1994) Chronic alcoholics have substantial glial pathology in the forebrain and diencephalon. *Alcohol and Alcoholism* suppl. 2: 253–57.

Guerri, C., Pascual, M., and Renau-Piqueras, J. (2001) Glia and fetal alcohol syndrome. *Neurotoxicology* 22: 593–99.

Johnston, I. N., et al. (2004) A role for proinflammatory cytokines and fractalkine in analgesia, tolerance, and subsequent pain facilitation induced by chronic intrathecal morphine. *J. Neurosci.* 24: 7353–65.

Korbo, L. (1999) Glial cell loss in the hippocampus of alcoholics. *Alcohol. Clin. Exp. Res.* 23: 164–68.

Lancaster, F. E. (1994) Gender differences in the brain: implications for the study of human alcoholism. *Alcohol. Clin. Exp. Res.* 28: 740–46.

Riikonen, J., Jaatinen, P., Rintala, J., Porsti, I., Karjala, K., and Hervonen, A. (2002) Intermittent ethanol exposure increases the number of cerebellar microglia. *Alcohol and Alcoholism* 37: 421–26.

Tarnowska-Dziduszko, E., Bertrand, E., Szpak, G. M. (1995) Morphological changes in the corpus callosum in chronic alcoholism. *Folia Neuropathol.* 33: 25–29.

Watkins, L. R., Hutchinson, M. R., Johnston, I. N., and Maier, S. F. (2005) Glia: novel counterregulators of opioid analgesia. *Trends Neurosci.* 28: 661–69.

Watts, L. T., Rathinam, M. L., Schenker, S., and Henderson, G. I. (2005) Astrocytes protect neurons from ethanol-induced oxidative stress and apoptotic death. *J. Neurosci. Res.* 80: 655–66.

Welch-Carre, E. (2005) The neurodevelopment consequences of prenatal alcohol exposure. *Adv. Neonatal Care* 5: 217–29.

Wickramasinghe, S. N. (1987) Neuroglial and neuroblastoma cell lines are capable of metabolizing ethanol via an alcohol-dehydrogenase-independent pathway. *Alcohol. Clin. Exp. Res.* 11: 234–37.

Wilson, M. A., and Molliver, M. E. (1994) Microglial response to degeneration of serotonergic axon terminals. *Glia* 11: 18–34.

Chapter 11: Mother and Child

Alvarez-Buylla, A., Garcia-Verdugo, J. M., and Tramontin, A. D. (2001) A unified hypothesis on the lineage of neural stem cells. *Nat. Rev. Neurosci.* 2: 287–93.

Amateau, S. K., and McCarthy, M. M. (2002) Sexual differentiation of astrocyte morphology in the developing rat preoptic area. *J. Neuroendocrinol.* 14: 904–10.

Bergles, D. E., Roberts, J. D. B., Somogyi, P., and Jahr, C. E. (2000) Glutamatergic synapses on oligodendrocyte precursor cells in the hippocampus. *Nature* 405: 187–91.

Blondel, O., et al. (2000) A glial-derived signal regulating neuronal differentiation. *J. Neurosci.* 20: 8012–20.

Boehler, M. D., Wheeler, B. C., and Brewer, G. J. (2007) Added astroglia promote greater synapse density and higher activity in neuronal networks. *Neuron Glia Biology* 3: 127–40.

Boulanger, L. M. (2004) MHC class I in activity-dependent structural and functional plasticity. *Neuron Glia Biology* 1: 283–89.

Cashion, A. B., Smith, M. J., and Wise, P. M. (2003) The morphometry of astrocytes in the rostral preoptic area exhibits a diurnal rhythm on proestrus: relationship to the luteinizing hormone surge and effects of age. *Endocrinology* 144: 274–80.

Chen, Y., Ai, Y., Slevin, J. R., Maley, B. E., and Gash, D. M. (2005) Progenitor proliferation in the adult hippocampus and substantia nigra induced by glial cell line–derived neurotrophic factor. *Exp. Neurol.* 196: 87–95.

Cohen, J. E., and Fields, R. D. (2008) Activity-dependent neuron-glial signaling by ATP and leukemia-inhibitory factor promotes hippocampal glial development. *Neuron Glia Biology* 4: 43–55.

Couzin, J. (2009) Celebration and concern over U.S. trial of embryonic stem cells. *Science* 323: 568.

Fields, R. D. (2001) Development of the vertebrate nervous system, in *The Encyclopedia of the Neurological Sciences.* Academic Press, San Diego, pp. 643–51.

———. (2007) The spaces between. *Scientific American Mind* October/November 18: 8.

Gritti, A., and Bonfanti, L. (2007) Neuronal-glial interactions in central nervous system neurogenesis: the neural stem cell perspective. *Neuron Glia Biology* 3: 309–23.

Herrera, A. A., et al. (2000) The role of perisynaptic Schwann cells in development of neuromuscular junction in the frog (*Xenopus laevis*). *J. Neurobiol.* 45: 237–54.

Ihrie, R. A., and Alvarez-Buylla, A. (2008) Neural stem cells disguised as astrocytes, in *Astrocytes in (Patho)Physiology of the Nervous System*, V. Parpura and P. G. Haydon, eds. Springer, New York.

Jahromi, B. S., et al. (1992) Transmitter release increases intracellular calcium in perisynaptic Schwann cells in situ. *Neuron* 8: 1069–77.

Káradóttir, R., Hamilton, N., Bakiri, Y., and Attwell, D. (2008) Spiking and nonspiking classes of oligodendrocyte precursor glia in CNS white matter. *Nat. Neurosci.* 11: 450–56.

Lemke, G. (2001) Glial control of neuronal development. *Annu. Rev. Neurosci.* 24: 87–105.

LePore, A., et al. (2004) Differential fate of multipotent and lineage-restricted neural precursors following transplantation into the adult CNS. *Neuron Glia Biology* 1: 113–26.

Lu, B., and Chang, J. H. (2004) Regulation of neurogenesis by neurotrophins: implications in hippocampus-dependent memory. *Neuron Glia Biology* 1: 377–84.

McCarthy, M. M., Amateau, S. K., and Mong, J. A. (2002) Steroid modulation of astrocytes in the neonatal brain: implications for adult reproductive function. *Biol. Reprod.* 67: 691–98.

Mennerick, S., Benz, A., and Zorumski, C. F. (1996) Components of glial responses to exogenous and synaptic glutamate in rat hippocampal microcultures. *J. Neurosci.* 16: 55–64.

Morest, D. K., and Silver, J. (2003) Precursors of neurons, neuroglia, and ependymal cells in the CNS: What are they? Where are they from? How do they get where they are going? *Glia* 43: 6–18.

Orzhekhovskaya, N. S. (2001) Neuron-glial relationships in various fields of the frontal area of the brain in children at different stages of life. *Neurosci. Behav. Physiol.* 31: 355–58.

Pfrieger, F. W. (2002) The role of glia in the development of synaptic contacts, ch. 2 in *The Tripartite Synapse.* Volterra, A., Magistretti, P. J., and Haydon, P. G., eds. Oxford University Press, New York.

Pfrieger, F. W., and Barres, B. A. (1997) Synaptic efficacy enhanced by glial cells in vitro. *Science* 277: 1648–57.

Svenningsen, A. F., Colman, D. R., and Pedraza, L. (2004) Satellite cells of dorsal root ganglia are multipotential glial precursors. *Neuron Glia Biology* 1: 85–93.

Ullian, E. M., Sapperstein, S. K., Christopherson, K. S., and Barres, B. A. (2001) Control of synapse number by glia. *Science* 291: 657–61.

Zhu, X., Hill, R. A., and Nishiyama, A. (2008) NG2 cells generate oligodendrocytes and gray matter astrocytes in the spinal cord. *Neuron Glia Biology* 4: 19–26.

Chapter 12: Aging

Abramov, A. Y., et al. (2003) Changes in intracellular calcium and glutathione in astrocytes as the primary mechanism of amyloid neurotoxicity. *J. Neurosci.* 23: 5088–95.

Abramov, A. Y., Canevari, L., Duchen, M. R. (2004) Calcium signals induced by amyloid beta peptide and their consequences in neurons and astrocytes in culture. *Biochem. Biophys. Acta* 1742: 81–87.

Akiyama, H. (2000) Inflammation and Alzheimer's disease. *Neurobiol. Aging* 21: 383–421.

Alzheimer, A. (1907) Über eine eigenartige Erkrankung der Hirnrinde. Allg. Zeitschr. Psychiatrie Psych.-Gerichtliche Med. 64: 146–48. (For English translation see Stelzmann, et al. [1995], below.)

Assis-Nascimento, P., et al. (2007) Beta-amyloid toxicity in embryonic rat astrocytes. *Neurochem. Res.* 32: 1476–82.

Avshalumov, M. V., MacGregor, D. C., Sehgal, L. M., and Rice, M. E. (2004) The glial antioxidant network and neuronal ascorbate: protective yet permissive for H_2O_2 signaling. *Neuron Glia Biology* 1: 365–76.

Barger, S. W., and Basile, A. S. (2001) Activation of microglia by secreted amyloid precursor protein evokes release of glutamate by cystine exchange and attenuates synaptic function. *J. Neurochem.* 76: 846–54.

Boelen, E., et al. (2007) Inflammatory responses following *Chlamydia pneumoniae* infection of glial cells. *Eur. J. Neurosci.* 25: 753–60.

Bowman, B. H., et al. (1996) Human APOE protein localized in brains of transgenic mice. *Neurosci. Lett.* 219: 57–59.

Carre, J. L., Abalain, J. H., Sarlieve, L. L., Floch, H. H. (2001) Ontogeny of steroid metabolizing enzymes in rat oligodendrocytes. *J. Steroid Biochem. Mol. Biol.* 78: 89–95.

Carrer, H. F., and Cambiasso, M. J. (2002) Sexual differentiation of the brain: genes, estrogen, and neurotrophic factors. *Cell. Mol. Neurobiol.* 22: 479–500.

Cascio, C., et al. (2000) Pathways of dehydroepiandrosterone formation in rat brain glia. *J. Steroid Biochem. Mol. Biol.* 75: 177–86.

Celotti, F., Melcangi, R., and Martini, L. (1992) The 5-alpha reductase in the brain: molecular aspects and relation to brain function. *Front. Neuroendocrinol.* 13: 163–215.

Chan, J. R., et al. (1998) Glucocorticoids and progestins signal the initiation and enhance the rate of myelin formation. *Proc. Natl. Acad. Sci. USA* 95: 10459–64.

Chowen, J. A., et al. (1995) Sexual dimorphism and sex steroid modulation of glial fibrillary acidic protein (GFAP) mRNA and immunoreactive levels in the rat hypothalamus. *Neuroscience* 69: 519–32.

Conejo, N. M., et al. (2005) Influence of gonadal steroids on the glial fibrillary acidic protein–immunoreactive astrocyte population in young rat hippocampus. *J. Neurosci. Res.* 79: 488–94.

Das, S., and Potter, H. (1995) Expression of Alzheimer amyloid-promoting factor antichymotrypsin is reduced in human astrocytes by IL1. *Neuron* 14: 447–56.

Day, J. R., et al. (1990) Castration enhances expression of glial fibrillary acidic protein and sulfated glycoprotein-2 in the intact and lesion-altered hippocampus of the adult male rat. *Mol. Endocrinol.* 4: 1995–2002.

Del Cerro, S., et al. (1995) Neuroactive steroids regulate astroglial morphology in hippocampal cultures from adult rats. *Glia* 14, 65–71.

de Vellis, J., ed. (2002) *Neuroglia in the Aging Brain.* Humana Press, Totowa, N.J.

Drew, P. D., Chavis, J. A., Bhatt, R. (2003) Sex steroid regulation of microglial cell activation: relevance to multiple sclerosis. *Ann. N.Y. Acad. Sci.* 1007: 329–34.

Dudas, B., Hanin, I., Rose, M., and Wulfert, E. (2004) Protection against inflammatory neurodegeneration and glial cell death by 7-beta-hydroxy-epiandrosterone, a novel neurosteroid. *Neurobiol. Dis.* 15: 262–68.

Fernandez-Galaz, M. C., et al. (1999) Diurnal oscillation in glial fibrillary acidic protein in the perisuprachiasmatic area and its relationship to the luteinizing hormone surge in the female rat. *Neuroendocrinology* 70, 368–76.

Fernandez-Tome, P., Brera, B., Arevalo, M. A., and de Ceballos, M. L. (2004) Beta-amyloid 25-35 inhibits glutamate uptake in cultured neurons and astrocytes: modulation of uptake as a survival mechanism. *Neurobiol. Dis.* 15: 580–89.

Foy, M., Baudry, M., and Thompson, R. (2004) Estrogen and hippocampal synaptic plasticity. *Neuron Glia Biology* 1: 327–38.

Garcia-Estrada, J., Luquin, S., Fernandez, A. M., and Garcia-Sequra, L. M. (1999) Dehydroepiandrosterone, pregnenolone and sex steroids down-regulate reactive astroglia in the male rat brain after a penetrating brain injury. *Int. J. Dev. Neurosci.* 17: 145–51.

Garcia-Ovejero, D., Azcoitia, I., Doncarlos, L. L., Melcangi, R. C., and Garcia-Sequra, L. M. (2005) Glia-neuron crosstalk in the neuroprotective mechanisms of sex steroid hormones. *Brain Res. Brain Res. Rev.* 48: 273–86.

Garcia-Sequra, L. M., and McCarthy, M. M. (2004) Minireview: role of glia in neuroendocrine function. *Endocrinology* 145: 1082–86.

Gerard, H. C. (2005) The load of *Chlamydia pneumoniae* in Alzheimer's brain varies with APOE genotype. *Microb. Pathog.* 39: 19–26.

Gonzalez-Perez, O., Ramos-Remus, C., Garcia-Estrada, J., and Luquin, S. (2001) Prednisone induces anxiety and glial cerebral changes in rats. *J. Rheumatol.* 28: 2529–34.

Hajos, F., et al. (2000) Ovarian cycle-related changes of glial fibrillary acidic protein (GFAP) immunoreactivity in the rat interpeduncular nucleus. *Brain Res.* 862: 43–48.

Hartlage-Rubsamen, M. (2003) Astrocyte expression of Alzheimer disease beta secretase (BSCE1) is stimulus-dependent. *Glia* 41: 169–79.

Hoyk, Z., Parducz, A., and Garcia-Segura, L. M. (2004) Dehydroepiandrosterone regulates astroglia reaction to denervation of olfactory glomeruli. *Glia* 48: 207–16.

Hung, A. J., et al. (2003) Estrogen, synaptic plasticity and hypothalamic reproductive aging. *Exp. Gerontol.* 38: 53–59.

Jantzen, P. T., et al. (2002) Microglial activation and beta-amyloid deposit reduction caused by nitric oxide releasing non-steroidal anti-inflammatory drug in amyloid precursor protein plus preseniline-1 transgenic mice. *J. Neurosci.* 22: 2246–54.

Jellinck, P. H., et al. (2006) Dehydroepiandrosterone (DHEA) metabolism in the brain: identification by liquid chromatography/mass spectrometry of the delta-4-isomer of DHEA and related steroids formed from androstenedione by mouse BV2 microglia. *J. Steroid Biochem. Mol. Biol.* 98: 41–47.

Kimura, N., et al. (2006) Amyloid beta up-regulates brain-derived neurotrophic factor production from astrocytes: rescue from amyloid beta-related neuritic degeneration. *J. Neurosci. Res.* 84: 782–89.

Klintsova, A., Levy, W. G., and Desmond, N. L. (1995) Astrocytic volume fluctuates in the hippocampal CA1 region across the estrous cycle. *Brain Res.* 690: 269–74.

Landis, G. N., and Tower, J. (2005) Superoxide dismutase evolution and life span regulation. *Mech. Ageing Dev.* 126: 365–79.

Leib, K., Engels, S., and Fiebich, B. L. (2003) Inhibition of LPS-induced iNOS and NO synthesis in primary rat microglial cells. *Neurochem. Int.* 42: 131–37.

Liang, Z. (2002) Effects of estrogen treatment on glutamate uptake in cultured human astrocytes derived from cortex of Alzheimer's disease patients. *J. Neurochem.* 80: 807–14.

Marin-Hussteg, M., Muggironi, M., Raban, D., Skoff, R. P., and Casaccia-Bonnefil, P. (2004) Oligodendrocyte progenitor proliferation and maturation is differentially regulated by male and female sex steroid hormones. *Dev. Neurosci.* 26: 245–54.

Martin, L. J., Pardo, C. A., Cork, L. C., and Price, D. L. (1994) Synaptic pathology and glial responses to neuronal injury precede the formation of senile plaques and amyloid deposits in the aging cerebral cortex. *Am. J. Pathol.* 145: 1258–81.

Masliah, E. (2000) Abnormal glutamate transporter function in mutant amyloid precursor protein transgenic mice. *Exp. Neurol.* 163: 381–87.

McCarthy, M. M., Todd, B. J., and Amateau, S. K. (2003) Estradiol modulation of astrocytes and the establishment of sex differences in the brain. *Ann. N.Y. Acad. Sci.* 1007: 283–97.

Melcangi, R. C., et al. (1999) Steroid metabolism and effects in central and peripheral glial cells. *J. Neurobiol.* 40, 471–83.

Melcangi, R. C., Magnaghi, V., Galbiati, M., and Martini, L. (2001) Glial cells: a target for steroid hormones. *Prog. Brain Res.* 132: 31–40.

Miller, K. R., and Streit, W. J. (2007) The effects of aging, injury and disease on microglial function: a case for cellular senescence. *Neuron Glia Biology* 3: 245–53.

Mong, J. A., Roberts, R. C., Kelly, J. J., and McCarthy, M. M. (2001) Gonadal steroids reduce the density of axospinous synapses in the developing rat arcuate nucleus: an electron microscopy analysis. *J. Comp. Neurol.* 432: 259–67.

Mouton, P. R., et al. (2002) Age and gender effects on microglia and astrocyte numbers in brains of mice. *Brain Res.* 22: 30–35.

Park, J. H., et al. (2006) Subcutaneous Nogo receptor removes brain amyloid-beta and improves memory in Alzheimer's transgenic mice. *J. Neurosci.* 26: 13279–86.

Ramirez, B. G., et al. (2005) Prevention of Alzheimer's disease pathology by cannabinoids: neuroprotection mediated by blockade of microglial activation. *J. Neurosci.* 25: 1904–13.

Schenk, D., et al. (1999) Immunization with amyloid-beta attenuates Alzheimer-disease-like pathology in PD APP mouse. *Nature* 400: 173–77.

Schoen, S. W., and Kreutzberg, G. W. (1994) Synaptic 5-nucleotidase activity reflects lesion-induced sprouting within the adult rat dentate gyrus. *Exp. Neurol.* 127: 106–18.

Schubert, P., and Ferroni, S. (2005) Role of microglia and astrocytes in Alzheimer's disease, in *The Role of Glia in Neurotoxicity*, second ed., Michael Aschner and Lucio G. Costa, eds. CRC Press, New York, pp. 299–311.

Schubert, P., Ogata, T., Marchini, C., and Ferroni, S. (2001) Glia-related pathomechanisms in Alzheimer's disease: a therapeutic target? *Mech. Ageing Dev.* 123: 47–57.

Shenk, D. (2006) The memory hole. *New York Times,* November 3.

Simard, A. R., Soulet, D., Gowing, G., Julien, J. P., and Rivest, S. (2006) Bone marrow–derived microglia play a critical role in restricting senile plaque formation in Alzheimer's disease. *Neuron* 49: 489–501.

Simard, M., et al. (1999) Glucocorticoids—potent modulators of astrocytic calcium signaling. *Glia* 28: 1–12.

Simard, M., Arcuino, G., Takano, T., Liu, Q. S., and Nedergaard, M. (2003) Signaling at the gliovascular interface. *J. Neurosci.* 23: 9254–62.

Sinchak, K., et al. (2003) Estrogen induces de novo progesterone synthesis in astrocytes. *Dev. Neurosci.* 5: 343–48.

Small, D. H., Mok, S. S., and Bornstein, J. C. (2001) Alzheimer's disease and Abeta toxicity: from top to bottom. *Nature Rev. Neurosci.* 2: 595–98.

Stelzmann, R. A., Schitzlein, H. H., and Murtagh, F. R. (1995) An English translation of Alzheimer's 1907 paper, "Über eine eigenartige Erkankung der Hirnrinde." *Clin. Anat.* 8: 429–31.

Stone, D. J., et al. (1998) Bidirectional transcriptional regulation of glial fibrillary acidic protein by estradiol in vivo and in vitro. *Endocrinology* 139: 3203–9.

Streit, W. J. (2002) Microglia as neuroprotective immunocompetent cells of the CNS. *Glia* 40: 133–39.

Svenningsen, A. F., and Kanje, M. (1999) Estrogen and progesterone stimulate Schwann cell proliferation in a sex and age-dependent manner. *J. Neurosci. Res.* 57: 124–30.

Takano, T., et al. (2007) Two-photon imaging of astrocyte Ca^{2+} signaling and the microvasculature in experimental mice models of Alzheimer's disease. *Ann. N.Y. Acad. Sci.* 1097: 40–50.

Takata, K., et al. (2007) Microglia transplantation increases amyloid-beta clearance in Alzheimer model rats. *FEBS Lett.* 6: 475–78.

Tanapat, P., Hastings, N. B., Reeves, A. J., and Gould, E. (1999) Estrogen stimulates a transient increase in the number of new neurons in the dentate gyrus of the adult female rat. *J. Neurosci.* 19: 5792–5801.

von Bernhardi, R., et al. (2007) Pro-inflammatory conditions promote neuronal damage mediated by amyloid precursor protein and decrease its phagocytosis and degradation by microglial cells in culture. *Neurobiol. Dis.* 26: 153–64.

Wang, M. J., Huang, H. M., Chen, H. L., Kuo, J. S., and Jeng, K. C. (2001) Dehydroepiandrosterone inhibits lipopolysaccharide-induced nitric oxide production in BV-2 microglia. *J. Neurochem.* 77: 830–38.

Webster, S. D., et al. (2001) Antibody-mediated phagocytosis of the amyloid-beta peptide in microglia is differentially modulated by C1q. *J. Immunol.* 166: 7496–503.

Wisniewski, H. H., and Wegiel, J. (1991) Spatial relationships between astrocytes and classical plaque components. *Neurobiol. Aging* 12: 593–600.

Yin, K. J., et al. (2006) Matrix metalloproteinases expressed by astrocytes mediate extracellular amyloid-beta peptide catabolism. *J. Neurosci.* 26: 10939–48.

Zilka, N., Ferencik, M., and Hulin, I. (2006) Neuroinflammation in Alzheimer's disease: protector or promoter? *Bratisl. Lek. Listy* 107: 374–83.

Zwain, I. H., Arroyo, A., Amato, P., and Yen, S. S. (2002) A role for hypothalamic astrocytes in dehydroepiandrosterone and estradiol regulation of gonadotropin-releasing hormone (GnRH) release by GnRH neurons. *Neuroendocrinology* 75: 375–83.

PART III:
GLIA IN THOUGHT AND MEMORY
Chapter 13: The Mind of the Other Brain

Amzica, F. (2002) In vivo electrophysiological evidences for cortical neuron-glia interactions during slow (<1 Hz) and paroxysmal sleep oscillations. *J. Physiol. Paris* 96: 209–19.

Amzica, F., and Massimini, M. (2002) Glial and neuronal interactions during slow wave and paroxysmal activities in the neocortex. *Cerebral Cortex* 12: 1101–13.

Amzica, F., Massimini, M., and Manfridi, A. (2002) Spatial buffering during slow and paroxysmal sleep oscillations in cortical networks of glial cells in vivo. *J. Neurosci.* 22: 1042–53.

Amzica, F., and Steriade, M. (2000) Neuronal and glial membrane potentials during sleep and paroxysmal oscillations in the neocortex. *J. Neurosci.* 20: 6648–65.

Blanco, E., et al. (2006) Astroglial distribution and sexual differences in neural metabolism in mammallary bodies. *Neurosci. Lett.* 27: 82–86.

Carey, B. (2008) H. M., an unforgettable amnesiac, dies at 82. *New York Times,* December 4.

Crunelli, V. (2002) Novel neuronal and astrocytic mechanisms in thalamocortical loop dynamics. *Philos. Trans. R. Soc. Lond. B, Biol. Sci.* 357: 1675–93.

Green, J. D., and Maxwell, D. S. (1961) Hippocampal electrical activity I. Morphological aspects. *Neurophysiol.* 13: 837–46.

Hatton, G. I. (1997) Function-related plasticity in hypothalamus. *Annu. Rev. Neurosci.* 20: 375–97.

———. (2004) Morphological plasticity of astroglial/neuronal interactions: functional implications, in *Glial Neuronal Signaling*, G. I. Hatton and V. Parpura, eds. Kluwer Academic Publishers, Norwell, MA, pp. 99–124.

Hatton, G. I., Perimutter, L. S., Salm, A. K., and Tweedle, C. D. (1984) Dynamic neuron-glial interactions in hypothalamus and pituitary: implications for control of hormone synthesis and release. *Peptides* 5 suppl. 1: 121–38.

Hsu, J. C., Lee, Y. S., Chang, C. N., Ling, E. A., and Lan, C. T. (2003) Sleep deprivation prior to transient global cerebral ischemia attenuates glial reaction in the rat hippocampal formation. *Brain Res.* 984: 170–81.

Iino, M., et al. (2001) Glia-synapse interaction through Ca^{2+}-permeable AMPA receptors in Bergmann glia. *Science* 292: 926–29.

Jackson, F. R., and Haydon, P. G. (2008) Glial regulation of neurotransmission and behavior in *Drosophila. Neuron Glia Biology* 4: 11–17.

Oliet, S. H., et al. (2008) Neuron-glia interactions in the rat supraoptic nucleus. *Prog. Brain Res.* 170: 109–71.

Panatier, A., and Oliet, S. H. R. (2006) Neuron-glia interactions in the hypothalamus. *Neuron Glia Biology* 2: 51–58.

Sawaguchi, T., et al. (2003) Clinicopathological correlation between brainstem gliosis using GFAP as a marker and sleep apnea in the sudden infant death syndrome. *Early Hum. Dev.* 75 suppl: S3–11.

Sawaguchi, T., et al. (2002) Association between sleep apnea and reactive astrocytes in brainstems of victims of SIDS and in control infants. *Forensic. Sci. Int.* 130 Suppl: S40–46.

Slezak, M., Pfrieger, F. W., and Soltys, Z. (2006) Synaptic plasticity, astrocytes and morphological homeostasis. *J. Physiol. Paris* 99: 84–91.

Sul, J. Y., Orosz, G., Givens, R. S., and Haydon, P. G. (2004) Astrocytic connectivity in the hippocampus. *Neuron Glia Biology* 1: 3–11.

Theodosis, D. T. (2002) Oxytocin-secreting neurons: a physiological model of morphological neuronal and glial plasticity in the adult hypothalamus. *Front. Neuroendocrinol.* 23: 101–35.

Theodosis, D. T., and Poulain, D. A. (1984) Evidence for structural plasticity in the supraoptic nucleus of the rat hypothalamus in relation to gestation and lactation. *Neuroscience* 11: 183–93.

Tweedle, C. D., and Hatton, G. I. (1977) Ultrastructural changes in rat hypothalamic neurosecretory cells and their associated glia during minimal dehydration and rehydration. *Cell Tissue Res.* 181: 59–72.

Wang, Y. F., and Hatton, G. I. (2009) Astrocytic plasticity and patterned oxytocin neuronal activity: dynamic interactions. *J. Neurosci.* 29: 1743–54.

Chapter 14: Memory and Brain Power beyond Neurons

Bullock, T. H., Bennett, M. V. L., Johnston, D., Josephson, R., Marder, E., and Fields, R. D. (2005) The neuron doctrine, redux. *Science* 310: 791–93.

Bushong, E. A., Martone, M. E., Jones, Y. Z., and Ellisman, M. H. (2002) Protoplasmic astrocytes in CA1 stratum radiatum occupy separate anatomical domains. *J. Neurosci.* 22: 183–92.

Chadwick, D. J., and Good, J., eds. (2006) *Purinergic Signaling in Neuron-Glia Interactions*. Novartis Symposium 276, John Wiley and Sons, Chichester, UK.

Fields, R. D. (2004) The other half of the brain. *Scientific American* April 290: 54–61.

———. (2004) Volume transmission in activity-dependent regulation of myelinating glia. *Neurochem. Int.* 45: 503–9.

———. (2005) Making memories stick. *Scientific American* February 292: 74–81.

———. (2005) Erasing memories. *Scientific American Mind* November 16: 28–37.

———. (2006) Nerve impulses regulate myelination through purinergic signalling. *Novartis Found. Symp.* 276: 148–58; discussion on pp. 158–61, 233–37, 275–81.

———. (2007) Sex and the secret nerve. *Scientific American Mind* 18: 20–27.

———. (2008) Purinergic signaling in astrocyte function and interactions with neurons. in *Astrocytes in Patholophysiology of the Nervous System*, V. Parpura and P. G. Haydon, eds. Springer Press, New York.

———. (2008) White matter matters. *Scientific American* 298: 54–56.

———. (2008) Activity-dependent myelination, in *Beyond the Synapse*, R. D. Fields, ed. Cambridge University Press, Cambridge, UK, pp. 36–42.

———. (2008) Oligodendrocytes changing the rules: action potentials in glia and oligodendrocytes controlling action potentials. *Neuroscientist* 14: 540–43.

———. (2008) White matter in learning, cognition, and psychiatric disorders. *Trends Neurosci.* 31: 361–70.

Fields, R. D., and Bukalo, O. (2008) Signaling to the nucleus in long-term memory, in *Beyond the Synapse*, R. D. Fields, ed. Cambridge University Press, Cambridge, UK, pp. 169–78.

Grosjean, Y., Grillet, M., Augustin, H., Ferveur, J.-F., and Featherstone, D. E. (2008) A glial amino-acid transporter controls synapse strength and courtship in Drosophila. *Nat. Neurosci.* 11: 54–61.

Haber, M., and Murai, K. K. (2006) Reshaping neuron-glial communication at hippocampal synapses. *Neuron Glia Biology* 2: 59–66.

Ishibashi, T., Dakin, K. A., Stevens, B., Lee, P. R., Kozlov, S. V., Stewart, C. L., and Fields, R. D. (2006) Astrocytes promote myelination in response to electrical impulses. *Neuron* 49: 823–32.

Itoh, K., Ozaki, M., Stevens, B., and Fields, R. D. (1997) Activity-dependent regulation of N-cadherin in DRG neurons: differential regulation of N-cadherin, NCAM, and L1 by distinct patterns of action potentials. *J. Neurobiol.* 33: 735–48.

Itoh, K., Stevens, B., Schachner, M., and Fields, R. D. (1995) Regulation of the neural cell adhesion molecule L1 by specific patterns of neural impulses. *Science* 270: 1369–72.

Jackson, F. R., and Haydon, P. G. (2008) Glial cell regulation of neurotransmission and behavior in Drosophila. *Neuron Glia Biology* (2008) 4: 11–17.

Markham, J. A., and Greenough, W. T. (2004) Experience-driven brain plasticity: beyond the synapse. *Neuron Glia Biology* 1: 351–63.

Stevens, B., and Fields, R. D. (2000) Action potentials regulate Schwann cell proliferation and development. *Science* 287: 2267–71.

Stevens, B., Ishibashi, T., Chen, J.-F., and Fields, R. D. (2004) Adenosine: an activity-dependent axonal signal regulating MAP kinase and proliferation in developing Schwann cells. *Neuron Glia Biology* 1: 23–34.

Stevens, B., Porta, S., Haak, L. L., Gallo, V., and Fields, R. D. (2002) Adenosine: A neuron-glial transmitter promoting myelination in the CNS in response to action potentials. *Neuron* 36: 855–68.

Stevens, B., Tanner, S., and Fields, R. D. (1998) Control of myelination by specific patterns of neural impulses. *J. Neurosci.* 15: 9303–11.

Suh, J., Jackson, F. R. (2007) *Drosophila* Ebony activity is required in glia for the circadian regulation of locomotor activity. *Neuron* 55: 435–47.

Zalc, B., and Fields, R. D. (2000) Do action potentials regulate myelination? *Neuroscientist* 6: 5–13.

Chapter 15: Thinking beyond Synapses

Barr, C. L., et al. (2001) Linkage study of polymorphisms in the gene for myelin oligodendrocyte glycoprotein located on chromosome 6p and attention deficit hyperactivity disorder. *Am. J. Med. Genet.* 105: 250–54.

Bengtsson S. L., et al. (2005) Extensive piano practicing has regionally specific effects on white matter development. *Nat. Neurosci.* 8: 1148–50.

Brenneman, M. M., et al. (2008) Nogo-A inhibition induces recovery from neglect in rats. *Behav. Brain Res.* 187: 262–72.

Diamond, M. C., Krech, D., and Rosenweig, M. R. (1964) The effects of an enriched environment on the histology of the rat cerebral cortex. *J. Comp. Neurol.* 123: 111–19.

Dupree, J., et al. (2004) Oligodendrocytes assist in the maintenance of sodium channel clusters independent of the myelin sheath. *Neuron Glia Biology* 1: 179–92.

Fields, R. D. (2005) Myelination: an overlooked mechanism of synaptic plasticity. *Neuroscientist* 11: 528–31.

———. (2008) White matter in learning, cognition, and psychiatric disorders. *Trends Neurosci.* 31: 361–70.

Fields, R. D., ed. (2008) *Beyond the Synapse.* Cambridge University Press, New York.

Fields, R. D., and Burnstock, G. (2006) Purinergic signaling in neuron-glia inter-actions. *Nat. Rev. Neurosci.* 7: 423–36.

Fields, R. D., and Itoh, K. (1996) Neural cell adhesion molecules in activity-dependent development and synaptic plasticity. *Trends Neurosci.* 19: 473–80.

Fields, R. D., and Stevens-Graham, B. (2002) New insights into neuron-glia communication. *Science* 298: 556–62.

Gruber, S. A., and Yurgelun-Todd, D. A. (2006) Neurobiology and the law: a role in juvenile justice? *Ohio State Journal of Criminal Law* 3: 321–40.

Hu, F., and Strittmatter, S. M. (2004) Regulating axon growth within the postnatal central nervous system. *Semin. Perinatol.* 28: 371–78.

Jones, T. A., and Greenough, W. T. (2002) Behavioral experience-dependent plasticity of glial-neuronal interactions, ch. 19 in *Glial Neuronal Signaling,* G. I. Hatton and V. Parpura, eds. Kluwer Academic Publishers, Norwell, MA.

Lee, J. K., Kim, J. E., Sivula, M., and Strittmatter, S. M. (2004) Nogo receptor antagonism promotes stroke recovery by enhancing axonal plasticity. *J. Neurosci.* 24: 6209–17.

Maier, I. C., and Schwab, M. E. (2006) Sprouting, regeneration and circuit formation in the injured spinal cord: factors and activity. *Philos. Trans. R. Soc. Lond. B, Biol. Sci.* 361: 1611–34.

McGee, A. W., et al. (2005) Experience-driven plasticity of visual cortex limited by myelin and Nogo receptor. *Science* 309: 2222–26.

Nishiyama, H., Knopfel, T., Endo, S., and Itohara, S. (2002) Glial protein S100B modulates long-term neuronal synaptic plasticity. *Proc. Natl. Acad. Sci. USA* 99: 4037–42.

Noppeney, U., Friston, K. J., Ashburner, J., Frackowiak, R., and Price, C. J. (2005) Early visual deprivation induces structural plasticity in gray and white matter. *Current Biol.* 15: R488–90.

Papadopoulos, C. M., et al. (2006) Dendritic plasticity in the adult rat following middle cerebral artery occlusion and Nogo-A neutralization. *Cereb. Cortex* 16: 529–36.

Roth, A. D., Ivanova, A., and Colman, D. R. (2006) New observations on the compact myelin proteome. *Neuron Glia Biology* 2: 15–21.

Schmithorst, V. J., Wilke, M., Dardzinski, B. J., and Holland, S. K. (2005) Cognitive functions correlate with white matter architecture in a normal pediatric population: diffusion tensor MRI study. *Hum. Brain Mapp.* 26: 139–47.

Yamazaki, Y., et al. (2007) Modulatory effects of oligodendrocytes on the conduction velocity of action potentials along axons in the alveus of the rat hippocampal CA1 region. *Neuron Glia Biology* 3: 325–34.

Chapter 16: Into the Future

Chadwick, D. J., and Good, J. (2006) *Purinergic Signalling in Neuron-Glia Interactions.* John Wiley and Sons, Ltd., Chichester, UK.

Fields, R. D. (2004) The other half of the brain. *Scientific American* April 290: 54–61.

———. (2006) Beyond the neuron doctrine. *Scientific American Mind* June/July 17: 20–27.

———. (2008) White matter in learning, cognition, and psychiatric disorders. *Trends Neurosci.* 31: 361–70.

———. (2008) White matter matters. *Scientific American* March 298: 42–49.

Fields, R. D., ed. (2008) *Beyond the Synapse: Cell-Cell Signaling in Synaptic Plasticity.* Cambridge University Press, Cambridge, UK.

Fields, R. D., and Stevens-Graham, B. (2002) New insights into neuron-glia communication. *Science* 298: 556–62.

Kettenmann, H., and Ransom, B. R., eds. (2005) *Neuroglia.* Oxford University Press, New York.

Parpura, V., and Haydon, P. G., eds. (2008) *Astrocytes in (Patho)Physiology of the Nervous System.* Springer, New York.

INDEX

FIGURE CREDITS

The author and publisher gratefully acknowledge the following sources for permission to reprint figures.

Figure 1. Reprinted with permission from Diamond, M.C., Scheibel, A.B., Murphy, G.M., Jr., and Harvey, T. (1985), On the brain of a scientist: Albert Einstein. *Experimental Neurology* 88: 198–204.

Figure 2. Courtesy Jonathan E. Cohen, Children's National Medical Center, Washington, D.C.

Figure 3. Courtesy Jonathan E. Cohen, Children's National Medical Center, Washington, D.C.

Figure 4. Santiago Ramón y Cajal, 1885, public domain.

Figure 5. Santiago Ramón y Cajal, 1899, public domain.

Figure 6. From Ramón y Cajal, S., *Histologie du Système Nerveux de l'Homme et des Vertèbres*. Paris, Maloine, 1909–1911.

Figure 7. From *J. Comp. Neurol.* 30. Volume dedicated to Professor Camillo Golgi, December, 1918, Wistar Institute of Anatomy.

Figure 8. From Camillo Golgi, "The Neuron Doctrine—Theory and Facts." The Nobel Lectures, 1906. The Nobel Foundation. Reprinted with permission.

Figure 9. From Ramón y Cajal, S., *Degeneration and Regeneration of the Nervous System*. Oxford University Press, London, 1928.

Figure 10. National Library of Medicine photo archives image #20885, Munich, J.F. Lehmann, 1907.

Figure 11. From Stevens, B., and Fields, R.D. (2000), Response of Schwann cells to action potentials in development. *Science* 287: 2267–2271. See also video of this figure at http://www.sciencemag.org/feature/data/1046675.dtl.

Figure 12. Reprinted with permission from Fields, R.D. (2008), White Matter Matters. *Scientific American* March 298: 54–61; and courtesy of Jen Christiansen, artist. Tractography by Derek Jones, Cardiff University.

Figure 13. Reprinted with permission of the artists, Alan Hoofring, NIH Medical Arts, and Jen Christiansen, *Scientific American*. From Fields, R.D. (2008), White Matter Matters. *Scientific American* March 298: 54–61.

Figure 14. Reprinted with permission of the illustrator Jeff Johnson, Hybrid Medical Animation. From Fields, R.D. (2004), The Other Half of the Brain. *Scientific American* April 290: 54–61.

Figure 15. Reprinted with permission of the illustrator Jeff Johnson, Hybrid Medical Animation. From Fields, R.D. (2004) The Other Half of the Brain. *Scientific American* April 290: 54–61.

Figure 16. From Nansen, F. (1897), *Farthest North Being the Record of a Voyage of Exploration of the Ship "Fram" 1893–1896 and of a Fifteen Months' Sleigh Journey by Dr. Nansen and Lieut. Johansen.* Harper and Brothers, New York.

Figure 17. From Nansen, F. (1897), *Farthest North: Being the Record of a Voyage of Exploration of the Ship "Fram" 1893–1896 and of a Fifteen Months' Sleigh Journey by Dr. Nansen and Lieut. Johansen.* Harper and Brothers, New York.

Figure 18. From Fields, R.D. (2006), Nerve impulses regulate myelination through purinergic signalling. Novartis Foundation Symp. 276: 148–174. John Wiley and Sons, Ltd, Chichester, UK.

Figure 19. From Johnson, A.B. (1996), Alexander disease. In *Handbook of Clinical Neurology,* Moser, H.W. (ed.), pp. 701–710. Elsevier Science, Amsterdam. Reprinted with permission.

Figure 20. From Fig. 1 in Fields, R.D., and Stevens-Graham, B. (2002), New Insights into Neuron-Glia Communication. *Science* 298: 556–562. Courtesy Kelly Fields.

Figure 21. From Spielmeyer, W. (1922), *Histopathologie des Nervensystems.* Verlag Julius Springer, Berlin.

Figure 22. From Río-Hortega, P. del, in Penfield, W., ed. (1932), *Cytology and Cellular Pathology of the Nervous System.* P.B. Hoeber, Inc, New York.

Figure 23. Reprinted with permission of the illustrator Jeff Johnson, Hybrid Medical Animation. From Fields, R.D. (2004), The Other Half of the Brain. *Scientific American* April 290: 54–61.

Figure 24. From Río-Hortega, Pío del (1920), La microglia y su transformación en células en basoncito y cuerpos gránulo-adiposos. *Trab. Lab. Invest. Biol. Madrid.* 18:37–82.

Figure 25. Reprinted with permission of the illustrator Jeff Johnson, Hybrid Medical Animation. From Fields, R.D. (2004), The Other Half of the Brain. *Scientific American* April 290: 54–61.

Figure 26. From Fields, R.D., and Stevens-Graham, B. (2002), New Insights into Neuron-Glia Communication. *Science* 298: 556–562. Electron micrograph courtesy Mark Ellisman and Thomas Deerinck, University of California, San Diego.

Figure 27. Reprinted with permission from Hockaday, D.C., et al., (2005), Imaging glioma extent with I311-TM-601. *J. Nuc. Med.* 46:580–586.

Figure 28. Courtesy Dr. Nitin N. Bhatia. Reprinted from Bhatia, N.N., and Gupta, R. (2009), Orthopaedic surgeons and spinal cord injury patients. American Academy of Orthopaedic Surgeons. http://www.aaos.org/news/aaosnow/feb09/research1.asp.

Figure 29. From Fields, R.D., Neale, E.A., and Nelson, P.G. (1990), Effects of patterned electrical activity on neurite outgrowth from mouse sensory neurons. *J. Neurosci.* 10: 2950–2964.

Figure 30. Reprinted with permission from Kulbatski, I., et al., (2008), Glial precursor cell transplantation therapy for neurotrauma and multiple sclerosis. *Progress in Histochemistry and Cytochemistry* 43: 123–176.

Figure 31. Gajdusek, C.D. (1957). Gajdusek Collection, History of Medicine Division, National Library of Medicine, NIH.

Figure 32. Gajdusek, C.D. (1957). Gajdusek Collection, History of Medicine Division, National Library of Medicine, NIH.

Figure 33. Reprinted with permission from Stanley B. Prusiner, "Prions." The Nobel Prize Lecture, 1997. The Nobel Foundation.

Figure 34. Gajdusek, C.D. (1980). Gajdusek Collection, History of Medicine Division, National Library of Medicine, NIH.

Figure 35. Friedrich Schiller University of Jena. University Archives. Germany.

Figure 36. Photo by the author (2006).

Figure 37. Friedrich Schiller University of Jena. University Archives. Germany.

Figure 38. Friedrich Schiller University of Jena. University Archives. Germany.

Figure 39. National Aeronautics and Space Administration, Kennedy Space Center, Florida.

Figure 40. Reprinted with permission of the artist, Lydia Kibiuk. From Brooks, P.J., Cheng, T.-F., and Cooper, L. (2008), Do all of the neurologic diseases in patients with DNA repair gene mutations result from the accumulation of DNA damage? *DNA Repair* 7: 834–848. Inset: Reprinted with permission from Simard, M., et al. (2003), Signaling at the gliovascular interface. *J. Neurosci.* 23: 1529–2401.

Figure 41. Reprinted with permission of the artist, Lydia Kibiuk. From Bruce S. McEwen, *The Hostage Brain*. Rockefeller University Press, New York, 1994.

Figure 42. From Miller, K.R., and Streit, W.J. (2007), The effects of aging, injury and disease on microglial function: a case for cellular senescence. *Neuron Glia Biol.* 3: 245–253. Reprinted with permission.

Figure 43. Courtesy of Varda Lev-Ram (2001).

Figure 44. Courtesy of Ulrika Wilhelmsson, University of Gothenburg, and Eric Bushong and Mark Ellisman, University of California, San Diego. Unpublished image from their research published in 2004 in *J. Neurosci.* 24: 5016–5021.

Figure 45. From Fields, R.D., and Stevens-Graham, B. (2002), New Insights into Neuron-Glia Communication. *Science* 298: 556–562. See also video of this figure at http://www.sciencemag.org/cgi/content/full/sci;298/5593/556/DC1.

Figure 46. From W.A. Locy (1915), *Biology and Its Makers,* third edition. Henry Holt and Company, New York.